Sears's
Anatomy and Physiology for Nurses

Sears's
Anatomy and Physiology for Nurses

R.S. Winwood

MB FRCP

J.L. Smith

SRN, SCM, RNT, Dip N. (Lond.) DANS (Man.)

Sixth edition

Edward Arnold

A division of Hodder & Stoughton

LONDON   BALTIMORE   MELBOURNE   AUCKLAND

© 1985 R. S. Winwood and E. Sears

First published in Great Britain 1941
Reprinted 1942, 1943 (twice), 1944 (twice), 1945, 1947, 1948, 1950
Second Edition 1951
Reprinted 1952, 1953, 1954, 1955, 1956, 1957
Third Edition 1958
Reprinted 1959, 1960, 1961, 1962, 1964
Fourth Edition 1965
Reprinted 1967, 1969, 1971, 1972
Fifth Edition 1974
Revised Reprint 1976
Reprinted 1978, 1980, 1982, 1984
Sixth Edition 1985
Reprinted 1988, 1989

British Library Cataloguing in Publication Data

Sears, W. Gordon
    Sears' anatomy and physiology for nurses.—6th ed.
    1. Human physiology
    I. Title II. Winwood, R.S. III. Smith, J.L.
    IV. Sears, W. Gordon. Anatomy and physiology for nurses and
    students of human biology
    612'.0024613 RT69

    ISBN 0-7131-4463-7

Typeset by Colset Private Ltd, Singapore
Printed and bound in Great Britain for Edward Arnold, the
educational, academic and medical publishing division of Hodder and
Stoughton Limited, 41 Bedford Square, London WC1B 3DQ by
Butler & Tanner Ltd, Frome and London.

Preface to the Sixth Edition

It is with sadness that we record the death after a short illness of Dr Sears on the 29th April 1983. He wrote the first edition of this book over forty-four years ago and his teaching and personal charm earned him the respect and love of many grateful members of the Nursing profession over the years. He shared in the planning of this latest edition of 'Anatomy and Physiology' and, whilst entrusting it to the current authors, maintained a caring interest in the revision. We hope that in continuing the life of this book we shall maintain its value and the popularity it has justly enjoyed.

Social, scientific and technological advances have led to much more being expected of a nurse in educational background, professional training, knowledge, expertise and versatility. The authors, being mindful of this, have gone into considerable detail where they thought it was required.

Important new facts and concepts have been included in this edition and, in particular, the complex and rapidly growing subject of immunity has been tackled.

The work, as before, is intended to be a basic textbook for nurses and other students of Human Biology, to be supplemented by consulting appropriate reference works when necessary.

This book itself may be used for reference and therefore contains in places more detail than may be needed for the purpose of passing examinations.

The authors hope that the reader will enjoy using this book and benefit from doing so.

R.S. Winwood and J.L. Smith
London 1985

Contents

1

Introduction

Life is, perhaps, the most mysterious fact in the Universe and it is not unreasonable that humans have devoted much study to this phenomenon. One of the results of their labours is the science of Biology. In its broad sense, this subject embraces all living matter, both animal and vegetable, in all its forms, both visible and microscopic.

The study of the simplest forms of life contributes to the better understanding of those which have attained a more complicated and advanced degree of development in the scale of Nature.

From the earliest concepts of the subject, careful study and the application of logical thinking, backed by evidence supported by ever-growing scientific techniques, has provided an enormous amount of knowledge. Much of this knowledge is so advanced and so specialized that it can only be appreciated by the few and even they would be the first to admit that such knowledge is incomplete and always capable of further expansion.

Human Biology may be studied as a pure science. On the other hand, for doctors, nurses and many other workers it is the practical application of this knowledge to the understanding of disease and the general well-being of the human race that is of greatest importance.

In order to attain the necessary understanding, some familiarity with science in general is essential and, in order to apply it to full advantage, there must be further appreciation of the workings of the human mind and the development of one of the greatest of human attributes, namely sympathy.

To return to the basic aspects of Human Biology, this has numerous subdivisions which include Anatomy, Physiology and Biochemistry. However, 'the divisions of the sciences are like the branches of a tree that join in one trunk' (Francis Bacon) and they are, therefore, more or less closely related to one another.

Anatomy is the study of the parts of the body, their form, position and relationship to each other.

This knowledge has been obtained by careful dissection and further expanded by the detailed study of the structure of the various tissues under the microscope (Histology).

A greater understanding of the subject has been obtained by studying the anatomy of other members of the animal kingdom (Comparative Anatomy), development from conception to birth (Embryology), and a general consideration of Evolution.

In recent years it has been increasingly possible to gain knowledge of the internal anatomy of the living subject by means of non-invasive *imaging techniques.*

Ultrasonography uses the echoes of ultrasound waves to build up a picture of the contents of the abdomen in which a fetus or any abnormal masses may be studied.

Echocardiography uses the same principles to obtain structural and functional information about the heart. With this technique, the opening and closing of the valves and the wall-motion may be seen and measured.

Computerized tomography (CT scanning) produces sharp clear images of normal and abnormal structures by distinguishing between their relative densities. Cross sections of the head and trunk may be studied at any level. The anatomical display may be helpful in teaching but is primarily used for diagnostic purposes. Abscesses, cysts and tumours may be detected and the exact anatomical position for a needle biopsy determined. Similarly, the investigation may be helpful in planning radiotherapy.

Nuclear magnetic resonance (NMR) is the latest major non-invasive imaging technique to be developed but its availability is currently limited by its cost.

Many techniques are available for the study of physiology and its alteration by disease. The *electroencephalogram (EEG)* is a well-known example of one which is non-invasive. It provides information on brain function and dysfunction and this

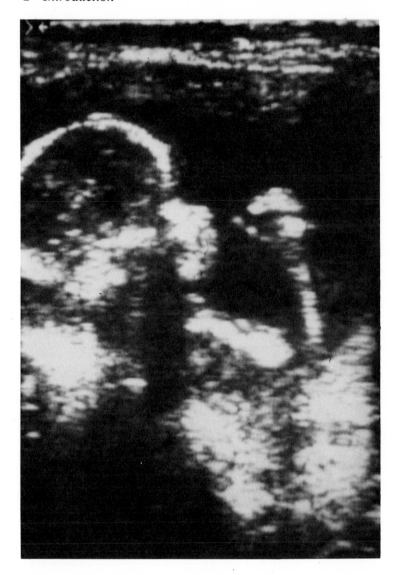

Fig. 1.1 Ultrasound image of a fetus in the womb (*in utero*) with hand to mouth.

method of investigation is frequently complementary to a CT scan.

Physiology is the study of the functions of the body as a whole and of the individual structures and organs contained therein. Some of this is reasonably simple; some involves complicated chemical, physical and electrical details.

Every living structure, whether it be animal or vegetable, is derived, so far as we know, from another living structure. It has the power of growth and reproduction, and its life is dependent upon its ability to absorb non-living material which it builds up into the framework of its own body.

Before considering living matter, it is necessary to go back a step further and ascertain the nature of the **chemical substances** of which it is composed, and which are, therefore, found in the human body as a whole.

Broadly speaking, there are two types of matter: elements and compounds. The latter may be divided into inorganic and organic.

An **element** is a substance which contains only one kind of matter. The following are the most important elements found in the human body: carbon, hydrogen, nitrogen, oxygen, sulphur, phosphorus, chlorine, iodine, sodium, potassium, magnesium, calcium and iron. Of these, oxygen

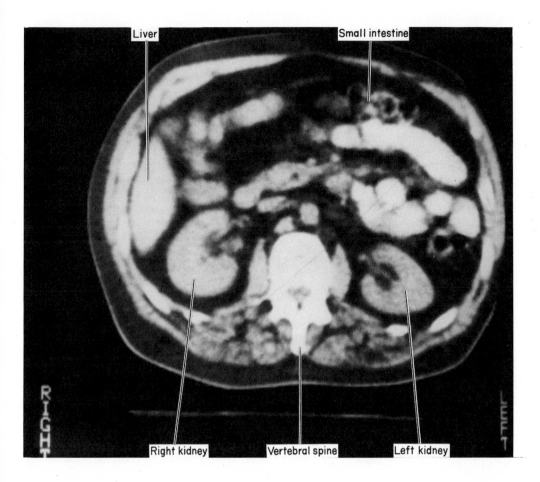

Fig. 1.2 CT scan at the level of the kidneys.

and nitrogen commonly occur in their uncombined natural state. The others are found combined with one another in the form of compounds.

A chemical **compound** is a combination of two or more elements in fixed proportions forming an entirely new substance in which the individual elements apparently lose their identity, thereby differing from a simple mixture. Every part of a compound has exactly the same composition and properties as every other part.

Inorganic compounds are relatively simple combinations of the elements found especially in non-living matter such as minerals, water and salts.

The essential feature of **organic compounds** is the presence of the element **carbon**, usually combined with hydrogen and oxygen. In addition, nitrogen and other elements may also be included and form compounds of a highly complicated nature which are found specially in living matter.

The main organic compounds found in the body are

carbohydrates
fats
proteins.

All three of these contain carbon, hydrogen and oxygen but proteins also contain nitrogen and other elements.

Going back one stage further in the structure of matter, and in order to understand some of the principles which must be considered in Physiology, it is necessary to have some knowledge of the atomic theory which was propounded by Dalton nearly one hundred and eighty years ago. This has been the basis of chemical science ever since.

Stated simply this implies the following.

(1) The basis of all matter is the atom.

(2) If further subdivided, an atom consists of protons, neutrons and electrons. The protons and electrons each carry a unit charge of electricity. That of the former is positive and that of the latter negative. The neutrons, as their name implies, are electrically inactive.

(3) Every atom consists of a central particle or nucleus around which constantly revolve in their own orbit one, a few or many smaller electrons. On an astronomical scale these are rather like planets revolving around the sun.

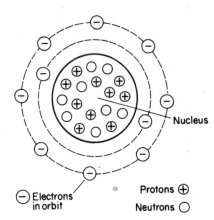

(4) The nucleus consists of a compressed mass of protons and neutrons. Because of the protons, it is positively charged with electricity. In order to render the atom as a whole electrically neutral it has an appropriate number of circulating negative electrons.

(5) For example, the atom of hydrogen carries one electron; that of carbon, six; nitrogen, seven; oxygen, eight; sodium, eleven; chlorine, seventeen, and so on.

(6) Under ordinary circumstances the atoms of each element (except for its radioactive isotopes) are stable. In 1919 Rutherford succeeded in splitting the atom and his work led, step by step, to the modern science of Nuclear Physics.

(7) The atoms of most elements have the property of combining with atoms of other elements to form the molecules of new compounds.

(8) This power can best be visualized by imagining that the atom of each element has one or more hooks or bonds which can link up with a similar hook or hooks of another atom or atoms:

For example: hydrogen, sodium and chlorine have one hook;

oxygen, calcium and sulphur, two;

nitrogen, three;

carbon, four; and so on.

Thus, one atom of sodium can combine with one atom of chlorine to form one molecule of the compound, sodium chloride or common salt. Using standard chemical symbols this might be expressed thus:

and one atom of oxygen can combine with two of hydrogen:

Nitrogen having three 'hooks' can link up with three atoms of hydrogen to make a molecule of the compound ammonia.

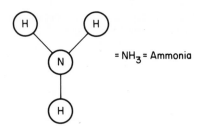

(9) In other words, the atoms of elements, by union with those of other elements, form the molecules of chemical compounds which have the combined mass of the individual elements which compose them. Therefore in scientific terms an atom is said to have its own atomic weight and a compound its molecular weight.

The atomic weight of an element is the average weight of an atom of that element in relation to the weight of an atom of hydrogen, which is taken as 1: for example carbon, 12; iron, 56; lead, 207.

Because the nucleus of some elements does not always have the same number of neutrons but still has the identical number of protons and electrons, there is a slight variation in the weights of individual atoms of a particular element. Those atoms which vary in weight from the standard are called **isotopes**. Many of these are unstable and are radioactive, discharging their nuclei (alpha particles) at high velocity.

Such radioactive isotopes introduced into the body can be detected and traced by a Geiger counter. Thus radioactive iodine given to an individual may be traced to the thyroid gland. In larger doses radioactive isotopes, for example radium, may be used to destroy the abnormal cells which occur in cancer.

The following table lists the elements present in the body and their respective symbols.

carbon (C)	chlorine (C1)
hydrogen (H)	iodine (I)
nitrogen (N)	sodium (Na)
oxygen (O)	potassium (K)
sulphur (S)	magnesium (Mg)
phosphorus(P)	calcium (Ca)
	iron (Fe)

Many other elements are present in only minute amounts and are known as **trace elements**. Some of these are essential to life; examples are cobalt, copper, manganese, molybdenum and possibly selenium. Chromium may protect arteries from atherosclerosis. Some trace elements serve no useful purpose in the body but are contaminants.

(10) Returning to the example of sodium chloride, it will be recalled that the atom of sodium has eleven electrons and that of chlorine seventeen. In effecting the combination to form a compound there is a rearrangement of the electrons in such a way that the sodium atoms become positively charged with electricity and the chlorine atoms become negatively charged. Such electrically charged atoms are called **ions**.

(11) When such compounds are dissolved in water some, but not all, of the molecules of the compound become *ionized* into their individual electrically charged atoms. Some of these atoms will carry a positive charge and the others a negative charge. In the case of sodium chloride, the sodium ions are positive ($+$) and the chlorine ions negative ($-$).

Atoms in this state of ionization, because they carry an electric charge, are referred to as electrolytes, and the solution containing them, an electrolyte solution. It is in this form that many salts circulate in the water contained in the blood and tissue fluids of the body.

(12) Another aspect of this subject is the 'acid/alkaline' reaction of body fluids. The hydrogen ions (H^+) have a positive charge and are acid, while the hydroxyl ions (OH^-) are negative and cause alkalinity. If the H^+ ions and the OH^- ions are equally balanced the reaction of the fluid will be neutral:

$$H^+ = OH^- = H_2O \text{ (or water which is neutral)}$$

For practical purposes a scale has been devised and numbered from 1 to 14. The central figure of 7 is taken to represent neutrality and the hydrogen ion concentration is indicated by the symbol pH.

Fluids having a pH of 1 to 7 are acid and those with a pH exceeding 7 are alkaline.

It must be understood that this is a scale only used to measure very small degrees of acidity and alkalinity. For example, the pH of the blood is kept constant at 7.4, i.e. very slightly alkaline but never acid. That of urine is usually slightly acid with a pH of 5 to 6.

One of the most important functions of the various salts present in the body is to keep the pH of the blood constant and there is a continuous interchange of various positive and negative ions in the tissues to maintain this equilibrium. Any excess of acid hydrogen ions is excreted in the urine by the kidney cells.

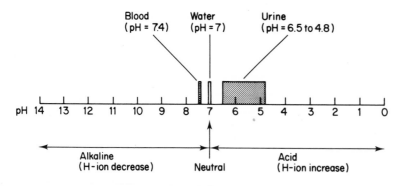

Fig. 1.4   Diagram illustrating hydrogen ion concentration (pH) in the body.

On the other hand, one of the most important end-products of carbohydrate and fat metabolism is carbon dioxide (CO_2). When this is dissolved in water it forms a weak acid (carbonic acid):

$$CO_2 + \quad H_2O \quad \rightleftharpoons H_2CO_3 \quad \rightleftharpoons H^+ \quad + \quad HCO_3^-$$

carbon   water   carbonic   hydrogen   bicarbonate
dioxide   acid   ion   ion

This, when ionized, also liberates acid H^+ in considerable quantity which would tend to lower the pH of the blood towards the neutral figure of 7. However, the respiratory centre in the medulla of the brain is particularly sensitive to any change in the pH of the blood and immediately causes an increase in the rate and depth of breathing which is maintained until the excess of carbon dioxide (and at the same time the excess of hydrogen ions in the blood) is removed.

There are also certain salts together with the blood plasma which themselves are neutral but which have the power of reacting with hydrogen ions without becoming acid. These are called buffer substances which also help to maintain the pH of the blood at a constant level.

2

Cells and Tissues

Cells are the smallest units of living matter in the animal kingdom; the most primitive forms of animal life are single-celled organisms, such as the amoeba and paramecium, which are able to organize the whole range of their activities within the walls of one cell.

A human is a multicellular animal, the human body consisting of trillions of cells of many different types. Masses of similar cells are organized together to form tissues and organs, each of which have special functions within the body. Thus kidney cells are different from liver cells, which in turn are different from brain and blood cells.

However, despite the differences between the types of cells, they all have some basic characteristics in common which indicate signs of life, whether we consider a single-celled organism or a highly developed multicellular animal such as a human.

Characteristics of cells

(*1*) *Irritability* The cell has the ability to detect, and respond to changes in its environment.

(*2*) *Nutrition* The cell is capable of absorbing fluids and dissolved substances directly through the cell membrane, and these can be used by the cell for growth and repair, or to provide energy and heat.

(*3*) *Respiration* The cell has the ability to use oxygen combined with food substances to form carbon dioxide and water, whilst releasing energy for intracellular activity.

(*4*) *Excretion* The cell is able to discharge unused and waste materials through the cell membrane.

(*5*) *Growth and reproduction* The cell has the ability to increase in size, and when it reaches the limit of its growth it reproduces by dividing into two smaller cells.

(*6*) *Movement* Some cells have the power of movement.

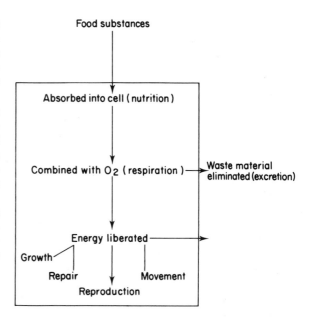

Fig. 2.1 Characteristics of a living cell in action (diagrammatic representation).

Life involves a process of combustion, or the burning of fuel supplied in the form of food. One substance, oxygen, is necessary for combustion of any kind. Its utilization in this way is known as the process of *oxidation*.

A candle consists principally of carbon and hydrogen. When this burns it consumes oxygen and forms carbonic acid (carbon dioxide and water). At the same time it liberates energy in the form of light and heat. The protoplasm of living cells is more complex in structure than a candle, but so far as its carbon and hydrogen are concerned the final products of combustion are the same. A living cell assimilates nutrients in the form of carbohydrates and fats, and these react with oxygen to form carbon dioxide and water with the release of energy, which is then available for cellular activities. The carbon dioxide and water are excreted by the cell as waste products.

Cells are also subject to wear and tear, and in order to repair themselves, and to have the power of growth, protein is necessary. This is assimilated from outside sources as amino acids. In utilizing protein, nitrogenous waste products are formed and excreted.

The chemical reactions by which the cell is able to produce energy, utilize protein and perform cellular functions are made possible by the presence of enzymes.

Enzymes are protein compounds which cause, or accelerate, chemical changes, without themselves changing or being used up. Many thousands of different types of enzymes exist in the human body, each type highly specific for one particular chemical reaction. Some enzymes are produced by cells for use outside the cell, such as those of the digestive system, whilst others are produced solely for use inside the cell. An example of an intracellular enzyme is carbonic anhydrase (present inside red blood cells), which increases the ability of the red blood cells to carry the waste product carbonic acid to the lungs for excretion.

When cells are damaged or destroyed, the intracellular enzymes burst out of the cells and may be found in the tissue fluid and the blood. Since different types of cells produce specialized enzymes for their individual functions it is useful to measure the presence of these enzymes by special blood tests. For example, following damage to cardiac muscle cells (as when a myocardial infarction occurs), the enzymes spill out of the cells and measurement of the serum concentrations of these enzymes assists in confirming a diagnosis of myocardial infarction, and in estimating the extent of the damage.

Structure of the cell

All cells are made of **protoplasm**, a viscid granular substance resembling the white of an egg, which consists of water, electrolytes, proteins, lipids and carbohydrates. The protoplasm forms the **cytoplasm** and the **nucleus**. Surrounding the cytoplasm is the **cell membrane**, a thin, elastic and highly complex structure composed of proteins and lipids. It is a semi-permeable membrane, containing 'pores' that allow the passage of water, oxygen, carbon dioxide and some solutes in and out of the cell, and

plays a vital part in maintaining the homeostatic balance of the cell.

Substances can pass through the cell membrane in a variety of ways, as outlined below.

(a) By diffusion (Fig. 2.2)

Molecules of gas and liquids are in constant motion, and tend to spread from a region of high concentration to a region of lower concentration until a uniform mixture is produced. Water, gases and some solutes diffuse easily through a permeable membrane; when the concentration of molecules is higher on one side than the other the molecules will move through the permeable membrane until both the concentrations are equal. Large molecules diffuse more slowly than small ones.

(b) By osmosis (Fig. 2.3)

Osmosis is the movement of *water* through a semi-permeable membrane, from a solution of low concentration to one of higher concentration (i.e. the reverse of diffusion). Molecules in solution tend to hold or attract water, this 'drawing' power is known as osmotic pressure. Osmotic pressure is determined by the number of molecules dissolved in a solution, thus the greater the number of molecules in solution the greater its 'drawing' power for water.

(c) By active transport (Fig. 2.4)

Some molecules are unable to pass through the cell membrane alone, and these molecules are linked to a special *carrier* substance in the cell membrane, often under the control of enzymes, and transported through the cell membrane to be released on the other side. The carriers are highly specific and usually only respond to one type of molecule.

(d) By pinocytosis (Fig. 2.5)

This is a mechanism by which the cell membrane indents and actually engulfs the substance to be absorbed into the cell.

(e) By phagocytosis (Fig. 2.6)

This process is similar to pinocytosis, the major difference being that the cell extends pseudopodia to surround and engulf larger particles such as bacteria and foreign material.

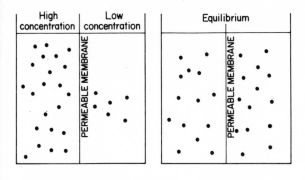

Fig. 2.2 Diffusion. Molecules move from a region of high concentration to a region of low concentration.

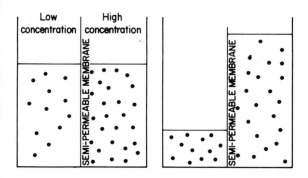

Fig. 2.3 Osmosis. Water moves from a region of low concentration to a region of high concentration.

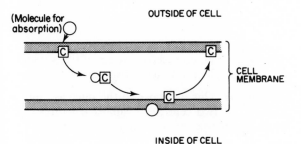

Fig. 2.4 Active transport. A molecule from absorption becomes attached to a carrier (C) for transportation through the cell membrane.

Fig. 2.5 Pinocytosis. A molecule is engulfed by the cell.

Fig. 2.6 Phagocytosis. The cell extends pseudopodia to engulf a molecule.

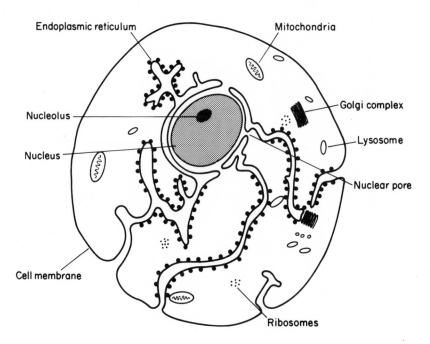

Fig. 2.7 A cell and its detailed structure.

The world of the cell is rich and intricate in detail and the **cytoplasm** contains a number of structures, called organelles, which are concerned with a variety of cellular functions. These include:

Mitochondria Small granular structures containing enzymes which extract energy from nutrients and oxygen, in the form of ATP (adenosine triphosphate). This energy is made available for cellular activity.

Endoplasmic reticulum A complicated network of tubules and vesicles connected with the nucleus and the cell membrane. It transfers substances from one part of the cell to another. On its outer surface are ribosomes.

Ribosomes Granules which contain RNA (ribonucleic acid). They are responsible for protein synthesis in the cell.

Lysosomes These contain enzymes that digest and remove particles which are useless, or may be harmful to the cell.

Golgi body This consists of a series of fine vesicles, and is especially prominent in secretory cells.

Centrosome This lies close to the nucleus, and is made up of two centrioles, small structures that play a major role in initiating cell division.

Also within the cytoplasm is the nucleus, which is enclosed by a nuclear membrane. The **nucleus** is the control centre of the cell and contains a special type of protein, called nucleoprotein. It controls both the chemical reactions that occur in the cell and the reproduction of the cell.

The nucleoprotein actually consists of a number of minute threads, the **chromosomes**. When the cell is in a non-dividing state the chromosomes are scattered throughout the nucleus, and appear as a mass of darkly-staining material called **chromatin**.

All human cells contain 23 pairs of chromosomes (i.e. 46 chromosomes) and each chromosome consists of a chain of smaller units, rather like a string of beads. These small units are called **genes**, and each gene has a very specific location on a particular chromosome.

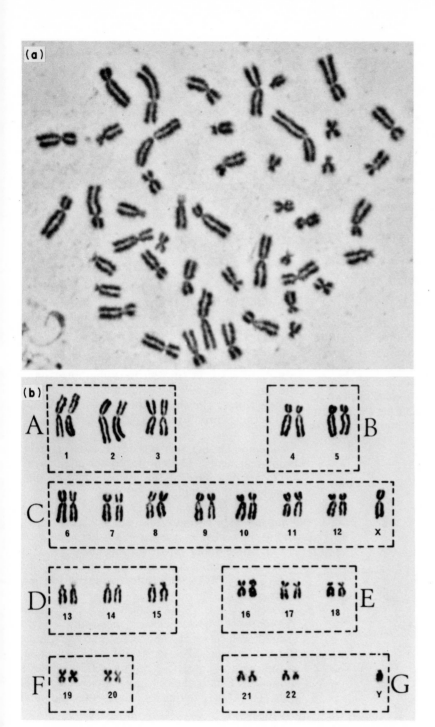

Fig. 2.8 **(a)** The 46 chromosomes from a single male cell undergoing mitosis. The chromosomes have doubled but are still held together by their centromeres. **(b)** The same 46 doubled chromosomes as in **(a)**. The chromosomes are arranged in decreasing order of size and numbered from 1 to 22. The X and Y are not numbered. The letters A to G show the various groupings. (By courtesy of Dr S. Walker, Cytogenetics Unit, Nuffield Wing, School of Medicine, University of Liverpool.)

Genes are made of a complex protein compound, *DNA* (deoxyribonucleic acid), and carry all the hereditary information of the cell. It is the genes which pass on the characteristics of the parent organism. Not only are the more obvious features such as colour of hair and eyes, height and body shape so transmitted, but also the factors which influence an individual's blood group, and in some instances, certain congenital defects and hereditary diseases.

Thus DNA is capable of self-replication. It can also make RNA, which transmits encoded information from the genes to the cell's proteins.

The structure of DNA is highly complicated, and can be affected by some outside influences, for instance viruses can enter cells and substitute their own type of DNA for that of the cell. Further, the structure of DNA can be altered by exposure to X-rays and atomic radiation.

By using special techniques, chromosomes can be arranged in groups and in the human have been given letters and numbers to distinguish the pairs. The sex chromosomes, called X and Y, can be distinguished from the others, which are known as somatic (body) chromosomes. A female has two X chromosomes (XX), whereas a male has an X and a Y chromosome (XY). The sex of a child clearly depends on whether it inherits an X or a Y chromosome from its father.

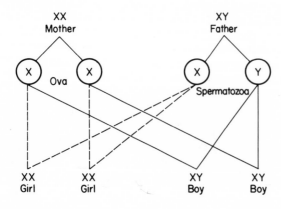

Fig. 2.9 From this diagram it is seen that the chances of having a boy or girl child are 50:50.

Cell division

New cells are necessary for the growth of an organism, and for the replacement of worn-out or damaged cells. However the actual regulation of cell growth and reproduction is still largely a mystery. Some cells, such as those of the blood-forming tissues of the bone marrow, grow and reproduce continuously, whilst other cells, such as muscle cells, do not reproduce for many years. A few cells, (i.e. nerve cells) are unable to replace those that die.

The process of cell division which produces two new 'daughter' cells identical to the 'parent' cell is called mitosis.

Mitosis

This involves a series of changes in which there is a rearrangement of the centrioles and chromosomes so that each of the two new cells has a nucleus with 23 pairs of chromosomes.

Prophase This is the first stage of mitosis. The centrosome divides and the centrioles migrate to opposite poles of the cell, but remain attached by fine spindle fibres. The chromatin material forms into well-defined chromosomes.

Metaphase During this phase the nuclear membrane fragments. The chromosomes align themselves at the centre of the nucleus, and become attached to the spindle fibres.

Anaphase The chromosomes are tightly stretched, and as the centrioles are drawn further apart each chromosome splits into two chromosomes. The separated chromosomes move towards opposite poles of the cell. The centrioles divide to form new centrosomes.

Telophase A new nuclear membrane forms around each set of chromosomes, and the spindle fibres disappear. The cytoplasm and cell membrane begin to constrict until the cell resembles a dumb-cell. Finally the cell splits into two identical cells. The clearly defined chromosomes fade and become a mass of scattered threads again.

Tissue fluid

The fluid inside the cells is called the intracellular fluid, whilst that outside the cells is called the extra-

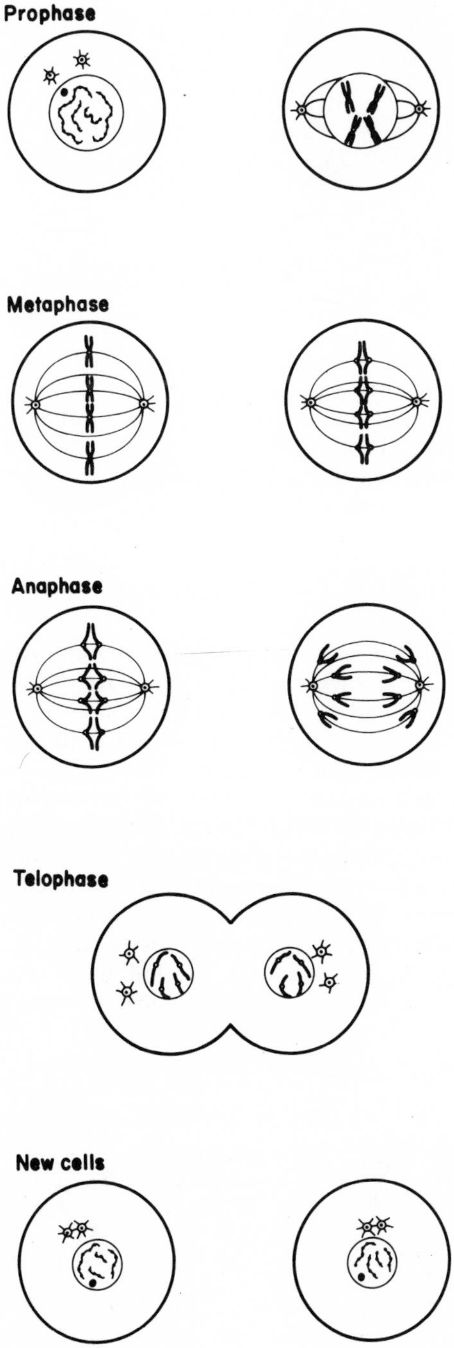

Prophase

Metaphase

Anaphase

Telophase

New cells

Fig. 2.10 Cell division.

cellular fluid, and includes the blood plasma and the tissue fluid which surrounds the cells. The normal functioning of the cells depends on the maintenance of a relatively stable environment, and the extracellular fluid circulates through the spaces between the cells and also exchanges freely with the blood plasma through the capillary walls. In this way the tissue fluid acts as a sort of 'middle-man' between the blood and the tissues, supplying food and oxygen to the cells, and removing waste products from cellular activity.

Tissues

The cells of the human body differ in appearance according to the particular type of tissue to which they belong, and the specialized functions which these tissues perform.

The following are the principal kinds of tissues found in the body. Each consists of individual types of cells which can be seen and recognized under a microscope.

Epithelium

Epithelium is a tissue composed of cells generally arranged to form a membrane or lining, covering either an internal or external body surface.

Simple squamous epithelium (also known as pavement epithelium) consists of a single layer of flattened cells. It is found in the alveoli of the lungs, lining the interior of the heart and blood vessels, and the lymphatic vessels. It forms a smooth, flat membrane, which when lining an internal surface, is sometimes called endothelium.

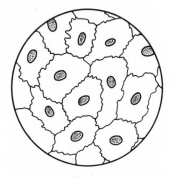

Fig. 2.11 Simple squamous epithelium.

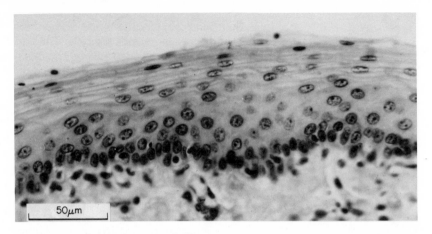

Fig. 2.12 Stratified squamous epithelium from the pharynx.

Stratified squamous epithelium is composed of similar cells arranged in layers. The surface cells are flattened like the simple variety, but the deeper ones are rounded.

The epidermis of the skin consists of stratified epithelium, the outer layers of which become hard and horny because they contain a substance called **keratin**. Stratified epithelium is also found in the mouth, pharynx, oesophagus, vagina and part of the urethra. Its structure is specially adapted to the wear and tear experienced by body surfaces. The superficial layers are constantly being shed and replaced by growth from the deeper layers.

Transitional epithelium is composed of cells which provide a watertight surface, but are also capable of expansion. It is found lining parts of the urinary tract, where it is able to withstand the action of urine with which it is in constant contact.

Columnar epithelium consists of cylindrical-shaped cells, one layer thick, and is found in the secretory glands of the body, such as the salivary glands and the breasts. The stomach and intestines are also lined with columnar epithelium, where some of the cells are responsible for the absorption of fluids and foodstuffs, whilst others secrete a thick, sticky, protective substance called mucus.

Ciliated columnar epithelium is a special form of columnar epithelium. The free surface of each cell is surmounted by fine hair-like processes or cilia. The cilia bend rapidly to one side and then straighten again; this whipping movement sweeps onwards in one direction any substance or fluid in contact with the surface of the cell. Ciliated epithelium is found in the respiratory system, and lines the nasal cavities, the trachea and bronchi. The movement of the cilia conveys mucus, dust, etc. from the deeper parts of the lungs towards the exterior. It is also found in the Fallopian (uterine) tubes, where it assists the ovum on its passage to the uterus.

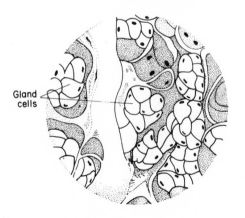

Gland cells

Fig. 2.13 Glandular epithelium from a salivary gland (submandibular gland).

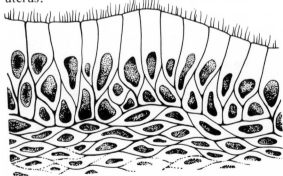

Fig. 2.14 Ciliated columnar epithelium from the trachea.

Connective tissue

The connective tissues are the most widespread and abundant tissues in the body, and exist in more varied forms than the other basic types. Fibres and fibrous tissue cells may be present, and there is a variable amount of supporting substance between the cells. This may be jelly-like, as in the umbilical cord at birth, and in the vitreous humour of the eye, firm as in hyaline cartilage, or hard from the presence of lime salts as in bone.

Fibrous tissue There are two types. Ordinary firm **white fibrous tissue**, which does not stretch, consists of bundles of white fibres containing a few cells; tendons and ligaments consist of this material, also the dura mater, the outer layer of the pericardium, the fascia and fibrous covering of organs. Gelatine can be extracted from this tissue, and it also contains a substance called collagen.

The second variety is **yellow elastic tissue** and consists of yellow fibres which stretch. This is found in the walls of arteries, in the bronchi and alveoli of the lungs and in a few special ligaments in the spine where, on account of their elasticity, they help to maintain the erect posture.

Areolar tissue may be described as the general packing and supporting tissue of the body. It is found under the skin and mucous membranes, and surrounding blood-vessels and nerves. It is a loosely woven tissue containing white fibres, various cells and a gelatinous substance between the cells.

Adipose tissue is a variety of areolar tissue, the bulk of which is made up of globules of fat contained in thin membranous envelopes. It is found in all parts of the body, especially where fat is deposited or stored, such as under the skin and around the eyes, heart and kidneys, and provides protection and insulation.

Brown fat This is a type of fat that contains large numbers of mitochondria in its cells, which are concerned with thermogenesis (heat production). The adult human has almost no brown fat, but infants have a small amount which may play an important part in maintaining normal body temperature in newborn babies.

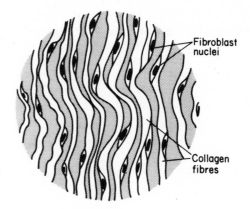

Fig. 2.15 White fibrous tissue.

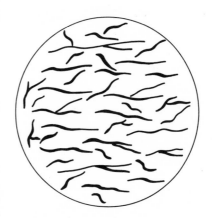

Fig. 2.16 Yellow elastic tissue.

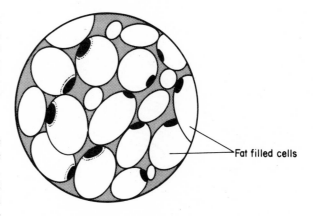

Fig. 2.17 Adipose tissue, showing fat cells.

Cartilage is a firm flexible tissue found mainly in connection with the skeleton. The outer surface of cartilage is covered by a fibrous membrane, the perichondrium, which is supplied with blood-vessels. No blood-vessels, however, enter the cartilage itself, which is nourished by tissue fluid. There are three types: hyaline cartilage, fibro-cartilage and elastic cartilage.

Hyaline cartilage is a bluish-white tissue with a smooth, glassy surface. It is found covering the ends of bones where they form joints (articular cartilage), in the costal cartilages, trachea and larynx.

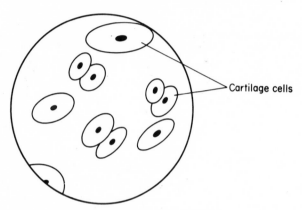

Fig. 2.18   Hyaline cartilage.

Fibro-cartilage contains white fibrous tissue. It is found in the intervertebral discs and the semi-lunar cartilages of the knee joint, where great strength combined with a certain amount of elasticity is required.

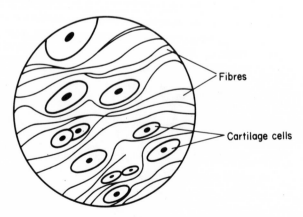

Fig. 2.19   Fibrocartilage.

Elastic cartilage contains yellow elastic fibres and is found in the epiglottis and pinna of the ear.

Bone   This is the hardest of all the connective tissues. The property of hardness is due to the impregnation of its ground substance with calcium salts. There are two types of bone: compact bone and cancellous bone.

Compact bone is the surface layer of bone; it is a strong, dense substance resembling ivory. Its microscopic structure consists of many units called Haversian canals.

Cancellous bone   Although of the same microscopic structure as compact bone, it appears as a spongy, porous tissue containing red or yellow bone marrow.

The periosteum is a tough membrane of fibrous tissue containing blood-vessels which covers the outer surface of bone, except where bone is covered with hyaline cartilage in the formation of joints.

(The details of the microscopic structure and the development of bone are discussed in Chapter 4.)

Blood is a connective tissue and is discussed in Chapter 8.

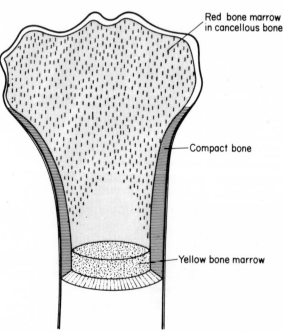

Fig. 2.20   Section of a long bone.

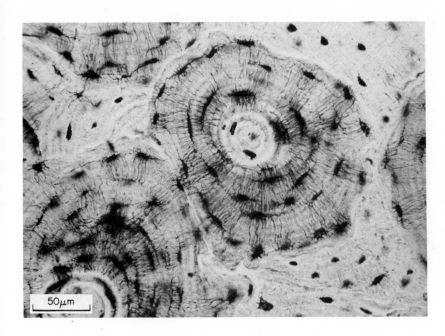

Fig. 2.21 Transverse section of compact bone (microscopic).

Muscular tissue

The muscles are structures which give the power of movement. Muscles are composed of thousands of elongated cells, called **muscle fibres**, each containing a small nucleus. Bundles of muscle fibres lie side by side like the threads in a skein of wool, and are bound together by a thin membrane of connective tissue.

Muscle cells have the power of contraction, in the process of which each one becomes shorter and thicker when it contracts; this can be demonstrated by contracting the biceps muscle of the arm when the fingers are raised to touch the shoulder.

Three types of muscular tissue are found in the body:

(*a*) voluntary muscle (striated), found in the muscles attached to the skeleton;

(*b*) involuntary muscle (plain), present in various internal organs and structures;

(*c*) cardiac muscle, a special type found only in the heart.

Voluntary muscle

As its name implies, skeletal muscle is under the control of the will and, from its microscopic structure, is sometimes referred to as striped or striated muscle (Figs 2.22, 2.23). All the muscles attached to the skeleton are of this type and their functions are to move the bones at their respective joints and to help in maintaining the posture of the limbs and body as a whole.

The voluntary muscles are connected to the cells of the motor cortex of the brain by nerve fibres which pass down the spinal cord, from which a second set of fibres from the anterior horn cells run in the peripheral nerves to the muscles. These motor nerve fibres terminate in special structures in the muscle fibres known as motor end-plates. It is through these end-plates that the impulse to contract is conveyed to the muscle cells.

Involuntary or plain muscle

The cells found in involuntary muscle differ in appearance from those of voluntary muscle (Fig. 2.24); in particular, they do not show the characteristic striation or striping of the latter.

Involuntary muscle is found in the internal organs and structures of the body such as the stomach, intestines, bladder, uterus, bronchi and blood-vessels, and is therefore, sometimes called visceral muscle. It is usually arranged in a series of layers, i.e. circular and longitudinal, thus forming part of the wall of the organ concerned.

It cannot be consciously controlled and its nerve supply comes from the involuntary or autonomic nervous system.

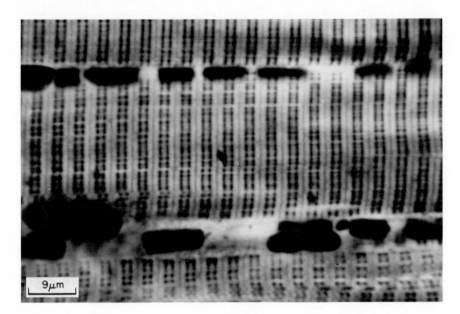

Fig. 2.22   Section of striped muscle.

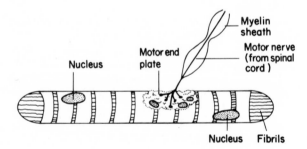

Fig. 2.23   Diagram of a single striated muscle fibre.

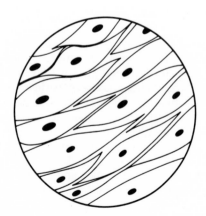

Fig. 2.24   Involuntary muscle fibres.

Cardiac muscle

This is a special type of muscle found only in the heart. Although it is an involuntary muscle it has a form of striation resembling that seen in striped muscle. It has the special property, not observed in the other varieties of muscle, of automatic rhythmic contraction which can occur independently of its nerve supply.

Cardiac muscle fibres are cylindrical with centrally placed nuclei. The fibres branch and connect with adjacent fibres at intercalated discs, and this allows an impulse to spread from one fibre to the next.

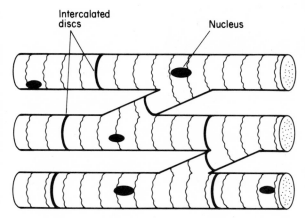

Fig. 2.25 Specialized muscle fibres of the heart.

Properties of muscle

(*1*) *The power of contraction* Voluntary muscles contract as a result of stimuli reaching them from the nervous system and many nerves have their endings in muscles. Other stimuli such as electricity, applied direct to the muscle or its nerves, will cause the muscle to contract. This response is used in Medicine to test nerves or muscles which may have been affected by disease.

(*2*) *Elasticity* Muscle tissue is elastic, and can be stretched by a weight. When the weight is removed the muscle returns to its normal length.

(*3*) *Fatigue* When a muscle contracts it uses energy. This energy is derived mainly from glucose stored in the muscle as glycogen and partly from glucose carried by the blood. The blood also conveys oxygen which the muscle uses to burn up the glucose with the formation of lactic acid, which in its turn is ultimately broken down into carbon dioxide and water. After a number of contractions the supply of glucose immediately available is used up and a certain amount of lactic acid accumulates. The muscle then becomes tired and is unable to contract with the same degree of efficiency. It requires rest in order to replenish its supply of glucose and to remove the lactic acid.

(*4*) *Muscular tone* Even when a muscle appears to be at rest it is always partially contracted, and therefore ready for immediate action. This state of partial contraction is called muscle tone, and is important in maintaining body posture. If a muscle lost its tone it would, on receipt of a stimulus, be compelled to take up the slack which would result in delayed and inefficient movement.

Involuntary muscle, however, retains its tone when all connections with the central nervous system have been severed.

Nervous tissue

Although the muscles are the structures which give the body the power of movement, almost every movement is governed by some portion of the nervous system, which acts as a medium between the brain and muscles.

Two distinct kinds of cell are found in the nervous system:

(a) the nerve cells, or neurons, with their processes or neurites which convey motor or sensory impulses (Fig. 2.26); and

(b) neuroglia, ependyma and Schwann cells; names covering several varieties of nonexcitable cells with numerous functions including mechanical support for the neurons.

Nerve cells vary considerably in size, some being the largest cells in the whole body. The nervous tissue containing the cell bodies of neurons is sometimes referred to as **grey matter**. Aggregations of nerve cell bodies are known as nuclei or ganglia and they are dark in appearance.

Each nerve cell contains a nucleus and highly specialized cytoplasm. From the body of the cell extend two types of neurite, the axon and dendrites.

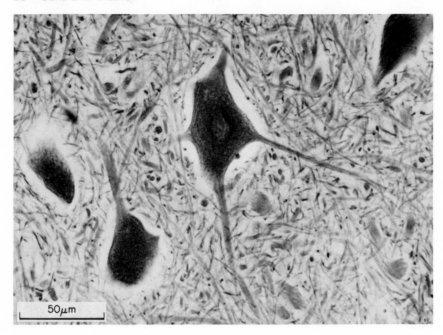

Fig. 2.26   A nerve cell in its supporting tissue.

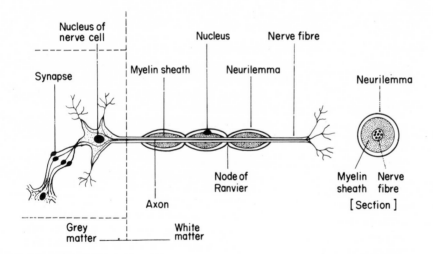

Fig. 2.27   Diagram of a neuron.

(a) The **axon** From the body of each nerve cell passes one main fibre known as an axon. It is along this fibre that the impulses from the nerve cell pass, in one direction only. An axon may be of considerable length: for example, the axons of certain cells in the lower part of the spinal cord extend in the nerves as far as the foot.

(b) **Dendrites** These small fibres extending from the body of the nerve cell receive impulses from other nerve cells. The name implies 'branching like a tree'. When impulses pass from one neuron to another they go from the axon of the first to the dendrites of the second. The point at which the axon and dendrites meet is called a **synapse**.

The appearance of tissue containing myelinated nerve fibres is light and hence it is referred to as **white matter**. The myelinated nerve fibres have a fatty sheath containing myelin, which accounts for the light appearance; those without such a sheath are called non-myelinated fibres. The myelin sheath acts as an insulation, and helps to speed up the passage of impulses along the axon.

Myelin may degenerate in the absence of vitamin B^{12}, as in the degeneration sometimes occurring in the spinal cord in pernicious anaemia, or may be destroyed by disease as in multiple sclerosis. The functional unit of the nervous system is the neuron and consists of a nerve cell, its axon and dendrites. The neuron may be compared to an electric battery. The nerve cell generates the impulse and the axon conveys it in the same way as wires lead the current from the battery.

Table 2.1

Tissues			Functions/Locations
(1) Epithelium	Squamous	Simple	Linings of organs (e.g. endothelium of heart)
		Stratified	Coverings (e.g. skin)
		Transitional	Linings of organs (e.g. bladder)
	Columnar	Glandular	Secretion (e.g. gastric juices)
			Absorption (e.g. walls of intestine)
		Ciliated	Movement of mucus (e.g. respiratory system)
(2) Connective tissue	Fibrous	White	Supporting (e.g. ligaments)
			Protecting – outer covering of organs (e.g. kidneys)
		Yellow elastic	In walls of arteries and bronchi of lungs
	Areolar		Packing and supporting (e.g. under skin)
	Adipose	Fat	Protection around organs (e.g. eyes, kidneys)
		Brown fat	Protection (e.g. thermogenesis)
	Cartilage	Hyaline	Protection (e.g. covering articular surfaces)
		Fibrocartilage	Supporting (e.g. intervertebral discs)
		Elastic	Supporting (e.g. epiglottis, pinna of ear)
	Bone	Compact	Support and protection (i.e. hard surface of bone)
		Cancellous	Provides lightness with strength (i.e. spongy bone); also contains red bone marrow
	Blood	Red cells	Carriage of oxygen and carbon dioxide
		White cells	Protection of body from infection
		Platelets	Coagulation of blood
(3) Muscle	Voluntary (striated)		Power of movement (i.e. muscles attached to skeleton)
	Involuntary (plain)		Present in internal organs (e.g. colon)
	Cardiac		Specialized muscle of heart
(4) Nerves	Nerve cells (neurons)		Generation of impulses
			Transmission of impulses
	Neuroglia		Support for nerve cells

Membranes

The term membrane is applied to any thin expansion of tissue which usually forms an enveloping or lining layer for various structures or organs. The most important types of membrane are:
- *(i)* mucous;
- *(ii)* serous; and
- *(iii)* synovial.

Mucous membranes

These are found lining the alimentary, respiratory and genito-urinary tracts. They are covered by epithelium of various types, for example stratified in the mouth, columnar in the stomach and intestines, ciliated in the respiratory tract.

A mucous membrane is usually supported by a sub-mucous layer of connective tissue. It has the following functions:
- *(a)* protection;
- *(b)* secretion of mucus which moistens its surface;
- *(c)* providing a surface for the absorption of foodstuffs in the alimentary canal.

Serous membranes

These have a surface layer of simple squamous epithelium and line the important closed cavities of the body together with the organs which lie within them. The term *parietal* layer is applied to the part of the membrane lining the cavity, whilst that which covers the organs is called the *visceral* layer.

The most important serous membranes are:

the **pericardium** covering the heart;

the **pleura** lining the thoracic cavity and covering the lungs;

the **peritoneum** lining the abdominal cavity and covering the abdominal organs.

The surfaces of the layers are smooth and moistened by serous fluid which enables the walls of the cavity and the organs contained within it to glide smoothly over each other without friction.

Synovial membranes

These are a specialized form of serous membrane which line the capsules of joints, the sheaths of tendons and the interior of bursa. They are moistened by synovial fluid and their function is to prevent friction. Serous fluid is thin and watery. Mucus is thicker and rather sticky. Synovial fluid is thicker than either of the other two. Each type of fluid is therefore adapted to the functions it has to perform.

Questions

1. What are the characteristics of cells?
2. What are enzymes?
3. What is (a) osmosis, (b) active transport, and (c) phagocytosis?
4. Describe mitosis.
5. Describe the different types of epithelium.
6. List the various types of connective tissue.
7. What are the properties of muscle?
8. What is (a) an axon, (b) a dendrite, and (c) a synapse?
9. Name the various types of tissue membrane and give an example of each.

3
Body Defences: Immunity

The body has a variety of defences against microbial invaders. Some of these immune mechanisms are *innate* and are present from birth; they are *non-specific* in that they counteract a wide variety of infective agents. In contrast, *acquired* immune mechanisms are *specific*, each being directed against a particular microorganism.

Innate immunity

There are major differences between species, and between strains within species, in susceptibility to infection. A dog, for example, stands no risk whatever of catching a cold from its master. Lesser determinants of innate immunity are individual genetic factors, age, sex, nutritional status and hormonal balance.

The intact **skin**, and particularly its outer horny layer, affords an effective *mechanical barrier* to the entry of microorganisms into the body. The **mucous membranes**, not having a horny layer, afford less protection but have mechanical mechanisms of their own. For example, the mucosa of the respiratory tract traps particles in its moisture and they are swept upwards, by the hair-like processes or cilia (performing mechanical cleansing), to reach the oropharynx, from which they are swallowed or expectorated. Unfortunately the cilia may be disabled by a variety of agents, including viruses which infect the respiratory tract. In the gastrointestinal tract, a layer of mucus traps organisms and peristaltic movement prevents stagnation and bacterial overgrowth. A *high flow rate* of secretions, as in the **urinary** and **biliary tracts**, discourages bacterial proliferation; such regions are usually sterile but may become infected if flow is impeded, by calculi or by prostatic obstruction of the bladder neck, for example. *Temperature* is also a factor in the innate immunity of an animal to some infective agents. Whilst warm-blooded animals are susceptible to tuberculosis, for example, cold-blooded animals are immune because the tubercle bacillus is temperature-dependent.

Chemical factors also protect against infection. Among these, the *fatty acids* on the surface of the skin have an antiseptic action. These are present in the secretions of the sebaceous glands and are also produced by *Proprionobacterium acnes*, which is a member of the normal microflora of the skin. The normal microflora also provides protection from pathogens in other ways. Large numbers of microorganisms live in a close harmonious relationship with humans and are known as commensals. The acidity of **sweat** (pH 5.5) also has a microbiocidal effect.

Lysozyme, an enzyme secreted by **macrophages**, is found in large quantities in human tears and helps to prevent bacterial conjunctivitis. It is found in relatively high concentration in most tissue fluids and in polymorphonuclear leukocytes, although it is not synthesized by these cells. Lysozyme disables or destroys many Gram-positive bacteria by chemically attacking their cell walls, breaking them up, a process known as lysis. The macrophages then remove damaged cells and bacteria from the tissues. Macrophages are large cells which are derived from the bone marrow but are distributed widely throughout the body and congregate at the site of an inflammatory response.

In the course of infection and inflammation, a variety of *basic polypeptides* released from damaged cells have a destructive effect on certain bacteria. The levels of a number of other 'acute phase' substances are increased in the blood in the inflammatory response but the roles of many of them are unclear.

A major defence mechanism is provided by a complex group of proteins, all of large molecular size, collectively known as *complement*. They form a series of enzymes which can be activated directly by some bacteria and indirectly, with the help of antibody, in other cases. Activated products of the complement system help in the first-line host defence against microorganisms, destroying them by lysis, and they mediate inflammatory changes. Complement can also lyse red cells (haemolysis) and indeed much of our knowledge of the complement system comes from studies of immune haemolysis.

The complement system is a triggered enzyme cascade system in which circulating inactive forms (proenzymes) are converted into active forms by their predecessors in the cascade. The complement system can be activated in two ways, known as the classical pathway and the alternate pathway. The classical pathway is typically initiated by complexes of antibody and antigen whereas the alternate pathway can be activated in the absence of antibody. Three stages are involved in the activation of the system, namely the recognition stage, the activation stage and the membrane attack stage.

The components of the complement system are known by their numbers, C1 to C9 and, of these components, C3 is the most abundant. The main sources of complement components are the intestinal epithelium, the macrophages, the liver and the spleen.

When attacked by a virus, many cells rapidly respond by producing *interferon*. This is a family of proteins which block the replication (multiplication) of viruses, both in the cells which produce the interferon and in other cells. Expressed simply, it interferes with the synthesis of new virus by the cells of the host. Interferons have two important actions:

(*a*) they alter the properties of the **cell membrane** against accessibility by viruses; and

(*b*) they activate cellular genes to produce **intracellular enzymes** which interact with double-stranded RNA to inhibit protein synthesis in the virus.

Because of their protective properties, lysozyme and interferons have been described as 'natural antibiotics.' They provide rapid protection against invaders, lysozyme against bacteria and interferon against viruses.

Phagocytes

Certain cells in the body have the power of engulfing bacteria and other foreign materials. They are known as **phagocytic cells** and are of two types, the circulating **polymorphonuclear leukocytes** of the blood and the *mononuclear* phagocytic cells known as **macrophages**. The latter are distributed throughout the body and are themselves of two kinds, namely those which circulate in the blood (monocytes) and those which are fixed in the tissues (e.g. histiocytes in the connective tissues). These actively phagocytic cells are collectively referred to as the *mononuclear phagocytic system*. The phagocytes either digest the material which they engulf or store it out of harm's way so that it can no longer act as an irritant. They quickly clear the blood of bacteria and other particulate matter.

The macrophages form an important link between the innate and acquired immune systems.

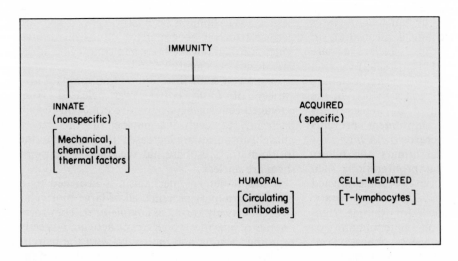

Fig. 3.1   Immunity.

They pass on to the lymphoid cells some of the antigens, or their products, but at the same time retain some of them so that the lymphoid cells are not overwhelmed.

The normal function of phagocytes depends upon their ability to *recognize*, *bind* and *ingest* microorganisms and other particles. This is, in part, non-specific but, in addition, bacteria coated with antibody and complement adhere to the phagocyte because the phagocyte cell membrane has **receptors** for a part of the antibody molecule and a component of the complement system. Antibodies can greatly enhance the phagocytosis and even the intracellular digestion of bacteria. An antibody which coats the surface of a bacterium or other particle (antigen), and makes it more susceptible to phagocytosis, is known as an opsonizing antibody or opsonin. IgM antibodies (see later) are particularly effective opsonizing agents. Clearly, the complement also acts as an opsonin in this situation.

Generally, polymorphonuclear neutrophils (see p. 118) are more effective than monocytes or macrophages in the rapid killing of bacteria. However, macrophages are more versatile than neutrophils and are essential for the elimination of various chronic infective agents.

Acquired immunity

All higher animals have recognition mechanisms that enable them to discriminate between their own cells and the cells of other individuals and other species, that is between **self** and **non-self**. The cell surface molecules or antigens which are recognized as foreign are often carbohydrates or glycoproteins. The molecules which recognize them are called **antibodies** and are proteins.

Each antibody recognizes and binds with one particular **antigen** (or a very similar one); this is known as *antibody specificity*. The cell surface antigens of an individual are under genetic control, which means they are controlled by a gene complex. These antigens are important in the rejection of tissues and organs transplanted from another individual and are known as histocompatibility antigens or transplantation antigens. Closely linked genes control the level of antibody response to antigens.

The recognition mechanisms are concerned with maintaining the integrity of the individual. If a substance is recognized by its surface antigens as being foreign, it can be destroyed or inactivated. A failure of this process is referred to as a lack of immunity or immunological deficiency. On the other hand, an attack on the individual's own cells will not be mounted, because of *immune tolerance*, unless something confounds the immune process and results in auto-immune disease. The recognition mechanisms are obviously to the biological advantage of the individual but, ironically, may not be to the medical advantage of a patient requiring transplantation of tissues or organs from another individual with different cell surface antigens.

Acquired immunity is the second line of defence utilized against microorganisms which have broken through the first line of defence, namely the individual's innate immunity. The immune response to an antigen is of two kinds, *humoral* and *cell-mediated*. Both depend upon cells of the **lymphoid system**. This is a system which filters foreign antigens from the blood stream and lymph and within which, immune responses usually occur. The lymph nodes consist of a meshwork of reticular cells in which are imbedded large numbers of lymphocytes, macrophages and plasma cells. These cells are similarly found in the spleen, which acts as a filter for the blood stream, filtering out particulate material such as bacteria and effete red blood cells.

An **antigen** is a substance capable of stimulating the immune system to produce a response specifically directed at that substance and not at unrelated substances. A substance can only be an antigen if it is foreign to the animal which is exposed to it. A substance may be antigenic in itself or may only be antigenic if attached to a carrier molecule, usually a serum protein; such a substance, incapable alone of inducing an immune response, is called a **hapten**. Immunogenicity is related to particle size, larger particles being more potent antigens than small particles.

Nutritional and endocrine factors affect the health of the lymphoid system and therefore the adequacy of the acquired immune response. Humoral immunity may be more affected than cell-mediated immunity in poorly nourished people because of the number of long-lived recirculating lymphocytes, particularly in adults.

The two types of acquired immunity, humoral and cell-mediated, are subserved by two types of lymphocyte, **B-lymphocytes** and **T-lymphocytes** respectively. They look alike and cannot be distinguished from each other under the microscope. The B-lymphocytes, which are concerned with antibody production (humoral activity), derive their name from the bursa of Fabricius which is present in birds. Human beings do not have such a bursa and, in adults, the B cells emerge from the bone marrow, which happens to begin with the same letter.

Thymus-dependent or T-lymphocytes, are so named because their development and maturation is dependent upon the thymus gland and under the influence of thymic hormones. If the thymus gland is surgically removed (thymectomy) from an animal soon after birth, T-lymphocytes are virtually eliminated.

All lymphocytes are formed from stem cells in the bone marrow. These are pluripotent, which means that, under the influence of different stimulating chemical factors, they can differentiate into the various cells of the blood, namely red cells, platelets and leukocytes, including B-lymphocytes and T-lymphocytes. The T cells migrate to the thymus, where they undergo maturation.

Some antigens can stimulate B cells directly, causing them to differentiate into **plasma cells**, which are the main antibody-producing cells, but many types of antigen cannot do so. On the other hand, T cells have the ability to be stimulated, so that they recognize an antigen, but they do not differentiate into antibody-producing cells. If recognition does occur, they induce B-lymphocytes to form antibody and are therefore called **helper T-lymphocytes**. Immune processes have to be regulated (modulated), however, and **suppressor T-lymphocytes** are also necessary; they limit the extent and duration of antibody production by B-lymphocytes and also limit the activities of other T-cells.

Humoral immunity

B cells and the plasma cells, into which they differentiate, reside mainly in the lymph nodes and spleen. **Antibody** (immunoglobulin) is synthesized predominantly in the plasma cells and eventually discharged into the blood stream. Individual cells make only one class of immunoglobulin at any one time.

An **immunoglobulin** molecule is based on a structure of two pairs of 'heavy' and 'light' polypeptide chains joined by disulphide bonds. The unit, called a domain, at one end of each chain is

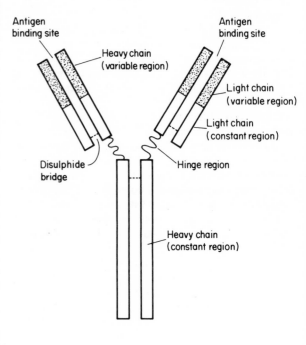

Fig. 3.2 An immunoglobulin molecule.

known as the *variable region* and this is the portion which recognizes and binds foreign substances. In other words, the antigen-binding site is located in this variable portion of the immunoglobulin molecule.

Five main classes of immunoglobulin (Ig) can be distinguished by the laboratory process known as *immuno-electrophoresis*. They are IgG, IgM, IgA, IgE and IgD. Seventy five percent of the immunoglobulin in serum is IgG, of which there are four subclasses with varying ability to bind complement. One of the functions of complement, which it is convenient to mention here, is that it attracts polymorphonuclear leukocytes to sites of antigen-antibody interaction, an important step, in the inflammatory process, known as *chemotaxis*.

IgM is a particularly efficient *agglutinating* antibody, judged by its ability to agglutinate bacteria. IgM antibodies are also particularly efficient *opsonizing* antibodies, coating particulate antigens, such as bacteria, and making them more susceptible to phagocytosis.

IgA is particularly concerned with *surface* protection, i.e. defence at mucosal surfaces. The IgA found in secretions differs from most of the serum IgA in having an attached 'secretory piece' which facilitates its transport from the lamina propria into the mucous secretions.

IgE antibodies have a strong affinity for tissues, to which they attach themselves. They are *reaginic* antibodies, which means that they are responsible for various forms of hypersensitivity reaction.

Immunoglobulin synthesis occurs in the red pulp of the spleen and in the cortex of the lymph nodes. In the early stages of an immune response, the B cells secrete IgM antibody but later they switch to the synthesis of IgG antibody. Otherwise, an immunoglobulin secreting cell synthesizes only one class of immunoglobulin. Furthermore, different families or clones of cells are genetically programmed each to make antibody of only one specificity. This is what is meant by acquired immunity being specific and contrasts with the nonspecific mechanisms of innate immunity.

Apart from antibodies which bring about the destruction of microorganisms, antibodies are formed which are able to neutralize bacterial toxins, such as the exotoxins of diphtheria, cholera and tetanus. To be effective, the neutralizing antibody must be produced at a faster rate than the toxin. This is likely to occur if the individual has been previously exposed to the organism or its toxin. An example of previous exposure would be by immunization with tetanus toxoid. An immuno-logical **memory** of the toxin enables the previously sensitized cell to mount a rapid secondary immune response. The primary immune response is that which results from first meeting an antigen, the antibody appearing in the blood about two weeks later. A subsequent dose of antigen results in a much quicker and greater rise in the level of antibody in the serum. This secondary response, which is due to a rapid mobilization of antibody-forming cells, depends upon the immune system retaining a memory of the antigen. Immunological memory is an essential component of immunity, of both the humoral and cell-based types. The lifetime persistence of immunological memory implies that the memory B and T cells are long-lived.

Macrophages

Macrophages have an early role in the immune response and are important cells in the body's resistance to infective disease. Among their functions, they

(*a*) *process antigens* and present them in a form which either stimulates lymphocytes or induces tolerance;

(*b*) later *act as phagocytic cells*, engulfing particulate antigens and immune complexes; and

(*c*) *secrete enzymes, complement components and prostaglandins*, thus generating some of the mediators of inflammatory reactions.

Despite its undisputed importance in the host resistance, it has to be remembered that the effectiveness of the macrophage is dependent upon the lymphocyte response.

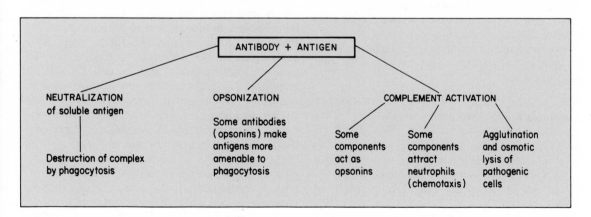

Fig. 3.3 Effects of antibody/antigen interaction.

Cell-mediated immunity

Contact with an antigen results not only in the production of circulating antibody but also in the development of a separate cell-mediated response, which activates lymphocytes. Both types of response are specific for the particular antigen.

The cell-mediated response is initiated in different areas of the spleen and lymph nodes from those concerned with antibody production. The areas concerned are those under the control of the thymus gland, containing thymus-dependent lymphocytes. Antigens stimulate the cells to undergo *lymphocyte transformation*, which can be demonstrated using plant extracts (phytohaemagglutinins) as an *in vitro* test of the competence of the cell-mediated immune system.

Lymphocytes activated by exposure to antigen produce soluble products called **lymphokines**. These *recruit* lymphoctyes which have not previously been involved (uncommitted lymphocytes) and encourage the *retention* of these cells and of phagocytes at the site of inflammation. The other main function of the lymphokines is to *activate* the retained cells so that they can participate in the inflammatory response. Lymphokines act as chemical messengers, allowing communication between cells and acting on lymphocytes, macrophages, polymorphonuclear leukocytes and other non-lymphoid cells.

The cell-mediated immune system is under the influence of the thymus and the cells involved are known as thymus-dependent lymphocytes or T-cells. Certain rare immune deficiency states are found in children born without a thymus gland. These children lack lymphocytes in those areas of spleen and lymph nodes which are under thymic control and their response to viral infections is defective.

The precursors of T-cells undergo maturation in the thymus, where they are 'educated' in two important ways.

(*1*) They learn to recognize the 'self' markers or **histocompatibility antigens** which they will encounter in the tissues. These 'self' markers, which are at the cell surface, are the antigens which cause transplant rejection and they are also known as HLA or transplantation antigens. It is the thymus which 'teaches' the T-cells what to recognize as 'self.'

The genes controlling the histocompatibility antigens are known collectively as the major histocompatibility complex (MHC). The immune response (Ir) genes, which control the ability of immunocompetent cells to respond to antigen, are very closely related to the major histocompatibility complex. The MHC genes are in some way involved in cell co-operation, such as that between T and B lymphocytes.

(*2*) They acquire the ability to kill cells which bear foreign (non-self) antigens. In doing this they become **cytotoxic T-lymphocytes**. They can 'see' the foreign antigens on cells infected with viruses or fungi and on transplanted cells and proceed to kill them. Because they have been programmed to recognize 'self', they do not attack the body's own normal cells. Under certain circumstances, however, they do attack 'self' and this results in one type of auto-immune disease. Such cases include those where damage to the individual's cells has resulted in a breach of the cell membrane and exposure of constituents which are normally 'hidden.' The immune system may not recognize these as 'self' and may mount an attack against them. It may also attack cells if they bear an antigen which is similar to that of a foreign antigen which the individual has encountered.

The lymphoid immune system can respond to a wide range of antigens which are the markers of foreign molecules. The immune response

(a) *recruits* cells wherever they are needed;
(b) *stimulates* the body's non-specific defence mechanism of inflammation; and
(c) *creates a reserve of 'memory cells'* (T-lymphocytes).

There is a *recirculation* of lymphocytes between blood and lymphoid tissues. This enables many lymphocytes to have access to an antigen, so that there is a good opportunity for a lymphocyte, carrying an antibody receptor for a particular antigen, to make contact and initiate an immune response. The recirculating pool also provides the means for *recruitment* of lymphocytes by lymphoid tissues and the *replenishment* with lymphocytes when, for example, the spleen has been depleted of these cells by infection, trauma or X-rays.

Certain aspects of immunity have been given particular prominence in this simplified account. It must be remembered, however, that the immune response is a highly complex process involving finely balanced relationships between many cells of different types.

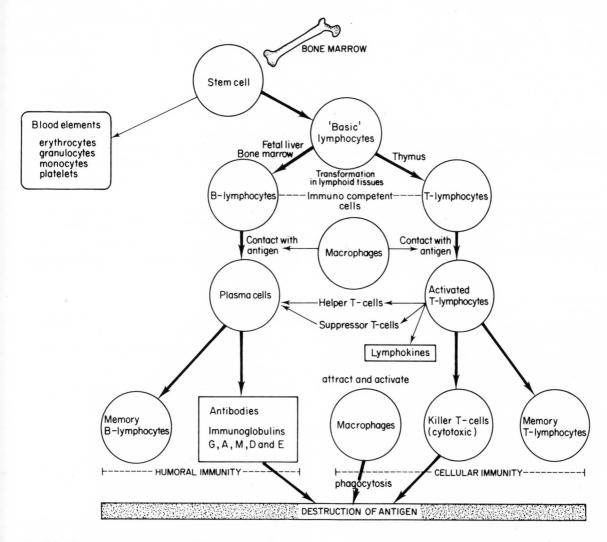

Fig. 3.4 The immune system: a simplified schematic diagram.

Questions

1. Explain what is meant by the term innate immunity? Give examples.
2. What are phagocytes? What are the two main varieties called and where in the body are they found?
3. What are (a) antigens (b) antibodies? Write brief notes on humoral immunity.
4. What are the functions of macrophages?
5. Describe some of the features of cell-mediated immunity.

4

The Body as a Whole

All the various cells and tissues which have been described form the basis of the separate systems of the body. A system may be defined as a group of structures or organs which carry out an essential fundamental function of the individual. Although, to some extent, each system works and can be considered on its own, the functions of the various systems are very closely connected and are dependent on each other. For example, the bones, joints, ligaments and muscles are all concerned with the function of movement which in turn is controlled by the activity of the nervous system. The vitality of the nervous system is dependent on an adequate circulation of blood and a supply of oxygen which enters the blood via the respiratory system.

It is essential to be familiar with certain **terms used in anatomical description**. The body is considered in the upright or erect position with the palms of the hands facing to the front and the toes pointing forwards. The following terms are then applied.

anterior (ventral)	towards the front of the body or limbs
posterior (dorsal)	towards the back of the body or limbs
median	in the middle
medial	the side nearest the mid-line of the body
lateral	the side farthest away from the mid-line
superior	above any point referred to
inferior	below any point referred to
plantar	belonging to the sole of the foot
palmar	belonging to the palm of the hand

The terms *internal* and *external* should only be used to describe the relationship to the inner and outer surfaces of the body and not as alternatives to *medial* and *lateral*.

In describing the limbs, the upper part nearest the trunk is called the *proximal* part; the lower portion farthest away from the trunk is called the *distal* part. The proximal part is the more central and the distal part is the more peripheral.

The terms for common movements are:

flexion	a bending at a joint so that two connected parts are approximated together
extension	a straightening out from a position of flexion
abduction	a drawing away from the median axis of the body
adduction	a bringing towards the median line of the body

The body as a whole is built up around the bony framework of the skeleton and consists of three main parts:

(1) the head and neck
(2) the trunk (a) the chest or thorax
(b) the abdomen and pelvis
(3) the limbs (a) the upper limbs
(b) the lower limbs.

The head is separated into two parts, the cranium or brain case and the face. The trunk is divided into the thorax or chest and the abdomen or belly. There are two pairs of limbs, the upper and lower, each divided into parts which roughly correspond.

Upper limbs: upper arm, elbow, forearm, wrist, hand, fingers
Lower limbs: thigh, knee, leg, ankle, foot, toes

The fingers and toes are referred to as digits, the bony parts of which, separated by joints, are called phalanges.

Covering the bones and giving rise to the general contour of the body are numerous muscles. Blood

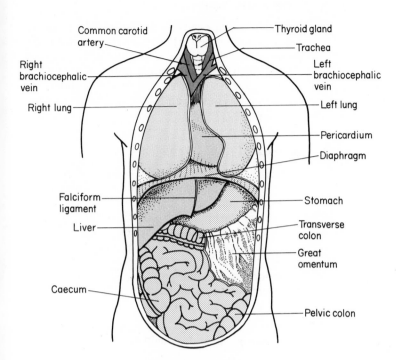

Fig. 4.1 The viscera of the body from the front.

vessels and nerves traverse all parts of the human frame.

The chief organs of the body are contained in special cavities:

(*1*) the brain in the cavity of the skull (cranium);

(*2*) the lungs and heart in the cavity of the thorax;

(*3*) the digestive organs in the abdominal cavity.

It is of importance to know the boundaries of these cavities and the structures which they contain. All those mentioned will be referred to later, but are included now for purposes of classification.

The skull

This consists of the cranium which protects the brain and the eyes, and the mandible or lower jaw which is hinged to it. The movements of the mandible are essential for the chewing of food and for the production of speech.

The thorax

This important cavity is situated in the upper part of the trunk and its walls consist of a bony framework supporting various muscles.

Boundaries of the thorax (Fig. 4.2)

Anterior: the sternum, costal cartilages and front ends of the ribs

Posterior: the thoracic or dorsal part of the vertebral column or backbone, consisting of twelve individual vertebrae and the intervertebral discs of cartilage between them

Lateral: the twelve ribs and the intercostal muscles

Superior: the root of the neck with its muscles and blood vessels

Inferior: the diaphragm – a large dome-shaped muscular structure, separating the cavity of the thorax from that of the abdomen, through which pass the oesophagus, aorta and inferior vena cava

Contents of the thorax

(*a*) The **lungs** occupy the greater part of the thoracic cavity except for a central portion behind the sternum which extends backwards to the vertebral column and is known as the mediastinum.

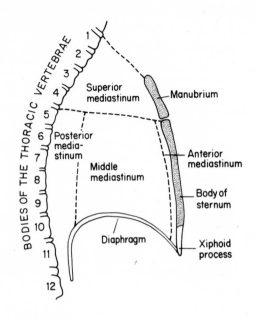

BODIES OF THE THORACIC VERTEBRAE

1
2
3
4
5
6
7
8
9
10
11
12

Superior mediastinum

Manubrium

Posterior media-stinum

Middle mediastinum

Anterior mediastinum

Body of sternum

Diaphragm

Xiphoid process

Fig. 4.2 Section of the thorax, showing the divisions of the mediastinum.

(*b*) The **heart** is situated in the mediastinum in the central part of the thoracic cavity and is enclosed in a fibrous bag or sac known as the pericardium.

(*c*) The **trachea** or windpipe enters the thorax through its upper opening from the neck and passes down in the posterior part of the mediastinum until it divides into the two main bronchi at the level of the fourth thoracic (dorsal) vertebra. A bronchus passes to each lung.

(*d*) The **oesophagus** or gullet also enters the superior opening of the thorax from the neck where it commences as a continuation of the pharynx. The oesophagus lies just in front of and to the left of the vertebral column and behind the trachea. It leaves the thorax by passing through an opening in the diaphragm and enters the abdominal cavity where it immediately joins the stomach.

Other structures in the thorax include the thymus gland, the phrenic and vagus nerves, the aorta, some important veins (the superior vena cava and inferior vena cava), the thoracic duct and lymphatic glands.

The abdomen

The abdominal cavity is the largest cavity in the body. It is described in two parts, the abdomen proper and the pelvic cavity. The latter is directly continuous with the rest of the abdomen but is situated in the space bounded by the sacrum behind, the ischium on each side and the pubic bones in front (see Fig. 5.12).

Boundaries of the abdominal cavity

Anterior: the muscles of the abdominal wall (the rectus, internal and external oblique and transversus on each side)

Posterior: the lumbar part of the vertebral column in the mid-line and the psoas, quadratus lumborum and iliacus muscles on either side

Superior: the diaphragm

Inferior: the abdominal cavity is continuous with the superior opening of the pelvic cavity

Contents of the abdominal cavity

(*a*) The stomach and intestines.
(*b*) The liver, gall-bladder and spleen.
(*c*) The pancreas, kidneys and suprarenal glands. The abdominal aorta and inferior vena cava. All of these structures lie outside the peritoneum and are situated on the posterior wall of the cavity.

Boundaries of the pelvic cavity

Anterior: the pubic bones and symphysis pubis
Posterior: the sacrum
Lateral: the ischium on each side
Superior: the pelvis is continuous with the abdominal cavity
Inferior: the muscles of the pelvic floor (levator ani, etc.). Through the pelvic floor pass the lower part of the bowel (rectum) in the hollow formed by the sacrum behind, the urethra and, in the female, the vagina towards the front.

Contents of the pelvic cavity

(*a*) The lower part of the large intestine (the sigmoid or pelvic colon and the rectum).
(*b*) The bladder.
(*c*) The female organs of reproduction (uterus,

ovaries and Fallopian tubes).

(*d*) Some loops of small intestine may also be present in the pelvis.

Systems of the body

As a matter of convenience the various systems of the body are generally described separately but it must be remembered that functionally they are all closely related and interwoven. The major systems are:

> The locomotor system (bones, muscles and joints)
> The nervous system (central and peripheral, somatic and autonomic)
> The skin and organs of special sense (sight, hearing, smell and taste)
> The cardiovascular system (heart, blood vessels and lymphatics)
> The blood (the haemopoietic system)
> The respiratory system
> The digestive system
> The endocrine system (ductless glands)
> The urinary system
> The reproductive system

Anatomically there is a close relationship between the urinary and reproductive systems. Hence they are frequently described together under the heading 'The Urogenital System'.

The nervous system receives and utilizes information from the sense organs and controls voluntary and involuntary (reflex) movements of muscles. The higher centres regulate thought, memory and social behaviour. The cardiovascular or circulatory system enables nutrients to be conveyed to and waste products removed from all parts of the body by means of the blood. The respiratory system provides for the intake of oxygen and discharge of carbon dioxide. The digestive system provides for the intake and breakdown of food into relatively simple chemical compounds suitable for absorption. The ductless glands influence metabolism, growth and reproduction. The urinary sytem removes waste products circulating in the blood and helps to keep its chemical and physical properties constant.

It should be noted that there is a variety of glands in various parts of the body. Broadly speaking there are exocrine glands which pour their secretions into the alimentary canal and the endocrine glands (ductless glands) whose secretions enter the blood stream. Lymph nodes, sometimes called lymph glands, are not really glands since they do not secrete hormones or enzymes but they produce lymphocytes and help to protect the body from the spread of infection. The ducts or lymphatic channels which connect these structures and ultimately enter the bloodstream convey tissue fluids which surround the various tissue cells and act as 'middle man' conveying nourishment to and removing waste products from them.

Questions

1. What is meant by the terms medial, lateral, median and inferior?
2. Describe the boundaries of the thoracic cavity.
3. List the contents of the thorax.
4. Describe the boundaries and contents of the abdominal cavity.
5. What do you understand by the term system? List the major systems of the body.

5
The Skeleton

The internal framework of the human body is its bony skeleton, which gives *support* and *protection* and provides levers for *locomotion* and other movements, such as those concerned with obtaining and eating food. There are animals which possess no backbone and even no rigid skeleton, either internal or external, but they are adapted to their own environment and humans to theirs.

The human being is the only fully upright vertebrate, the term applied to an animal with a backbone. Being upright, and needing to move from place to place or from one position to another, humans have special problems in relation to gravity if their antigravity mechanisms fail, as may happen in old age, for example.

The maintenance of posture and the performance of movements, simple or complicated, necessitates the activity of muscles and of the nervous system, which initiates, controls and co-ordinates movement. Often the movement has to be co-ordinated with the organs of special sense, especially the eyes.

Mobility is provided by joints between bones which, especially in the limbs, act as levers. The power to move the bones resides in the muscles which are under the direction and control of the central nervous system. The brain wills the movement (volition), and the message is carried to the appropriate muscles by motor nerve fibres, initially by upper motor neurones in the pyramidal tracts and then, via relay stations in the brain stem and spinal cord, by lower motor neurones in the cranial and spinal nerves. The active muscle (agonist) contracts whilst the antagonist relaxes. For example, in order to flex the arm and convey food to the mouth, the biceps contracts and the triceps relaxes and elongates just the right amount to permit the desired action. Movement is modulated by the extrapyramidal nuclei of the brain and co-ordinated by the cerebellum.

A distinction is made between bone and bones. **Bones** are organs, each of which is adapted for its own particular function. **Bone** is the tissue (p. 35) from which they are made. Bones and muscles are derived from the mesoderm, the middle layer of the embryo. Most bones pass through two stages, a first blastemal stage of condensation of tissue (mesenchyme) and a second, cartilaginous stage. Some bones such as those in the roof and sides of the skull, never go through a cartilaginous stage, calcification occurring directly in the fibro-cellular membrane which precedes the bone. Most bones, however, are modelled in cartilage before being mineralized; thus ossification is either intramembranous or, more commonly, intracartilaginous.

Bone is an active living tissue (a specialized type of connective tissue) with several major roles in the body. It is the substance from which bones (the organs) are made and is the body's reservoir of calcium. Apart from forming salts which strengthen bone, calcium is essential for normal cardiac and skeletal muscle contraction, nerve function and blood coagulation. Bone, the tissue, consists of a matrix of collagen fibres impregnated with mineral salts, mainly phosphates of calcium in the form of crystals of hydroxyapatites. Because bone has an extensive network of blood vessels, calcium salts are as easily removed from bone as they are deposited in it.

Bone is constantly being resorbed and reformed and there is a corresponding turnover of calcium which is greatest in infancy. The cells in bone which affect its formation and resorption are the osteoblasts, the osteocytes and the osteoclasts. **Osteoblasts** secrete the collagen which forms the matrix and are thus bone-forming cells. When they have surrounded themselves with bone (calcified matrix) they become **osteocytes**. These cells connect with each other and with osteoblasts deep in the bone by means of long protoplasmic processes which extend along the channels (canaliculi) which ramify through the bone. **Osteoclasts** erode bone, apparently by a phagocytic action (literally eating and digesting bone), and so bring about its resorption.

All of these cells are affected by the hormones which regulate bone structure and calcium metabolism. These hormones are 1,25-dihydroxycholecalciferol formed from Vitamin D, parathyroid hormone, calcitonin (secreted by the thyroid gland), glucocorticoids and growth hormone. 1,25-dihydroxycholecalciferol, also called 1,25-dihydroxy-D3, and parathyroid hormone both mobilize calcium from bone and increase its concentration in the serum. Calcitonin inhibits bone resorption and thereby tends to reduce serum calcium levels.

1,25-dihydroxycholecalciferol acts not only on bone but also on the small intestine, where it increases the absorption of calcium. Vitamin D deficiency therefore causes a state of calcium deficiency in which new bone fails to calcify and remains soft, resulting in rickets in children or osteomalacia in adults.

In cases of lead poisoning much of the lead is deposited in bones. Strontium-90, a product of atomic fission, is also deposited in bones. This is radioactive, and continues to be so for a long time after it has been absorbed.

A mature bone is composed of two kinds of bone tissue. **Compact bone** (Fig. 2.21) is the hard dense ivory-like bone which forms the shafts of long bones and the surface layers of flat bones. It is built up of units which are called osteons or Haversian systems. These are cylinders of bone, through the middle of which a minute circular canal, the Haversian canal, runs longitudinally, parallel with the surface of the bone. Blood vessels and lymphatics run in the Haversian canals and nourish the bone substance. In the bone tissue surrounding the Haversian canal are a number of small spaces, called lacunae, arranged in concentric rings, which contain the bone cells. The lacunae communicate with one another and with the central Haversian canal by the minute canals (canaliculi) in which the protoplasmic processes of the bone cells lie.

Trabecular (cancellous or spongy) bone has a microscopic structure similar to that of compact bone but, instead of being dense, it appears spongy, having more and larger spaces and less solid matter. Cancellous bone makes bones lighter without loss of strength. Compact bone is always arranged outside cancellous bone and forms the surfaces of bones. Cancellous bone is trabecular in its internal arrangement, with a criss-crossed pattern of small bars and beams, struts and trusses, determined by mechanical stresses and adding to the strength of the bone. A soft pulpy tissue, **bone marrow**, is found in the cylindrical cavities of long bones and in the spaces between the trabeculae and in the larger Haversian canals of all bones. In early life it is all red marrow (blood-forming) but after about the fifth year it is gradually replaced by yellow marrow (mostly fat cells) until, by the age of 20 to 25 years, red marrow remains only in the vertebrae, sternum, ribs, clavicles, scapulae, cranial bones and proximal ends of the femora and humeri.

Bone surfaces are covered with a tough sheet of fibrous tissue which is known as the **periosteum**. This clothes a bone all over except where the bone is covered by hyaline cartilage for the formation of a joint.

Bone receives its blood supply from two sources; the surface of the bone from the periosteum; the interior from an artery which enters the shaft, generally about its middle, through a canal known as the nutrient foramen.

In summary, bone has several important functions.

(*a*) It is the support tissue for the framework of the body.

(*b*) It is a mineral store, containing 97% of the total bodily content of calcium salts.

(*c*) It houses the blood-forming bone marrow.

Development of bone

In very early fetal life no actual bone is apparent, but the bones of the human skeleton are outlined by other forms of connective tissue, either cartilage or fibrous tissue (membrane). It will be seen later that the majority of bones in the skeleton are described as long bones and these are preceded by cartilage in the embryo; while certain flat bones, such as those of the skull, are developed in membrane.

First of all, therefore, the bones are represented by rods or blocks of cartilage, or sheets of membrane, which resemble in shape the mature bone. The next stage in development is the deposition of calcium salts in the cartilage or membrane (calcification). Later, bone cells or osteoblasts enter the calcified cartilage together with other cells which remove the cartilage. The bone cells proceed to lay down bone in the place of the cartilage which is absorbed. The area in which this process commences is called a **centre of ossification.**

At birth only a proportion of the skeleton is represented by actual bone. The remainder consists of ossifying cartilage, or membrane.

Growth of bones

Most bones go on growing in size for at least twenty years.

Centres of ossification, i.e. areas in which bone cells are replacing cartilage, are necessary for growth in length of a bone. In most long bones there are two such centres, one at each end of the bone. An additional centre exists for the shaft of the bone.

The main part or shaft of the bone is called the **diaphysis**. The portion at each end containing a centre of ossification is called the **epiphysis**. Each epiphysis is joined to the diaphysis by a layer of cartilage known as the epiphyseal cartilage. This is gradually replaced by bone, and by adult life has disappeared and the epiphysis and diaphysis are completely fused into a single bony structure, but so long as some epiphyseal cartilage remains, growth in the length of the bone is possible.

So far reference has only been made to growth in the length of a bone; increase in circumference takes place by new bone being laid down under the periosteum by osteoblasts situated in this position.

New bone may be formed from time to time in order to repair fractures, without any increase in size of the bone as a whole.

The three main factors which influence the normal growth of bone in children are:

(a) sex hormones (oestrogens and androgens);

(b) pituitary growth hormone;

(c) thyroid hormone.

Any defect in these hormones may cause stunting of growth.

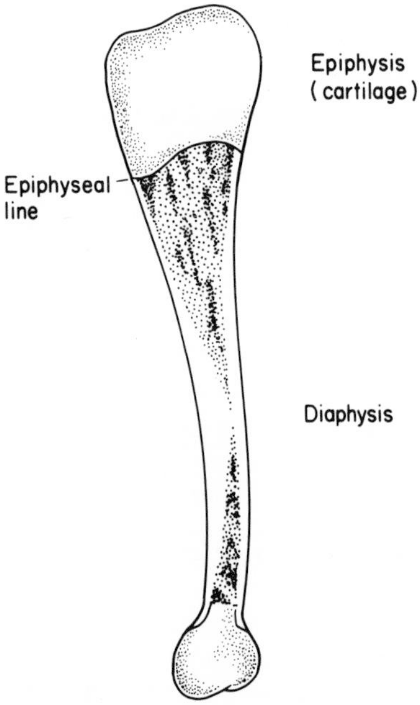

Fig. 5.1   Section of young bone before ossification of the epiphyses has started.

Practical considerations

In children, the bones do not contain quite as much calcium salts as in the adult, so that the bones are more elastic and instead of complete fractures resulting from injury, a partial fracture known as the greenstick variety sometimes occurs.

Injuries to the ends of bones in children often result in damage to the epiphysis, which may become displaced rather than the bony substance being broken.

Deficiency in the amount of calcium salts in rickets results in the bones becoming abnormally soft, so that bending with a corresponding deformity of the limbs may take place. Changes in the epiphysis are also apparent.

When a fracture of bone occurs, Nature attempts to repair it and the repair is most efficient when the surgeon can place the broken fragments in their natural position. The following process then occurs; the space between the broken ends is filled with blood clot, in which fibrous tissue develops; osteoblasts then migrate from the damaged ends of the bones and from the periosteum and gradually fill up the gap with new bone, known as callus; eventually the continuity of the bone is restored and the callus becomes hard and firm by the deposit of calcium salts in the damaged area. In a healthy individual the repair is usually complete in one to three months.

The skeleton as a whole

The skeleton is the framework of the body consisting of bones and, strictly speaking, the cartilages and ligaments which bind them together. It serves to support the soft structures which are grouped around it and to protect the organs of the body. It is so jointed that various parts move on each other and many bones act as levers for the muscles which are attached to them.

The *main functions of the skeleton* are:

(*1*) to act as a framework and to support the soft tissues;
(*2*) to enable free movement by the action of muscles (i.e. to combine stability with mobility);
(*3*) to protect delicate organs and structures;
(*4*) to form a store of calcium; and
(*5*) to provide for formation of blood cells in the bone marrow (p. 116).

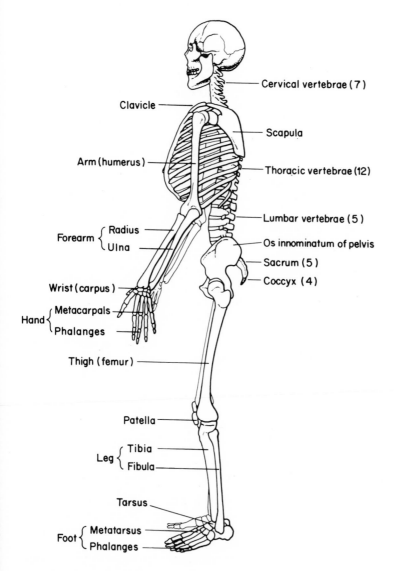

Cervical vertebrae (7)

Clavicle

Scapula

Arm (humerus)

Thoracic vertebrae (12)

Lumbar vertebrae (5)

Forearm { Radius / Ulna

Os innominatum of pelvis

Sacrum (5)

Coccyx (4)

Wrist (carpus)

Hand { Metacarpals / Phalanges

Thigh (femur)

Patella

Leg { Tibia / Fibula

Tarsus

Foot { Metatarsus / Phalanges

Fig. 5.2 The human skeleton.

Types of bone

The bones of the skeleton are classified according to their shape into long, short, flat and irregular bones.

(1) *Long bones* are found in the limbs and consist of an elongated shaft with two extremities. The bones of the arm, forearm, thigh and leg are typical examples. The shaft consists of a cylinder of compact bone containing yellow bone marrow. The extremities are formed by a thin outer shell of compact tissue with an interior network of spongy or cancellous bone containing red marrow.

(2) *Short bones* have no shaft but consist of smaller masses of spongy bone surrounded by a shell of compact bone. They are roughly box-like in shape. Examples are found in the small bones of the wrist (carpus) and ankle (tarsus).

(3) A *flat bone* consists of two layers of compact bone between which is a layer of cancellous bone. Examples are the scapula, innominate bone and bones of the skull.

(4) *Irregular bones* cannot be placed strictly in any of the previous categories and include the vertebrae and most of the bones of the face.

Sesamoid bones are small bones which are developed in the tendons around certain joints. The patella, in the front of the knee joint, is the largest and most important.

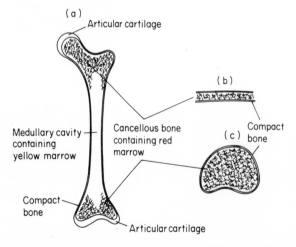

Fig. 5.3 Diagram illustrating the structure of **(a)** a long bone (femur), **(b)** a flat bone (skull) and **(c)** an irregular bone (body of vertebra).

Descriptive terms

The following terms are used in the description of bones.

Articulation A joint between two bones, e.g. the humerus articulates with the scapula and the femur with the tibia.

Border The edge separating two surfaces of a bone, e.g. the vertebral and axillary borders of the scapula separating the anterior from the posterior surfaces.

Condyle A rounded knuckle-shaped articular surface at the end of a bone, usually covered by cartilage, e.g. the condyles of the femur.

Epicondyle A small projection adjacent to a condyle and usually giving attachment to ligaments.

Crest An elevated ridge on a bone, e.g. the crest of the ilium and the crest of the tibia.

Facet A small articulating surface.

Foramen An opening or hole perforating a bone, e.g. the obturator foramen in the innominate bone and the nutrient foramen present in all bones.

Fossa A hollowed-out area or depression in the surface of a bone, e.g. the olecranon fossa and coronoid fossa of the humerus.

Process A projection from a bone, e.g. the spinous process of a vertebra.

Tubercle, tuberosity, trochanter All terms used to describe various types of process. A tubercle is a small rounded prominence, e.g. the tubercle of the tibia. A tuberosity is a larger protuberance of bone, e.g. the tuberosities of the humerus. A trochanter is a large, round eminence, e.g. the trochanter of the femur. The expanded proximal end of a long bone is often referred to as its head, e.g. the head of the femur, which articulates with the bony pelvis.

The human skeleton

The human skeleton, which contains approximately 200 individual bones, is made up of the following parts.

The skull, viz. the bones of the cranium, face and lower jaw.
The bones of the trunk, viz. the spinal column, ribs and sternum.
The bones of the limbs together with the shoulder and pelvic girdles.

The spinal column consists of 33 vertebrae and is divided for purposes of description into cervical (7 vertebrae), thoracic or dorsal (12 vertebrae), lumbar (5 vertebrae), sacral (5 vertebrae) and coccygeal (4 vertebrae) from above downwards. The five sacral vertebrae are fused together to form a single bone, the sacrum; the coccygeal are similarly joined to form the coccyx.

The ribs are twelve in number; they articulate behind with the thoracic vertebrae and, in front, the upper seven articulate with the sternum or breastbone.

The bones of the upper limb, which is attached to the shoulder girdle (clavicle and scapula), are those of the arm (humerus), forearm (radius and ulna), wrist (carpus), hand (metacarpals), and digits or fingers (phalanges).

Included in the lower limb, which is attached to the pelvic girdle, are the thigh bone (femur), the leg bones (tibia and fibula), the ankle bones (tarsus) and those of the foot (metatarsals) and toes (phalanges).

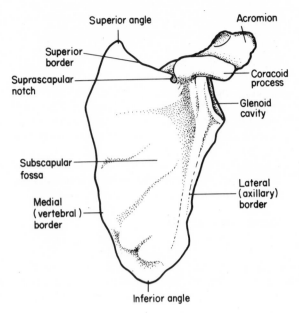

Fig. 5.4 Anterior surface of left scapula.

Bones of the shoulder girdle and upper limb

The scapula

The scapula or shoulder-blade is a large flat bone which forms a part of the shoulder-girdle and contributes to the wide range of movement of the upper limb, gliding, as it does, across the posterior aspect of the thorax. Parts of its roughly triangular outline can be seen or felt beneath the skin in the living subject.

The processes and thickened parts of the scapula contain cancellous bone and the remainder is a thin layer of compact bone.

The anterior (costal) surface of the scapula (Fig. 5.4) is slightly hollowed out in conformity with the ribs which it overlies.

The dorsal surface (Fig. 5.5) is slightly convex and is divided into two unequal parts by a large ridge known as the **spine** of the scapula. The smaller upper part is the **supraspinous fossa** (supra, L. above) and the larger area below is the **infraspinous fossa**. These fossae give attachments to the supraspinatus and infraspinatus muscle respectively.

A broad process, known as the **acromion**, projects forwards almost at right angles from the lateral end of the spine of the scapula. There is a

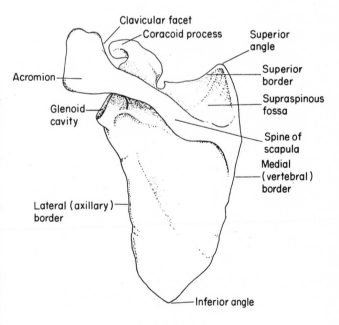

Fig. 5.5 Posterior surface of left scapula.

small facet on its medial border for articulation with the collar bone, or clavicle. The upper part of the spine gives attachment to the trapezius muscle of the neck and the lower portion to a part of the deltoid muscle, which abducts the arm.

The large irregular mass of bone which projects forwards from the outer end of the upper (superior) border of the scapula is the **coracoid process**. Attached to the coracoid process is the coracoclavicular ligament (which binds the clavicle to the process), the short head of biceps and pectoralis minor. These latter two muscles are concerned respectively with flexing the forearm on the upper arm and drawing the scapula forwards around the chest wall.

A small notch in the superior border of the scapula, the suprascapular notch, transmits the suprascapular nerve to the supraspinous fossa.

The head of the humerus articulates with the scapula at the glenoid cavity to form the shoulder joint. The glenoid cavity is a smooth, shallow pear-shaped cavity at the lateral angle of the scapula. The long head of the biceps muscle is attached just above and triceps just below the glenoid cavity. These two muscles act together in bending and straightening the elbow, the one contracting and, at the same time, the other lengthening an equal amount.

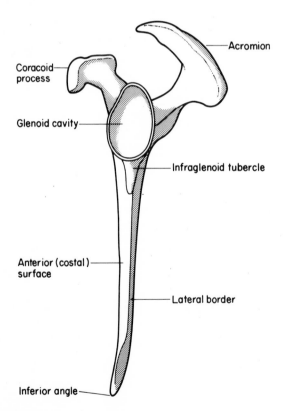

Fig. 5.6   Lateral aspect of the left scapula.

The scapula is so attached by muscles that it can glide across the surface of the thorax. If the shoulders are shrugged backwards and forwards a degree of transverse movement will be felt, and if the arms are raised above the head the inferior angle will be observed to move outwards towards the lower part of the axilla.

The scapula, being covered by muscles and lying flat against the chest wall, is not often exposed to injury and only a direct blow of considerable severity is likely to produce fracture.

The clavicle

The clavicle or collar bone is a long bone without a marrow cavity. It consists of cancellous bone within and a layer of compact bone externally. It is situated subcutaneously (just under the skin) at the root of the neck. Its lateral end articulates with the acromion of the scapula and its medial end with the manubrium sterni. It acts as a weight-bearing strut or brace for the shoulder and allows the arm to swing clear of the trunk. It also transmits some of the weight of the limb to the axial skeleton. The clavicle has a gently curved **shaft**, the medial two-thirds bowing forwards and the lateral third curving backwards. The bone is weakest at the junction of the two curves and breaks at this point when the collar bone is broken. Such a fracture is common but rarely results from direct trauma despite the exposed superficial position of the clavicle. The fracture is usually the result of indirect injury such as a fall on the outstretched hand or on the shoulder, the force being transmitted to the clavicle and snapping this slender bone. Deformity results from the lateral fragment of the clavicle being drawn downwards by the weight of the arm acting on it through the coracoclavicular ligament (see p. 74).

The humerus

The humerus (Fig. 5.8) is the long bone of the arm. It consists of a shaft and expanded upper and lower extremities. Medially, the upper extremity bears a rounded **head**, covered with hyaline articular cartilage, which forms a ball and socket joint (the shoulder joint) with the glenoid cavity of the scapula. The slight groove which surrounds the head of the humerus is called the **anatomical neck** of the humerus. This gives partial attachment to the capsular ligament of the shoulder joint. Laterally

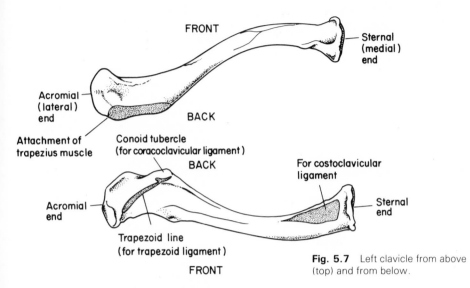

Fig. 5.7 Left clavicle from above (top) and from below.

the upper extremity of the humerus bears a prominence called the **greater tuberosity** and in front of this is the **lesser tuberosity**. A deep groove lying between the tuberosities and extending down the anterior surface of the upper shaft is known as the intertubercular sulcus. It is about 8 cm long and in it lies the long tendon of the biceps muscle. The tapering region where the upper extremity of the humerus joins the shaft is called the **surgical neck** of the humerus because it is a common site for fractures to occur. The axillary nerve and posterior humeral circumflex artery are close to it medially.

Nearly halfway down the outer side of the shaft is a roughened elevated area for attachment of the deltoid muscle and known as the **deltoid tuberosity**. On the posterior surface, passing from above downwards and outwards and having its lower end just below the deltoid tubercle, is a shallow groove, the **spiral groove** for the radial nerve.

The upper half of the shaft is cylindrical but the lower half is flattened. Protruberances from the medial and lateral borders of the lower extremity of the bone are known respectively as the **medial and lateral epicondyles**. The ulnar nerve passes on its way into the forearm in a shallow sulcus behind the medial epicondyle and a blow at this site, jarring the nerve against the epicondyle, gives rise to the expression 'funny bone'.

Between the epicondyles is an irregularly shaped area, a modified condyle, which has a smooth articular part shaped for articulation with the radius and ulna at the hinge of the elbow joint. The medial

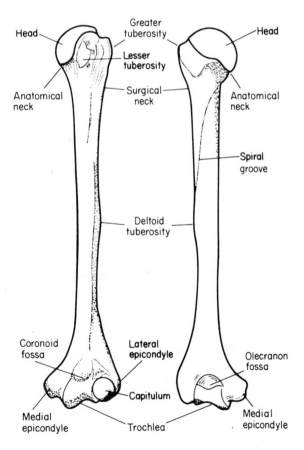

Fig. 5.8 Anterior and posterior aspects of the left humerus.

portion, a pulley-shaped surface called the **trochlea**, articulates with the trochlear notch of the ulna. The lateral portion, a rounded convex projection called the capitulum, articulates with the head of the radius.

Immediately above the articular surface deep depressions are present both on the anterior and posterior surfaces of the bone. On the anterior surface is the **coronoid fossa** into which the coronoid process of the ulna fits when the elbow is flexed. On the posterior surface is the larger **olecranon fossa** for the olecranon process of the ulna when the elbow is extended.

When the arm is at rest by the side of the body the head of the humerus is directed backwards and medially towards the glenoid cavity, which faces forwards and laterally. This rotated position of the humerus, with the posterior surface of its shaft facing laterally and backwards, has to be borne in mind when considering movements of the arm and forearm.

The humerus may be fractured at almost any level but is commonly fractured at one of three sites: the surgical neck, sometimes with damage to the axillary nerve; at about mid-shaft just below the attachment of the deltoid muscle, sometimes with damage to the radial nerve in its groove; and at the lower end, when the fracture may extend into the elbow joint or one of the epicondyles may be separated. A fractured medial epicondyle may damage the ulnar nerve. The ossifying medial epicondyle is relatively late in fusing (at about the twentieth year) with the shaft of the humerus and in X-ray films of young people it may therefore look fractured when it is not.

Bones of the forearm

The forearm contains two long bones, the **radius** on the lateral (outer) side and the **ulna** on the medial (inner) side. When the arm is placed in the position for anatomical description, namely with the palm of the hand facing forwards (p. 30), the radius and ulna are parallel, and the forearm and hand are said to be in the position of *supination*. The forearm can, however, be rotated so that the back of the hand is directed forwards. This position is called *pronation*, and the radius then lies partly across the ulna.

The radius

The radius is the lateral bone of the forearm. It is a long bone with a shaft and expanded extremities, the lower of which is much the wider of the two.

The smaller upper end of the radius has a circular **head** which articulates with the capitulum of the humerus and with the radial notch of the ulna, a **neck** and, medially, a **tuberosity** for the insertion of the tendon of the biceps muscle. The articular circumference of the head of the radius, is bound to the ulna by an annular ligament, within which it rotates during the movements of pronation and supination of the forearm.

The **shaft** has a sharp medial border facing the ulna to which it is attached by the interosseous membrane which stretches between the radius and ulna, dividing the forearm into anterior and posterior compartments. The former contains the flexor muscles of the wrists and fingers, the latter the extensor muscles.

The lower extremity of the radius presents an articular surface for the scaphoid and lunate bones of the wrist. At the lateral end of this surface is a downward projection of bone called the **styloid process** which can be felt through the skin on the lateral aspect of the wrist just above the base of the thumb. On the medial aspect of the extremity there is an articular surface for articulation with the ulna.

The ulna

The ulna is the medial bone of the forearm and is slightly longer than the radius. Like the radius, it is a long bone, with a shaft and upper and lower extremities, but whereas the lower end of the radius is the larger end, it is the upper end of the ulna which is expanded. This extremity is strong and hook-like, with a large C-shaped cavity, the **trochlear notch**, which articulates with the trochlear surface of the humerus. The large mass of bone forming the posterior wall of the trochlear notch and overhanging its upper part rather like a beak is the **olecranon process**, which fits into the olecranon fossa of the humerus. The posterior surface of the olecranon process can be felt through the skin and its upper limit forms the point of the elbow. The olecranon bursa separates the bone from the skin and may swell with excessive fluid if inflamed (bursitis). The floor of the trochlear notch projects forwards in a process known as the **coronoid process**, which fits into the coronoid fossa on the front of the humerus when the elbow is flexed. The articular surface of the trochlear notch is continued downwards on the lateral side of the ulna to form

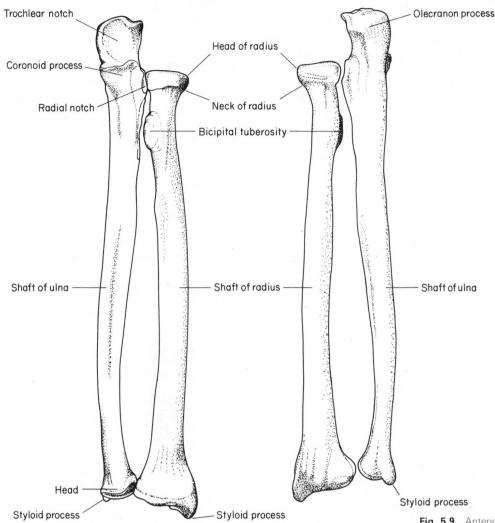

Trochlear notch
Coronoid process
Radial notch
Head of radius
Neck of radius
Bicipital tuberosity
Olecranon process
Shaft of ulna
Shaft of radius
Shaft of ulna
Head
Styloid process
Styloid process
Styloid process

Fig. 5.9 Anterior and posterior aspects of the left radius and ulna.

the radial notch, which articulates with the circumference of the head of the radius.

The **shaft** of the ulna narrows as it passes downwards but expands slightly at its lower end to form the **head** of the ulna, which can be seen projecting on the medial side of the posterior aspect of the wrist. A small downward process at the extreme lower end of the ulna, on its medial side, is called the **styloid process**; its tip is easily felt through the skin.

With the upper limb hanging straight down by the side, palm facing forwards (i.e. forearm fully extended and hand supinated), the forearm is angulated away laterally from the upper arm. The angle so formed is called the *carrying angle* and it results from the medial edge of the trochlea of the

humerus projecting lower than the lateral edge and from the obliquity of the superior articular surface of the coronoid process. This arrangement adds to the precision with which the hand using, say, a tool or weapon, can be controlled during or after extension of the elbow.

The bones of the forearm are a common site of fracture. Either may be involved in fractures occurring in the region of the elbow joint. Sometimes the olecranon process is separated from the rest of the ulna. Either or both bones may be broken across the shaft. A very common fracture occurs about an inch above the lower end of the radius and is known as Colles' fracture; the styloid process of the ulna is frequently torn off at the same time. This fracture usually results from a fall on the outstretched hand.

The bones of the wrist and hand

The carpus or wrist consists of eight bones arranged in two rows, proximal and distal, with four bones in each.

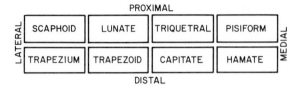

PROXIMAL

LATERAL	SCAPHOID	LUNATE	TRIQUETRAL	PISIFORM	MEDIAL
	TRAPEZIUM	TRAPEZOID	CAPITATE	HAMATE	

DISTAL

The carpal bones are irregularly shaped bones which articulate with one another and are held in position by ligaments. With the exception of the pisiform, the bones of the proximal row of the carpus articulate with the radius and the articular disc of the inferior radio-ulnar joint.

The palmar surface of the carpus forms a deep concavity, known as the carpal groove. A strong fibrous band or retinaculum bridges the bony margins of the groove and converts it into a tunnel. The *carpal tunnel* so formed transmits the flexor tendons and median nerve to the hand. This arrangement increases the strength of the wrist and the efficiency of the flexor muscles. However, if swelling or deformity due to pregnancy, disease or injury compromises the space in the tunnel, the bones and retinaculum being unable to 'give', pressure is exerted on the median nerve which is trapped in the tunnel, thereby damaging it and providing a common example of an *entrapment neuropathy*. The resulting pain in the hand (particularly at night), sensory loss and weakness and wasting of the short abduction muscle of the thumb (abductor pollicis brevis) is known as the *carpal tunnel syndrome*.

The **metacarpal bones**, or bones of the palm of the hand, are long bones, each with a base, a shaft and a head. The bases articulate with the distal row of carpal bones, and the heads with the proximal or first row of phalanges. The first metacarpal bone lies anteriorly to the others and is rotated so that its palmar surface faces medially. This makes it possible to oppose the thumb to the fingers and to obtain a good grip with the fingers on one side and the thumb on the other side of an object.

The **phalanges** (singular = phalanx). These also are long bones. The thumb has only two phalanges while the fingers have three, proximal, middle and distal, the proximal being the longest. A useful diagnostic clue in the X-rays of patients with acromegaly is that the distal phalanges appear to be 'tufted'.

The joints between the metacarpals and the phalanges are called the metacarpophalangeal joints, those between the phalanges themselves, the interphalangeal joints. The proximal interphalangeal joints are often swollen in patients with rheumatoid arthritis, making the fingers spindle-shaped.

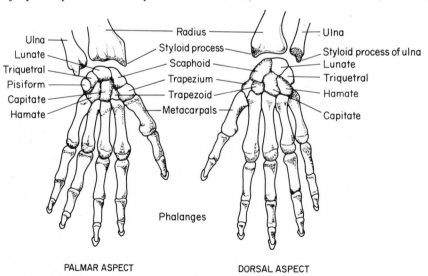

Fig. 5.10   Palmar and dorsal aspects of the left hand.

Bones of the pelvic girdle

The innominate bone

This is the name given to the hip bone. The two innominate bones join (articulate) together anteriorly at the pubic symphysis and are bridged posteriorly by the sacrum, together with which they form the bony pelvis or pelvic girdle. Each innominate bone is a large irregularly shaped bone which in the child consists of three parts separated by cartilage. In the adult these parts are fused together (i.e. united by bone) but their names are retained for descriptive purposes. The uppermost bone is called the ilium (not to be confused with the ileum, a part of the small intestine). The part situated in front is the pubis, and that posteriorly the ischium. All three bones unite and take part in the formation of the large cup-shaped cavity on the outer surface of the bone, known as the **acetabulum**, into which fits the head of the femur forming the hip joint.

The ilium

The ilium, which supports the flank, forms the upper, expanded and flat part of the innominate bone. The external (gluteal) and internal surfaces of the ilium are separated along the upper margin of the bone by the rough **iliac crest**, which can be felt underlying the flesh at the lower limit of the waist. This is the site from which bone marrow biopsies are frequently taken.

The gluteal surface is marked by three uneven ridges, the posterior, anterior and inferior gluteal lines. The gluteal muscles arise from this surface of the bone; gluteus maximus behind the posterior gluteal line, gluteus medius between the posterior and anterior lines, and gluteus minimus between the anterior and inferior lines.

The smooth hollow of the internal surface of the ilium is the iliac fossa. Behind and below it is the sacropelvic surface, which is roughened for the attachment of ligaments which bind the ilium to the sacrum and bears an ear-shaped facet where the sacrum articulates.

If the iliac crest is traced forwards, it will be found to end in the **anterior superior spine**, which gives attachment to the lateral end of the inguinal ligament. The spine can be easily felt in the living subject and is an important landmark in the examination of patients. The posterior end of the iliac crest ends in the **posterior superior spine**, which cannot be felt in the living subject but which may be marked by a dimple above the medial part of the buttock.

The anterior inferior spine and the posterior inferior spine are situated a little below the corresponding superior spines. Immediately below the posterior inferior spine is a large notch, the **greater sciatic notch**, which is usually much wider in the female than in the male. Through this notch passes the sciatic nerve, an important nerve which extends down the back of the leg. Damage to this nerve causes the painful condition known as sciatica.

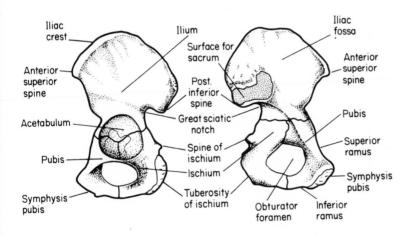

Fig. 5.11 The left innominate bone. External (left) and internal surfaces.

The pubis

The front portion of the innominate bone is called the pubis. It has a body and two rami. The **body** articulates with its fellow of the opposite side at the **pubic symphysis**. The bridge of bone which joins the upper part of the body of the pubis to the ilium, and takes part in the formation of the acetabulum, is called the **superior ramus**. The bridge which joins the lower part of the body to the ramus of the ischium is the **inferior ramus** of the pubis.

The free upper border of the body of the pubis is called the pubic crest. At its junction with the superior ramus is a process named the **pubic tubercle**, to which is attached the medial end of the inguinal ligament.

The urinary bladder lies behind the pubic symphysis and the bodies of the pubic bones and may therefore be injured in fractures of the pelvis involving the pubis. It is more likely to be ruptured if it is full at the time of injury. Rupture of the male urethra must be suspected if blood issues from the external urinary meatus following pelvic injury.

The ischium

The solid broad portion at the lower, posterior part of the innominate bone is the ischium. It consists of a body and a ramus. The **body** takes part in the formation of the acetabulum. The **ramus** of the ischium passes forwards to join with the inferior ramus of the pubis.

Below the acetabulum, the opening bounded by the pubis and the ischium is called the **obturator foramen**. It is large and oval in the male but is smaller and nearly triangular in the female. In life it is largely occupied by a fibrous sheet called the obturator membrane.

The inferior extremity of the body of the ischium is marked by the rough **ischial tuberosity**, which supports the body weight when sitting and provides attachment for the hamstring muscles. In life it is obscured by the gluteus maximus muscle when the hip is extended, as in the standing position, but can be easily identified by palpation when the joint is flexed.

The pelvis can be divided into 2 parts or segments. The upper part, formed mainly by the flanged parts of the two iliac bones, is called the greater pelvis or 'false pelvis.' The cavity of the greater pelvis is a part of the abdomen. The smaller segment below, known as the lesser pelvis or 'true pelvis', consists of the rest of the innominate bone (ischium and pubis), on each side and in front, and the sacrum behind. It bounds the pelvic cavity or canal, which is of obstetric importance. The upper or superior opening or aperture of the lesser pelvis is occupied by viscera and the lower or inferior aperture is largely closed in life by the muscular pelvic floor and its sphincters.

The boundary of the superior pelvic aperture, or *pelvic inlet*, is called the pelvic brim. Its antero-posterior, transverse and oblique diameters are measured for obstetric and anthropological reasons. Each of these three dimensions is greater on average in females than in males, the greatest difference being in the anteroposterior diameter.

The cavity of the lesser pelvis is deeper posteriorly, where it is bounded by the concave surface of the sacrum and coccyx, than it is anteriorly,

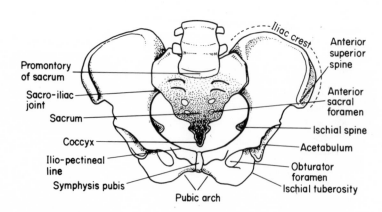

Fig. 5.12   The pelvis.

where it is bounded by the body of the pubis, the pubic rami and the pubic symphysis. On each side, the pelvic cavity is bounded by a smooth quadrangular area formed by the fused ilium and ischium. The pelvic cavity thus enclosed contains the bladder anteriorly, the rectum posteriorly and the uterus in between. Mid-cavity measurements are made, as for the pelvic brim, and again these are greater in the female.

The shape of the inferior pelvic aperture, or *pelvic outlet* is determined by three wide notches, the pubic arch in front, between the ischiopubic rami, and two large sciatic notches. The three diameters of the pelvic outlets corresponding to those of the pelvic inlet are all greater in the female.

Other pelvic measurements are also used in Obstetrics. What matters to the mother and her fetus is the comparison between her pelvic measurements (pelvimetry) and the measurements of the fetal head (cephalometry).

There are differences in shape between the male and female pelvis, the latter having a shape which facilitates the passage of the baby's head during childbirth. The axis of the pelvic cavity follows the curvature of the cavity, which is parallel to the profile of the sacrum and coccyx seen from the side (Fig. 5.13). The backward tilt of the sacrum and the anteroposterior diameter of the pelvic cavity are generally greater in the female. The pelvic axis and the disparity in depth between the anterior and posterior contours of the pelvic cavity are important to the passage of the fetus through the pelvic canal.

In the adult pelvis there are differences between the sexes in linear measurements, angles and shape, of various components of the pelvis. The greater prominence of the hip in the female is probably due to the differences in the ilium between the sexes.

Even in fetal life, the male and female pelves differ from each other, particularly in the suprapubic arch or angle, between the inferior pubic rami, which is greater in the female. Although the female pelvis is adapted for childbirth, the prime function of the pelvis in both sexes is *locomotor*. The male generally being the more muscular and more heavily built, has a pelvic architecture which is generally stouter than that of the female. The overall dimensions of the male pelvis are greater and its muscular and ligamentous markings are more obvious.

In forensic work involving the identification of human remains, the sex of the deceased can best be determined from examination of the pelvis. Some of its features, especially the iliac crest and the greater sciatic notch, also enable the bone age to be assessed to within a very few years, especially in young adults.

The bones of the lower limb

The femur or thigh bone

This is the longest and strongest bone in the skeleton, its length being associated with the striding gait of human beings and its strength being appropriate to the weight and muscular forces to which it is subjected. It has an almost cylindrical **shaft**, bowed forwards, and two extremities, the upper being a rounded head, a neck and two trochanters and the lower being in the form of a massive double 'knuckle' or condyle.

The **head** of the femur is hemispherical and its surface is covered with hyaline articular cartilage, except for a small roughened pit for attachment of the ligamentum teres, just below and behind its centre. It fits into the acetabulum of the innominate bone to form the hip joint.

The **neck** extends upwards and medially from the shaft at an angle of about 125° and terminates in

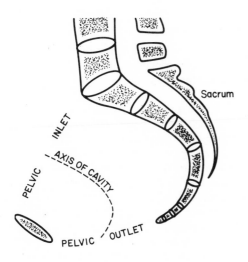

Fig. 5.13 Sargittal section through the female pelvis.

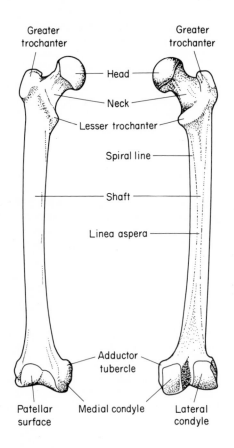

Greater trochanter

Greater trochanter

Head

Neck

Lesser trochanter

Spiral line

Shaft

Linea aspera

Adductor tubercle

Patellar surface

Medial condyle

Lateral condyle

Fig. 5.14   Anterior (left) and posterior views of right femur.

The **lesser trochanter** projects medially from the shaft just below the neck of the femur. To it is attached the psoas muscle. A ridge of bone joining the greater and lesser trochanters on the posterior aspect of the femur is called the intertrochanteric crest.

The **shaft** of the femur is almost cylindrical except towards its lower end, where it is flattened and expanded. Anteriorly and laterally the shaft is smooth but posteriorly a prominent broad rough ridge, the **linea aspera** (rough line) runs vertically, giving attachment to several muscles. In the lower third of the bone, the linea aspera divides into two smaller ridges, one of which passes to each of the condyles. The triangular area enclosed by these two ridges is called the popliteal surface of the femur. The shaft of the femur is surrounded by thick powerful muscles and cannot normally be felt through the skin.

The lower extremity of the femur consists of two (medial and lateral) **condyles**, separated behind by the intercondylar fossa, which articulate with the corresponding tuberosities of the tibia to form the knee joint. The articular cartilage covering the condyles continues on to the anterior surface of the bone and forms a surface for articulation with the patella or knee-cap. Immediately above the medial condyle is a small tubercle, the **adductor tubercle**, to which part of the adductor magnus muscle is attached.

The patella or knee-cap

The patella is a sesamoid bone developed in the tendon of the quadriceps muscle of the thigh. It is roughly triangular in shape, with its apex directed downwards.

That portion of the quadriceps tendon attaching the apex of the patella to the tuberosity of the tibia is called the patellar ligament. This is tapped by the physician, using a patellar hammer, to elicit the *knee jerk*, one of the reflexes commonly tested. The rough anterior surface of the patella can be palpated through the skin, from which it is separated by a bursa. Inflammation of this bursa, and of the suprapatellar bursa, gives rise to the condition known as housemaid's knee. The posterior surface of the patella is smooth and covered with articular cartilage; it articulates with the patellar surface of the femur, taking part in the formation of the knee joint.

the head. The angulation enables the lower limb to swing clear of the pelvis.

The neck of the femur is vulnerable to fracture as a result of tripping, in elderly people, especially women, owing to postmenopausal and senile osteoporotic changes. Twisting (medial rotation) of the thigh due to a similar accident in a person under the age of 16 years may result in a spiral fracture of the shaft of the femur.

The **greater trochanter** is a large process projecting upwards from the top of the shaft laterally to the neck, which it overhangs, creating a small hollow called the **trochanteric fossa**, into which is inserted the obturator externus muscle. The outer surface of the greater trochanter and the posterior surface of the upper part of the shaft provide insertion for the gluteal muscles.

The patella is quite commonly fractured, being (a) broken transversely into upper and lower portions or (b) shattered into fragments by a direct blow, causing a stellate (star-like) fracture.

Dislocation of the patella may result from fairly severe violence, frequently involving a twisting injury accompanied by a blow to the inner side of the knee. This is known as acute traumatic dislocation. Alternatively, recurrent dislocation may occur as a result of anatomical abnormalities. Partial dislocation or subluxation is not uncommon, especially in girls.

The tibia or shin bone

The tibia is the second longest bone in the body (the first being the femur) and is the medial and stronger of the two bones of the leg. It has a **shaft**, shaped like a prism in cross-section, and expanded extremities.

The upper extremity is composed of the **medial and lateral condyles**, which have weight-bearing surfaces, and a tuberosity. Each condyle has an articular surface for articulation with the corresponding condyle of the femur. Attached to the condyles, in life, are the medial and lateral menisci, which are semi-lunar (half-moon shaped) cartilages.

Between the articular surfaces of the two condyles is the roughened intercondylar area. This gives attachment to the horns of the menisci and to the cruciate (cross-shaped) ligaments.

On the under surface of the lateral condyle is a small articular facet for the head of the fibula.

The anterior margins of the tibial condyles can be easily felt when the knee is flexed, each forming the lower border of a depression at each side of the patella. The lower part of the **tibial tuberosity** is subcutaneous and therefore also easily felt.

The shaft of the tibia, being triangular in cross-section, has three borders and three surfaces. The anterior border runs down from the tibial tuberosity to the medial malleolus. It can be felt under the skin and with the exception of the lower quarter,

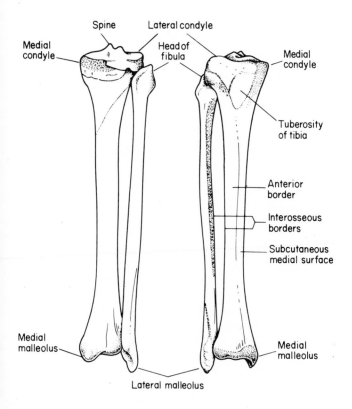

Fig. 5.15 Posterior (left) and anterior views of right tibia and fibula.

which is rounded and indistinct, it forms a sharp ridge (the 'shin'). It provides attachment for the deep fascia of the leg. The interosseous border gives attachment to the interosseous membrane which stretches between the tibia and the fibula. The medial border is sharp and distinct in the middle and rounded and ill-defined in its upper and lower fourths.

The medial surface of the tibia lies between the anterior and medial borders. It is smooth and situated directly beneath the skin throughout its length; hence it is sometimes known as the **subcutaneous surface**. When traced downwards, it is found to end in the **medial malleolus**. The lower part of the medial surface is crossed obliquely by the great saphenous vein, the profile of which can be seen in some people, especially those with varicose veins.

The posterior surface of the tibia is bounded by the interosseous and medial borders. It gives attachment to the popliteus, soleus and tibialis posterior muscles and to the long flexor muscle of the toes, flexor digitorum longus. The upper part, together with the similar area of the posterior aspect of the lower end of the femur, forms the floor of the popliteal fossa.

The lateral surface of the tibia gives attachment to the tibialis anterior muscle.

The lower extremity of the tibia is slightly expanded. Medially it projects downwards as the medial malleolus. Laterally is found the fibular notch, which is bound by ligaments to the lower end of the fibula. The inferior surface of the lower end of the tibia is smooth and articulates with the talus at the ankle joint.

The fibula

The fibula is the lateral of the two bones of the leg. Not having to bear weight, it is a slender long bone. It consists of a shaft with upper and lower extremities, named respectively the head and the lateral malleolus.

The **head** can be felt through the skin as a prominence on the posterolateral aspect of the knee. The common peroneal nerve runs across the neck of the bone just below the head of the fibula and in this superficial position it is vulnerable to injury. This may be direct trauma, such as a blow, or compression of the nerve against the bone by a tight knee bandage or leg plaster or as a result of prolonged sitting with the legs crossed. The result is *foot drop* with inability to dorsiflex or evert (turn outwards)

the foot. Sensation is impaired in the skin over the dorsum of the foot and the front and outer side of the leg.

The head of the fibula has a facet for articulation with the under surface of the lateral condyle of the tibia. (*Note*. This articulation takes no part in the formation of the knee joint.)

The lower end of the fibula is expanded to form the **lateral malleolus**, which projects downwards to a lower level than the tibia − a fact which students can easily check in themselves. On the medial surface of the lateral malleolus is a facet for articulation with the lateral surface of the talus in the ankle joint.

The lower ends of the tibia and fibula are firmly held together by ligaments and are also joined throughout their shafts by the interosseous membrane which divides the leg into anterior and posterior compartments.

Fracture of the lower ends of the tibia or fibula, or both, is very common and is sometimes called *Potts' fracture*. Fracture of the shaft of the tibia also occurs. A fracture of a condyle at the upper end may extend into the knee joint.

Bones of the foot

The framework of the foot consists of
  the tarsus
  the metatarsus
  the phalanges

The **tarsus** consists of a medial and a lateral series of bones.
(a) The medial series comprises the talus, the navicular and the three cuneiform bones.
(b) The lateral series are composed of the calcaneus and the cuboid.

If the general architecture of the foot is examined it will be seen that there are two arches − a longitudinal one and a transverse one. The **longitudinal arch** is most marked on the medial aspect of the foot and results from the fact that the talus is placed on top and, therefore, above the level of the calcaneus. The **transverse arch** is most marked at the level of the base (proximal end) of the metatarsus. These arches are very very important in walking and are maintained by strong ligaments aided by muscles. The arches can withstand the very considerable strain caused by an adult jumping from a height of several feet. If the arches become weakened and collapse, the condition known as 'flat foot' (pes planus) results. An abnormally high-arched foot is called pes cavus.

The **talus** is a very irregular bone having articular surfaces: (a) above, for the tibia, and on either side for the medial and lateral malleoli forming the ankle joint; (b) below, for the calcaneus; (c) in front for the navicular.

The **calcaneus**, also an irregular bone, and the largest of the tarsal bones, is situated below the talus. It projects backwards behind the talus, forming the heel, and to its posterior margin is attached the important tendo calcaneus, better known as the Achilles tendon. Projecting backwards, as it does, beyond the bones of the leg, it provides a short lever for the calf muscles acting through the tendo calcaneus. Rupture of the tendon is clearly a serious matter and necessitates surgical repair and immobilization of the ankle in plaster of Paris.

The calcaneus articulates in front with the cuboid.

The talus and calcaneus are both strong bones and much force is needed to break them. They are fractured usually as a result of falling from a height, depending among other factors upon the weight and muscularity of the victim but usually at least four metres in the case of an athletic man.

The **navicular bone** is a disc-like bone having a concave proximal surface, where it articulates with the talus, and a convex distal surface where it articulates with the three cuneiform bones. A tuberosity on the medial aspect of the bone provides the principal attachment for the tendon of the tibialis posterior muscle.

The **cuboid bone** articulates behind with the calcaneus and in front with the fourth and fifth metatarsals. Medially it is in contact with the navicular and lateral cuneiform bones.

The three (medial, intermediate and lateral) **cuneiform**, or wedge-shaped bones, are placed between the navicular bone behind and the three medial metatarsals in front. The arrangement of the cuneiform bones and the strong ligaments binding them are important factors in maintaining the transverse arch of the foot.

The **metatarsus** is composed of five **metatarsal bones** which connect the tarsus to the phalanges of the five toes. They are long bones, each with a proximal base, a shaft and a distal head. The bases of the first three articulate with the cuneiform bones, those of the lateral two with the cuboid. The first metatarsal (of the great toe) is shorter and stouter than the others. The fifth (little toe) has a rough eminence (tuberosity or styloid process) on the lateral side of its base and this can be seen and felt half-way along the lateral border of the foot; sometimes it is fractured when the foot is acutely inverted.

Pain beneath the three middle metatarsal heads is called *anterior metatarsalgia* and this may be associated with tender calluses under the ball of the foot. An untrained individual undertaking prolonged marching, hiking or running may sustain a stress fracture (*march fracture*) of a metatarsal.

The **phalanges** of the foot are also long bones and they correspond in number and general arrangement with the phalanges of the hand. The big toe, like the thumb, has two, a proximal and a distal phalanx. Each of the other toes has three phalanges, the additional one being a middle phalanx.

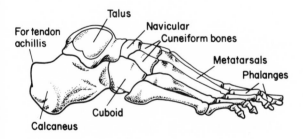

Fig. 5.16 Bones of the right foot — lateral view.

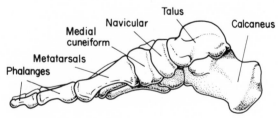

Fig. 5.17 Bones of the right foot — medial view.

The skull

The most obvious function of the skull is its protective one, the cranium protecting the brain from externally inflicted injury. The special sense organs are also afforded substantial protection. However, the skull is also involved in a diversity of other functions including the securing and devouring of food and the control of the cerebral circulation. The bony sockets (orbits) housing the eyes ensure a constant distance between these organs and provide a necessary pre-condition for binocular vision. The sockets also provide for rotation of the eyes, utilizing muscles attached to their walls. The fixation of the organs of balance, the labyrinths, within the skull ensures a fixed relationship between the six semicircular canals (three on each side) to each other and to the head, providing the basis for an orderly correlation between these sense organs (receptors) and their central nervous connections.

Strictly speaking the skull consists of the cranium and the mandible, although the term is commonly used for the cranium alone. The upper part of the cranium, forming the brain-box, is termed the calvaria and the remainder of the skull forms the facial skeleton.

Bones of the skull

The bones of the skull are divided into two groups.

(*a*) Those of the **calvaria** or brain-box, eight in number.

    1 frontal (forehead) bone
    2 parietal
    2 temporal
    1 ethmoid
    1 sphenoid
    1 occipital

(*b*) Those of the face, fourteen in number.

    2 maxillae (upper jaw)
    2 zygomatic (cheek) bones
    2 nasal
    2 lacrimal
    2 palatine
    2 inferior turbinate
    1 vomer
    1 mandible (lower jaw)

All the bones of the skull, except the lower jaw, are joined together by sutures or immovable joints. The lower jaw or temporomandibular joint is a

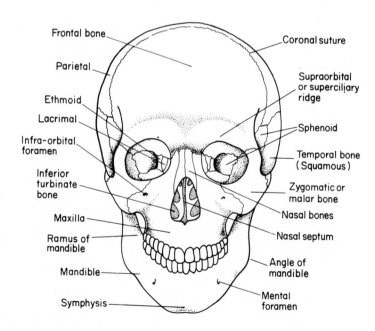

Fig. 5.18 The skull – anterior view.

condyloid joint (see p. 74) allowing both hinge-like and side-to-side movements.

Before discussing the individual bones, the skull as a whole must be examined. The top or vault can be removed from an anatomical specimen so that the interior can be examined. The following description is meant to be read with the skull at hand and with the aid of diagrams.

Anterior aspect (norma frontalis)

Looked at from the front, the upper part of the skull is formed by the frontal bone of the forehead. The remainder is formed by the facial bones, including the lower jaw.

The roofs and upper margins of the **orbits**, or eye-sockets, and the ridges underlying the eyebrows are formed by the frontal bone. The lateral wall of the orbit is formed by the zygomatic bone and the great wing of the sphenoid bone, while several bones, the lacrimal bone and parts of the maxilla, ethmoid and sphenoid bones, form the medial wall. The floor of the orbit is formed mainly by the maxilla. The orbit is cone-shaped, with its wide open base in front and its narrow apex behind. At its apex is a circular foramen for the passage of the optic nerve. A deep groove, the inferior orbital fissure, extends from the lateral side of this foramen into the floor of the orbit.

The portion of the maxilla which forms the medial margin of the orbit projects upwards to meet the frontal bone. Between these projections of the maxillae are the two nasal bones. Passing vertically downwards from these bones in the mid-line, the bony nasal septum can be seen dividing the nasal cavity into right and left halves.

On the side walls of the nasal cavity three bony projections are visible, namely the inferior, middle and superior nasal conchae.

Surrounding the opening of the nasal cavity and meeting in the mid-line below it are the right and left maxillae or upper jaws, which carry the upper set of teeth.

Maxillofacial injuries result from road traffic accidents, acts of violence and accidents in industry, sport and the home. The maxilla, which is less frequently fractured than the mandible, forms with the other bones of the region a mass capable of absorbing considerable violence, thus protecting the brain. Fractures of the middle third of the facial skeleton require prompt and skilled management (reduction and fixation) because of their great physiological and aesthetic importance. The surgeon's work is sometimes facilitated by a tracheostomy but this procedure is generally avoided unless required for associated injuries such as fractures of the larynx or a flail chest.

Lateral aspect

When viewed from the side the greater mass of the cranial bones can be seen behind the facial bones.

The central part of the lateral aspect is occupied by the temporal bone. A process from this bone passes forward to join the zygomatic (malar) bone, thereby enclosing a fossa (the temporal fossa) in which can be seen a portion of the sphenoid bone. This adjoins the frontal, parietal and temporal

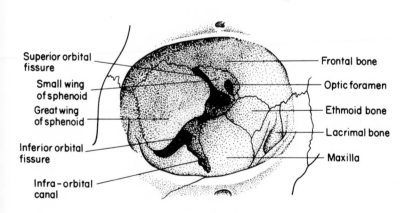

Fig. 5.19 Boundaries of the right orbit viewed from in front.

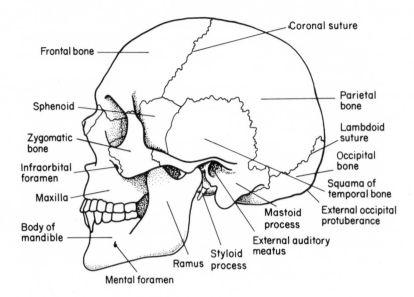

Fig. 5.20   The skull — lateral view.

bones, so that an irregular H-shaped arrangement of sutures can be seen. A small circle can be drawn enclosing the horizontal limb of the H and portions of all four bones. This area, which overlies the anterior branch of the middle meningeal artery, is called the *pterion* and its centre is an important landmark for the surgeon.

Obvious features of the lateral aspect of the skull are the external acoustic meatus and, behind it, the prominent mastoid process of the temporal bone. Posteriorly, the occipital bone will be seen.

Superior aspect (norma verticalis)

On looking at the skull from above, the frontal bone will be seen to occupy the front of the vault of the cranium and to be separated from the two parietal bones by a suture which runs transversely, the **coronal suture**. The two parietal bones together form the greater part of the vertex and are separated from each other in the mid-line by a suture placed at right angles to the coronal suture. This is the **sagittal suture**. The meeting point of the coronal and sagittal sutures is called the *bregma* and this marks the site of the anterior fontanelle, a gap filled by membrane in the fetal skull.

The lambdoid suture joins the parietal bones to the occipital bone.

Posterior aspect (norma occipitalis)

The complete **lambdoid suture** is seen in this view. It is sometimes complicated by a number of small irregular bones appropriately called sutural bones. The most prominent feature of the posterior aspect of the skull is the **external occipital protuberance**, which can be easily felt in the living subject at the upper end of the median furrow at the back of the neck. Ridges called the superior nuchal lines pass laterally from the protuberance and mark the boundary between the scalp and the neck.

Inferior aspect (norma basalis, base of the skull)

The lower jaw should be removed for study of the base of the skull. The anterior part is formed by the hard (bony) palate and the alveolar arches which carry the upper teeth. The hard palate is formed anteriorly by the two maxillae and posteriorly by the two palatine bones, all united together by a cruciform suture. Immediately behind the hard palate, the posterior openings of the nasal cavities (posterior nares) will be seen. Lateral to the maxilla and palate can be seen the fossa or hollow bounded laterally by the zygomatic bone and process of the temporal bone. This is the temporal fossa described in the lateral aspect. A part of the sphenoid bone can be seen here; behind and lateral to it is the

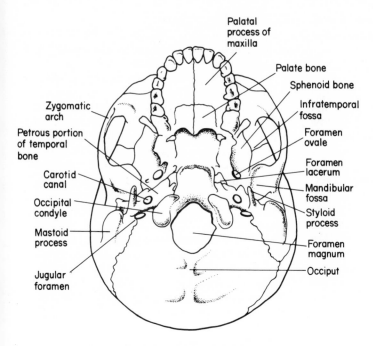

Palatal
process of
maxilla

Palate bone

Sphenoid bone

Zygomatic
arch

Infratemporal
fossa

Petrous portion
of temporal
bone

Foramen
ovale

Foramen
lacerum

Carotid
canal

Mandibular
fossa

Occipital
condyle

Styloid
process

Mastoid
process

Foramen
magnum

Jugular
foramen

Occiput

Fig. 5.21 Outline of skull as seen from below.

mandibular fossa for the articulation of the mandible or lower jaw.

The large oval opening further back in the midline is the **foramen magnum**, through which passes the spinal cord. This is situated in the occipital bone, a part of which projects forwards to meet the sphenoid bone close to the level of the posterior nares. Two smooth condyles, one on each side of the foramen magnum, articulate with the atlas vertebra. Behind the foramen magnum is the main portion of the occipital bone.

On either side of the occipital bone are portions of the temporal bone including the mastoid process and an irregular portion, the petrous portion, projecting forwards and medially towards the midline. A large round foramen in the petrous portion of the temporal bone leads upwards into the bone and is the opening into the carotid canal, which conveys the internal carotid artery on its course into the cranial cavity. The jugular foramen, a little further back, transmits the internal jugular vein, which returns blood from the brain, and other structures including three cranial nerves (IX, X and XI).

The interior of the skull

The skull cap (calva, cranial vault or vertex) having

been removed, it can be appreciated that the walls of the cranial cavity vary in thickness from region to region. Most of the bones have a tough outer table of compact bone separated from a thinner and more brittle inner table by the diploe, which is cancellous bone housing red marrow in its interstices.

The skull cap consists of portions of the frontal, parietal and occipital bones. Its internal surface shows numerous vascular markings. Anteriorly the frontal crest projects backwards in the midline. It gives attachment to the falx cerebri, the thick sheet of dura mater (one of the meninges) which separates the two cerebral hemispheres. It is grooved by the saggital sulcus which runs backwards in the median plane and accommodates the sagittal sinus. A number of irregular pits on each side of the sagittal sulcus are caused by arachnoid granulations, which are minute projections of the arachnoid mater, one of the membranous coverings (meninges) of the brain. Calcific nodules associated with the arachnoid granulations may sometimes be seen in the skull X-rays of elderly people and are not pathological, i.e. not due to disease.

The internal surface of the base of the skull is clearly divided into three compartments: the anterior, middle and posterior cranial fossae. Impressions for the cerebral gyri are conspicuous in the anterior and middle fossae.

The anterior cranial fossa
In the mid-line, in front, is a small vertical projection (the crista galli, for attachment of the falx cerebri), on either side of which there is a narrow plate of bone perforated by numerous minute holes and forming part of the roof of the nasal cavity. This is the cribriform plate of the ethmoid and the holes are for the passage of the olfactory nerves. On either side of these plates, a part of the frontal bone forms the roof of the orbits. The posterior margin of the anterior fossa is formed by a sharp edge of bone which is part of the lesser wing of the sphenoid. The medial extremity of this forms the anterior clinoid process, to which is attached the free border of the tentorium cerebelli, the double layer of dura mater which covers the cerebellum.

The middle cranial fossa
In the mid-line is the sella turcica (like a Turkish saddle) which is hollowed out as the hypophyseal fossa and in life contains the pituitary gland, or hypophysis cerebri. Enlargement of the fossa, as seen on X-ray pictures, suggests the presence of a pituitary tumour. The sella turcica is part of the body of the sphenoid bone which is traversed anteriorly by a groove, the sulcus chiasmatis, which runs from one optic canal to the other. Posterior to the hypophyseal fossa is the dorsum sellae, a plate of bone which projects upwards and forwards and ends on each side in a posterior clinoid process. Erosion of these processes, as seen on a skull X-ray, is taken as a sign of raised intracranial pressure.

The middle cranial fossa opens out laterally on each side into a deep, wide irregular hollow which supports the temporal lobe of the brain. It is bounded in front by the greater wing of the sphenoid bone and behind by the petrous part of the temporal bone, which contains the internal ear.

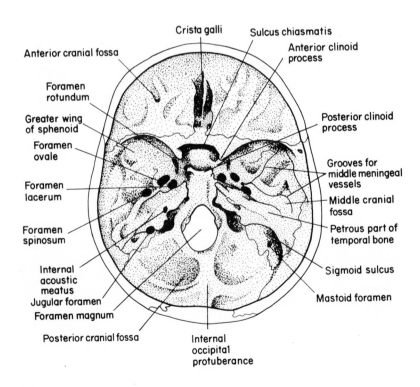

Fig. 5.22   Interior of skull.

The posterior cranial fossa

This is bounded in front by the petrous portions of the temporal bones, with a part of the occipital bone in the mid-line. Its most conspicuous feature is a large opening, the foramen magnum through which that part of the brain stem known as the medulla oblongata continues down into the spinal cord. Most of the posterior fossa is formed by the occipital bone and it contains the cerebellum, which functions primarily to co-ordinate and adjust movements and make them smooth. In front of the cerebellum lie the pons and medulla which, together with the cerebellum, make up the hindbrain.

Immediately behind and below the petrous portion of the temporal bone, on each side of the posterior fossa, is a deep S-shaped groove called the sigmoid sulcus. This contains the sigmoid sinus which drains venous blood into the internal jugular vein. The junction of these blood vessels lies in the jugular foramen which also transmits the ninth, tenth and eleventh cranial nerves. A number of other holes or foramina will be observed in the cranial fossae and these likewise transmit nerves or blood vessels or both.

Posteriorly in the posterior cranial fossa can be seen the internal occipital protuberance, in front of which lies the confluence of sinuses. On each side is a groove related to the commencement of the transverse sinus.

Having observed the general arrangement of the bones of the skull, some of them require individual study.

The frontal bone

This is an irregular flat bone which forms the forehead region, most of the anterior cranial fossa and the roofs of the orbits. In the forehead region, the bone is thick and consists of two compact laminae between which is sandwiched cancellous bone, except in the regions of the **frontal sinuses**, where the cancellous bone is absent. The frontal sinuses contain instead air which is in continuity with the nasal cavity via the frontonasal canal. The orbital part of the frontal bone consists entirely of compact bone.

On the anterior aspect of the frontal bone are two eminences, one on each side, called the frontal tubers or tuberosities. Below them are two curved **superciliary arches** joined across the midline by a smooth eminence known as the glabella. The super-ciliary arches originate as a result of the mechanical stresses imposed upon the frontal bone by the masticatory apparatus (jaws and associated muscles). Below the superciliary arches are the supraorbital margins which form the upper borders of the orbital openings. The **supraorbital notch**, at the junction of the lateral two-thirds and medial one-third of the supraorbital margin, transmits the supraorbital vessels and nerve. The doctor sometimes applies thumb pressure over this nerve to help in assessing a patient's level of consciousness. The supraorbital margin ends laterally in the **zygomatic process** which articulates with the zygomatic bone in a suture which can be felt through the skin as a slight depression in the living subject.

A portion of the frontal bone known as the nasal part projects downwards between the supraorbital margins and ends in a sharp **nasal spine**. Each nasal part articulates with one of the two nasal bones and supports the bridge of the nose.

The parietal bone

The right and left parietal bones, joined superiorly in the mid-line by the sagittal suture, form the greater part of the side walls and roof of the cranium. Each is a four-sided or irregularly quadrilateral bone. The anterior (frontal) border of each parietal bone articulates with the frontal bone at the coronal suture. The posterior margin articulates with the occipital bone at the lambdoid suture and the inferior margin articulates with the temporal bone. At birth, there are membranous intervals, known as fontanelles, in the skull at the four angles of the parietal bones (see also p. 63). The internal surface of the parietal bone is marked by impressions corresponding to the cerebral gyri, or convolutions of the brain, and furrows for the middle meningeal blood vessels. A shallow groove for the superior sagittal sinus runs along the superior or sagittal border of the bone.

The occipital bone

The occipital bone, situated at the back and base of the skull, forms the greater part of the floor of the posterior cranial fossa. It articulates in front with the two parietal bones, at the lambdoid suture, and on either side with the temporal bone. Its most distinctive feature is the large oval opening known as the **foramen magnum**, through which passes the lower end of the medulla oblongata.

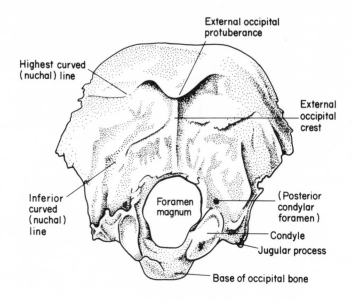

Fig. 5.23 Outer surface of occipital bone.

Behind the foramen magnum is an expanded plate known as the squamous part of the occipital bone. Its internal surface presents four fossae shaped to the two occipital lobes of the cerebrum and the hemispheres of the cerebellum. The **internal occipital protuberance** is easily identifiable at the junction of the four fossae. The sulcus of the superior sagittal runs upwards from this protuberance and the sulci of the transverse sinuses extend laterally from it. The internal occipital crest runs downwards from the protuberance and gives attachment to the falx cerebelli, a small sickle-shaped process of dura mater containing the occipital sinus. In front of the foramen magnum is the basilar part of the bone which unites with the sphenoid bone. The lateral (condylar) parts of the bone lie on either side of the foramen and on their inferior surfaces are the convex **occipital condyles** for articulation with the atlas vertebra.

The temporal bone

This bone consists of four parts:
    (*1*) the squamous part;
    (*2*) the petromastoid part;
    (*3*) the tympanic part;
    (*4*) the styloid process.

The **squamous part** of the temporal bone forms part of the side wall of the skull. It articulates above mainly with the parietal bone and in front with a part of the sphenoid bone. Behind and below it is continuous with the mastoid portion of the petromastoid part. In its lower part is a circular opening – the **external acoustic meatus** – which leads by a canal to the cavity of the middle ear. Immediately above the external acoustic meatus is the zygomatic process, which projects forwards to join the zygomatic bone.

The **mastoid portion** of the petromastoid part of the temporal bone projects downwards immediately behind the external acoustic meatus in a dense process of bone known as the **mastoid process**. This contains the mastoid air cells. Middle ear infection may occasionally spread into these cells, causing mastoiditis and, rarely, further spread may cause a brain abscess. The mastoid process and its air cells are closely related to the groove for the sigmoid sinus in the posterior cranial fossa. The lateral surface of the mastoid process gives attachment to the sternomastoid muscle, amongst others.

The **petrous portion** projects inwards and forwards at an angle, wedged between the sphenoid

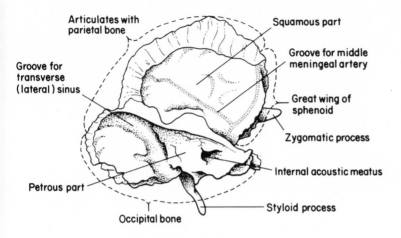

Fig. 5.24 Internal aspect of left temporal bone.

and occipital bones, and can be seen on the base of the skull from below and from the interior. As the ends of the petrous portions approach each other they are separated in the mid-line by the base of the occipital bone. Close to the anterior extremity is the **auditory canal** (not always easy to make out on the whole skull) or Eustachian tube, which connects the nasopharynx to the middle ear cavity.

The posterior surface of the petrous portion of the temporal bone forms the anterior wall of the posterior cranial fossa and is here grooved by the sigmoid sinus. A well marked foramen, the opening of the **internal acoustic meatus**, is seen near the centre of the posterior surface. It transmits the eighth cranial nerve which is also known as the vestibulocochlear nerve and subserves hearing and balance. The petrous portion of the temporal bone contains the cochlea, vestibule and semicircular canals.

The **tympanic part** of the temporal bone is fused internally with the petrous portion and posteriorly with the squamous part and the mastoid process. It forms the anterior wall, floor and lower part of the posterior wall of the external acoustic meatus, the remainder being formed by the squamous part of the temporal bone.

The **styloid process** projects from the under surface of the temporal bone. It is a long slender process which gives attachment to muscles and ligaments.

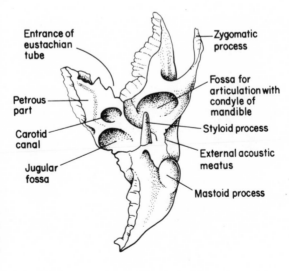

Fig. 5.25 Under-surface of left temporal bone.

The sphenoid bone

This bone, which takes part in the formation of the middle cranial fossa, is shaped somewhat like a bat. It has a body, situated in the midline, from which project outwards (a) two greater wings, and (b) two lesser wings.

Two pterygoid processes project downwards from the adjoining parts of the body and greater wings.

The **body** is hollow and contains two large air sinuses. Its upper surface is shaped like a Turkish saddle and is therefore known as the **sella turcica**. The deepest part of the hollow is the **hypophyseal fossa**, which houses the hypophysis cerebri or pituitary gland. Behind the sella turcica is the dorsum sellae from the upper angles of which the two **posterior clinoid processes** project, one on each side and give attachment to the tentorium cerebelli.

From the anterior part of the body the lesser wings spread out on either side to form the posterior margin of the anterior cranial fossa. From the sides of the body the greater wings spread outwards between the frontal and temporal bones, and close to their origin is a well-marked foramen (the foramen ovale) which transmits the mandibular branch of the trigeminal or Vth cranial nerve.

Developmental abnormalities of the sphenoid bone in the fetus can deform the bridge of the nose causing it to be depressed, as in achondroplasia, or abnormally broad with excessive separation of the orbits (hypertelorism), which may be associated with congenital heart disease.

The ethmoid bone

This is a box-shaped bone of delicate structure, hollowed out to contain the ethmoidal air cells. Its lateral walls help to form the medial wall of each orbit. Its roof is formed by the **cribriform plate** which can be seen between the two portions of the frontal bone in the anterior cranial fossa and is perforated by the olfactory nerves. The inferior surface enters into the formation of the roof of the nose.

From the cribriform plate, a **perpendicular plate** projects downwards to form the upper part of the nasal septum, dividing the cavity of the nose into right and left halves. The septal cartilage is attached to the lower border of the perpendicular plate, so that the nasal septum is bony in its upper part and cartilaginous in its lower anterior part. The nasal septum is usually deflected slightly to one or other side but excessive deflection may cause nasal obstruction and necessitate the surgical operation known as submucous resection (SMR) of the cartilaginous septum, which often appears on an ear, nose and throat (ENT) surgeon's operating list.

Laterally to the perpendicular plate, the superior and middle **nasal conchae** project from the side walls of the nasal cavities.

The ethmoidal air cells open into fissures (the superior and middle meati of the nose) below these conchaè. The inferior nasal conchae have a separate identity as bones and are not a part of the ethmoid or any other bone.

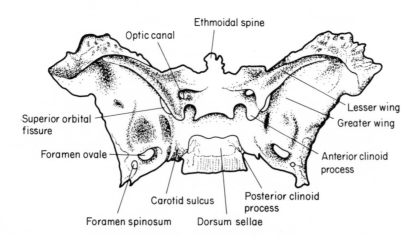

Fig. 5.26　The sphenoid bone viewed from above.

The maxilla

This is a large irregular bone which unites in the midline with its fellow to form the upper jaw. The upper part of the bone forms the greater part of the floor of the orbit; its lower part, the major portion of the hard palate. Laterally it articulates with the zygomatic bone. Medially it helps to make up the side wall of the nasal cavity and to it is attached the inferior concha.

The body of the maxilla is hollow and contains the important **maxillary sinus** or antrum, which communicates with the middle meatus of the nose through a small aperture. Each maxilla has an **alveolar process** containing the sockets for eight teeth. It will be noticed that the roots of the first and second molar teeth are closely related to the floor of the maxillary sinus.

Inflammation within the nose, resulting from a common cold, may spread to the maxillary sinus so that the cavity becomes filled with mucus or mucopus, causing discomfort in the area. If this does not drain spontaneously it may be necessary to perforate the medial wall of the sinus and wash it out; this is known as an antral washout. Similar inflammation (sinusitis) may affect the frontal sinus, causing a headache, and much less commonly, the ethmoid and sphenoid air cells.

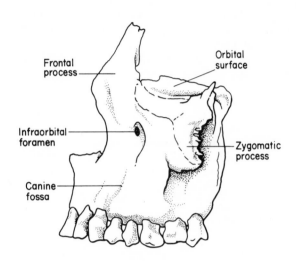

Fig. 5.27 Left maxilla — external or lateral view.

The zygomatic bone

The zygomatic bone (cheek bone) forms the prominence of the cheek, parts of the floor and lateral wall of the orbit and parts of the walls of the temporal and infratemporal fossae. It articulates medially with the maxilla, above with the frontal bone and posteriorly with a process from the temporal bone.

The lacrimal bones

These are two small bones which take part in the formation of the medial wall of the orbit (see Fig. 5.19).

The nasal bones

These are two small bones lying side by side between the frontal processes of the maxilla and forming the bridge of the nose.

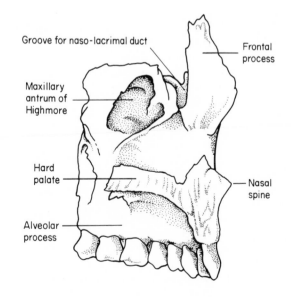

Fig. 5.28 Left maxilla — medial aspect.

The vomer

This is a thin, flat bone situated in the mid-line, where it rests on the hard palate and forms the lower and posterior part of the nasal septum. The rest of the septum is formed above by the perpendicular plate of the ethmoid bone and anteriorly by cartilage.

The palatine bones

These are irregular in shape and placed between the maxillae and sphenoid bone. Each takes part in the formation of the hard palate, the floor and lateral wall of the nasal cavity and the floor of the orbit.

The mandible or lower jaw

The mandible is the only moving bone in the skull. It consists of a curved horizontal **body**, which carries the teeth of the lower jaw, and two broad **rami** which project upwards almost vertically on each side posteriorly. The lower posterior corner of the ramus is called the angle of the jaw. On the inner surface of the ramus is a foramen for the entry of the inferior alveolar nerve and vessels into the mandibular canal, which re-emerges at the mental foramen on the lateral aspect of the bone. The ramus terminates above in two processes: the triangular **coronoid process** in front, and the convex **condylar process** of the mandible behind.

The coronoid process provides attachment for the temporalis muscle and the head of the condylar process articulates with the temporal bone to form the temporomandibular joint.

The mandible becomes smaller and undergoes changes in shape in old age as a consequence of the loss of teeth. Particularly striking is the reduced height (width) of the body of the lower jaw following loss of the teeth and associated absorption of adjacent bone. These edentulous atrophied mandibles are frequently fractured by relatively minor trauma. At the other extreme, mandibular fractures are rare in children with their more resilient lower jaws; loosening or breakage of a tooth is a more common result of trauma in this age group.

Generally, fractures of the mandible are due to violent trauma resulting from the blow of a fist, road traffic accidents, gunshot wounds or falling and hitting the ground (as in the 'parade ground fracture') or another hard stationary object. Such a fracture may cause deformity and *loss of function*, i.e. difficulty in eating and speaking.

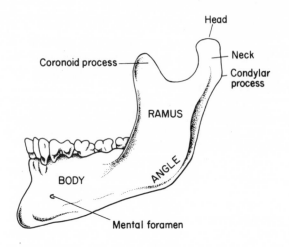

Fig. 5.29 Mandible – outer view of left half.

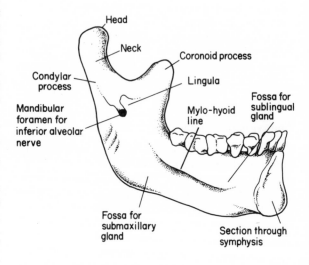

Fig. 5.30 Mandible – inner view of left half.

A severe blow to the point of the chin may result not only in mandibular fracture but the condylar heads may be displaced upwards and backwards to enter the middle cranial fossa, resulting in the leakage of cerebrospinal fluid from the ear (CSF otorrhoea).

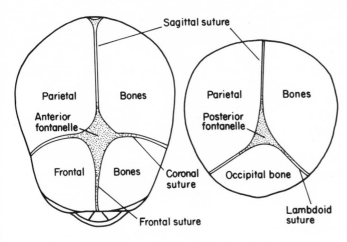

Fig. 5.31 The fetal skull from above (left) and from behind.

The fontanelles

It will be recalled that the sagittal suture separating the two parietal bones in the mid-line meets the frontal and the coronal suture in front and behind meets the lambdoid suture separating them from the occipital bone. At birth, however, the actual bones are not in contact. There is a diamond-shaped space covered only by fibrous tissue membrane between the frontal and two parietal bones in the mid-line: this is called the **anterior fontanelle**. It lies immediately above the superior longitudinal sinus.

The **posterior fontanelle** is situated at the area of junction between the sagittal and lambdoid sutures. It differs from the anterior fontanelle in being smaller and triangular and in closing a little earlier.

There are four other fontanelles which are small and of irregular shape. They are at the sphenoidal and mastoid angles of the parietal bones and are therefore named the sphenoidal and mastoid fontanelles (one pair of each).

The presence of the fontanelles and the width of the sutures permit the bones of the cranial vault, or vertex, to overlap slightly, thus facilitating the birth process, during which the skull is slowly compressed into a different shape, or moulded. During this process, the part of the scalp which lies more centrally in the birth canal may temporarily become swollen as a result of impeded venous drainage and oedema. The resulting swelling is known as a caput succedaneum.

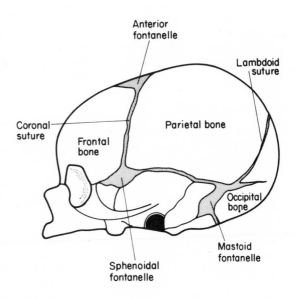

Fig. 5.32 The fetal skull − lateral aspect.

After birth the size of the fontanelles gradually lessens and the anterior fontanelle is usually closed by the age of eighteen months.

The sutures of the vault of the skull start to become obliterated in middle age.

The vertebral column

The cranium, vertebral column, ribs and sternum constitute the *axial skeleton*. The bones of the limbs and limb girdles make up the appendicular skeleton.

The vertebral column (spinal column, spine or backbone) is the central part of the skeleton which supports the head and encloses the spinal cord. Its construction combines *great strength* with a moderate degree of *mobility*. These features depend on the spine having a number of separate bones united by ligaments and by tough discs of fibrocartilage (the intervertebral discs) which act essentially as hydrostatic 'shock absorbers'. The nucleus of the disc is 85° water and it is bounded by a fibrous ring, the annulus. The bones give origin to a number of muscles.

The vertebral column is made up of 33 vertebrae which are grouped (from above downwards) as follows:

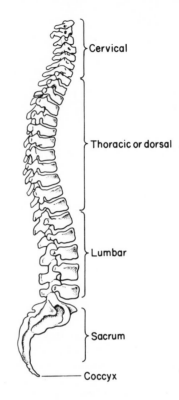

Fig. 5.33   The vertebral column.

7 cervical	discrete, true or movable vertebrae
12 thoracic	
5 lumbar	
5 sacral forming the sacrum	fused, false or fixed vertebrae
4 coccygeal forming the coccyx	

The spine as a whole

It will be noticed that the bones become increasingly larger as the column descends, reaching their maximum width at the upper part of the sacrum, only to become greatly reduced in size as it tapers off into the coccyx.

When looked at from the side the spine will be seen to have several curves.

(*1*) It is convex forwards in the cervical or neck region.

(*2*) The thoracic region is convex backwards. Excessive curvature is known as *kyphosis* (hump-back).

(*3*) The lumbar region is markedly convex forwards. When it is excessively convex the term *lordosis* is used. This term is often also applied to the normal cervical and lumbar curves but this is usually made plain in the text. For example, the normal lumbar lordosis of a pregnant woman acts as a support for the contracting uterus.

(*4*) The sacrum and coccyx form a marked forward concavity.

The three main curves in the vertebral column are developed from a single curve in the spine of an infant with the assumption of walking and the erect position.

It will be noticed also that the last lumbar vertebra joins the sacrum at a pronounced angle (the lumbosacral angle).

Lateral (sideways) curvature of the spine is abnormal and is known as *scoliosis*.

The vertebrae

The individual vertebrae are all built on the same plan, although there are certain variations in different parts of the spinal column and special vertebrae (atlas and axis), the sacrum and coccyx, which require separate description.

A **typical vertebra** consists of:
(*1*) a body;
(*2*) the vertebral or neural arch;
   (*a*) 2 pedicles,
   (*b*) 2 laminae,

(*c*) 2 transverse processes,
(*d*) 1 spine or spinous process,
(*e*) 4 articular processes.

The **body** is the solid box-shaped structure situated anteriorly which has slightly concave upper and lower surfaces. It is separated from the bodies of the vertebrae immediately above and below by the tough intervertebral discs of fibro-cartilage.

From either side of the posterior aspect of the body two short, stout bars of bone, the **pedicles**, project backwards. The pedicle has a notch on its upper surface and a similar notch on its lower surface. Each notch forms with its neighbour an intervertebral foramen. The intervertebral foramina transmit the segmental spinal nerves.

From the posterior ends of the pedicles, the two **laminae** are directed backwards and towards each other and meet in the mid-line behind. From the junction of the pedicles with the laminae, the **transverse processes** project outwards on each side of the bone.

Where the laminae unite in the mid-line behind, the **spinous process** or spine is formed and projects backwards and, in some parts of the column, downwards.

Two **articular processes** (zygapophyses) are situated on the upper and lower surface of each vertebra at the junction of the pedicles with the laminae close to the origin of the transverse processes.

The roughly circular opening enclosed by the body in front, the pedicles on either side and the laminae behind, through which passes the spinal cord, is called the **vertebral foramen** and it forms a part of the vertebral canal. This bony canal helps to protect the spinal cord from injury.

The cervical vertebrae

The lower five cervical vertebrae have the same general form, although the seventh has a particularly long spinous process, ending in a single tubercle, and large transverse processes to which cervical ribs are sometimes attached. The upper two vertebrae, the atlas and axis require separate description.

A typical cervical vertebra differs from those of the rest of the spinal column in having
(*a*) a smaller body,

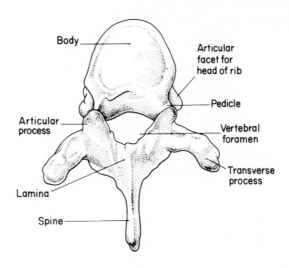

Fig. 5.34 A thoracic or dorsal vertebra seen from above.

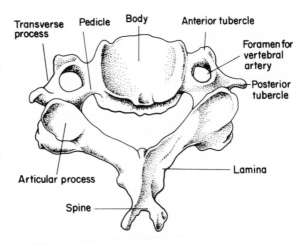

Fig. 5.35 A typical cervical vertebra viewed from above.

(*b*) an oblong shape, being broadest from side to side,
(*c*) a larger and roughly triangular vertebral foramen,
(*d*) a bifid (double-ended) spinous process, and
(*e*) a well-marked foramen in the transverse process for the passage of the vertebral artery.

The atlas

The first cervical vertebra, or atlas, is specially adapted, together with the axis, to carry the weight of the head and to facilitate its movements. The atlas is easily recognized because it has no body. Instead, it consists of a ring of bone enclosing a very large vertebral canal, on the anterior aspect of which is a small facet for articulation with the dens or odontoid process of the axis. The atlas pivots around the dens (at the atlanto-axial joint) when the head is rotated.

The upper surface of the atlas bears on each side a superior articular facet, on a thick lateral mass, for articulation with the occipital condyles of the skull. Nodding movements and lateral flexion are effected at the atlanto-occipital joints so formed. Unlike the other cervical vertebrae, the atlas has no spinous process.

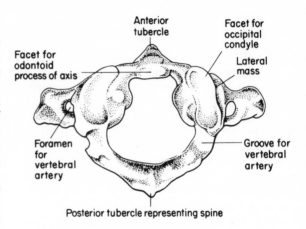

Fig. 5.36   The atlas viewed from above.

The axis

The second cervical vertebra, or axis, is characterized by a tooth-like process – the **dens** or odontoid process – which projects upwards from its body. This process actually represents and occupies the position of the missing body of the atlas. On its anterior surface is a small facet which articulates with the anterior arch of the atlas, an arrangement which permits the rotation of the head. The spinous process of the axis is large and strong; it takes the pull of muscles which extend, retract and rotate the head.

Dislocation, and less commonly fracture, of cervical vertebrae may occur as a result of a fall on the head with acute flexion of the neck or as a result of a forward jerk in a car or plane crash. Their intervertebral facets being relatively horizontal, *dislocation* of the cervical vertebrae can occur without fracture. In contrast, the thoracic and lumbar vertebrae nearly always fracture when dislocated forwards because their intervertebral facets are relatively vertical.

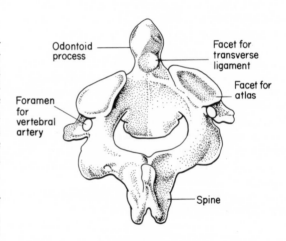

Fig. 5.37   The axis viewed from above and behind.

The thoracic vertebrae

Also known as the dorsal vertebrae, these carry the ribs and are twelve in number. Notable features are:

(*1*) A gradual increase in size from above downwards, reflecting the progressive increase in load.

(*2*) Bodies shaped like those of the cervical region in the case of the upper thoracic vertebrae and like those of the lumbar region in the case of the lower thoracic vertebrae. The fourth thoracic vertebra is heart-shaped. The bodies of the fifth to the eighth thoracic vertebrae are slightly flattened on their left side owing to the pressure of the descending aorta. An aortic aneurysm in this region will erode the bodies of these four vertebrae but leave their intervertebral discs unaffected.

(*3*) Costal facets; typically two on each side of the body, for the heads of the ribs and one on the tip of each transverse process, for articulation with the tubercles of the ribs.

(*4*) Long, markedly downward-sloping spinous processes.

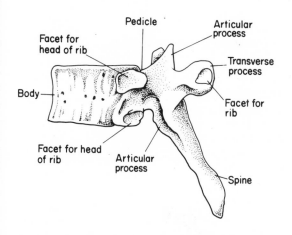

Fig. 5.38 A thoracic or dorsal vertebra seen from the side.

The lumbar vertebrae

(*1*) These are the largest vertebrae and are five in number.

(*2*) Their bodies are kidney-shaped.

(*3*) They have no costal facets.

(*4*) Their spinous processes are strong, quadrangular, broad and flat, and almost horizontal.

(*5*) The transverse process of the fifth lumbar vertebra is (*a*) massive and (*b*) connects with the whole of the lateral aspect of the pedicle and encroaches on the body of the vertebra. The transverse processes of the other lumbar vertebrae are thin and attach only to the junction of pedicle with lamina.

The sacrum

The sacrum is a large, triangular bone composed of five fused vertebrae, the individual parts of which can be discerned. It articulates above with the fifth lumbar vertebra (forming the lumbosacral angle), below with the coccyx and at the sides with the innominate bones. The dorsal surface of the sacrum is convex and the pelvic surface is concave, thereby increasing the capacity of the true pelvis.

The sacrum has a wide base, articulating with the fifth lumbar vertebra and a narrow blunted apex inferiorly, articulating with the coccyx. The transverse processes project from the central **body** and those of each side are fused together to form the lateral parts of the sacrum. Between the body and the lateral parts are the four anterior sacral foramina for the passage of nerves.

The upper margin of the body projects forwards and is called the **promontory** of the sacrum. The upper portion of the lateral part of the sacrum has an ear-shaped surface on its lateral aspect for articulation with the iliac part of the innominate bone, with which it forms the sacro-iliac joint. Behind this surface is a very rough area for the attachment of strong ligaments (see Figs 5.12 and 5.13).

On the convex posterior surface is the median sacral crest surmounted by three of four rudimentary spinous tubercles. On each side of the crest are four dorsal sacral foramina for nerves.

The neural canal is continued into the sacrum as the sacral canal and, at its lower end, opens on to the surface of the bone at the sacral hiatus, below the third or fourth spinous tubercle. The sacral canal contains the cauda equina, and the filum terminale and fifth sacral nerves emerge at the sacral hiatus. The inferior part or apex of the sacrum articulates with the coccyx.

Typically, the sacrum is wider in the female than in the male and its ventral concavity is deeper. The pelvic surface of the bone also faces downwards more in the female, thereby further increasing the size of the pelvic cavity and causing the sacrovertebral (lumbosacral) angle to be more prominent.

The coccyx (or tailbone)

This consists of four (sometimes three or five) vertebrae usually fused together so that their individual characteristics are not apparent, although the first coccygeal vertebra may be separate. The coccyx is triangular in shape, articulating at its base with the sacrum and tapering to its apex below. The apex or tip of the coccyx gives origin to the external anal sphincter. The dorsal surface of the coccyx and lower part of the sacrum gives origin to part of the gluteus maximus muscle of the buttock.

Ligaments of the vertebral column

The vertebrae are held together by strong ligaments which include:

(*1*) Those connecting the bodies. The tough **anterior and posterior longitudinal ligaments** (Fig. 5.39) run the whole length of the spine, joining the anterior and posterior aspects of the vertebral bodies respectively.

(*2*) Those connecting (*a*) the laminae and (*b*) the spinous processes.

(*a*) Those connecting the laminae are called the **ligamentum flava** and consist of elastic tissue.

(*b*) The tough **supraspinous ligaments** link the spinous processes. They are penetrated by the needle during the performance of lumbar puncture.

These, and to a lesser extent other ligaments, serve to support the vertebral column (spine) when it is in the fully flexed position.

(*3*) The vertebral bodies are also joined by the **intervertebral discs** of fibrocartilage, which are extremely strong and which account for a quarter of the length of the spine. Their atrophy in old age results in shrinkage in height. Each disc consists of a ring-like annulus fibrosus surrounding a gelatinous nucleus pulposus. The annulus fibrosus may rupture posteriorly, as a result of trauma or degeneration, and allow the nucleus pulposus to protrude backwards into the vertebral canal. This is known as a *prolapsed intervertebral disc* and it occurs most commonly between the fourth and fifth lumbar (L4/5) or fifth lumbar and first sacral (L5/S1) vertebrae. Prolapse at these sites exerts pressure on the roots of the fifth lumbar nerve and first sacral nerve, respectively. Referred pain (*sciatica*) is felt in the back of the lower limb, along the distribution of the sciatic nerve. Straight leg raising is painful and limited due to the traction exerted on the irritated and already stretched nerve root. The ankle jerk may be absent if the first sacral nerve (S1) is affected.

Movements of the spinal column

Apart from the movements of nodding and rotation of the head which, it has been seen, take place at the atlas and axis respectively, the movements of the spine as a whole are considerable although the actual range of movement between the individual vertebrae is small. They are:

(*1*) (*a*) flexion or bending forwards and (*b*) extension or bending backwards, in which the maximum movement takes place in the cervical and lumbar regions;

(*2*) lateral movement or bending from side to side, well marked in the neck and also possible in the dorsal region;

(*3*) rotation or twisting the spinal column as a whole around its long axis.

It is the elasticity of the intervertebral discs which makes all of these movements possible. The swing of the pelvis from side to side during walking is due to a rotatory movement at the lumbosacral articulation, together with similar movements of the lumbar intervertebral joints. This enables a patient to walk reasonably well even with fixed hip joints.

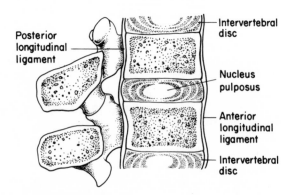

Fig. 5.39   Section of vertebral column showing ligaments and intervertebral discs.

NOTES: The following structures are on a level with various vertebrae:

structure	on a level with
oral pharynx	axis vertebra
opening of larynx	4th cervical vertebra
bifurcation of carotid artery	3rd cervical vertebra
upper margin of sternum	2nd thoracic vertebra
bifurcation of trachea	4th thoracic vertebra
heart	6th, 7th and 8th thoracic vertebrae
kidneys	12th thoracic, 1st and 2nd lumbar vertebrae
pancreas	1st and 2nd lumbar vertebrae
bifurcation of aorta	4th lumbar vertebra

A line joining the highest points of the iliac crests passes between the spines of the fourth and fifth lumbar vertebrae and is a landmark for lumbar puncture.

Fractures of the spine may be accompanied by damage to the spinal cord. Patients with spinal cord injuries suffer from retention of urine, which may be temporary or permanent. If infection is introduced at catheterization, ascending pyelonephritis may result and may be fatal.

Bones of the thorax

The skeletal framework of the thorax is formed (*a*) behind, by the thoracic or dorsal vertebrae; (*b*) anteriorly, by the sternum and costal cartilages; and (*c*) the remainder of the circumference, by the ribs. The thoracic cage thus formed communicates above with the root of the neck through the *'thoracic inlet'* and is separated below from the abdominal cavity by the diaphragm.

The sternum

This is an almost flat, dagger-shaped bone which has a slightly convex anterior and a slightly concave posterior surface. It is composed of highly vascular trabecular bone covered by a layer of compact bone. The spaces in its spongy interior contain red marrow and a sample of this may be aspirated through a wide-bore needle pushed through the thin cortical bone (*sternal puncture*). Such a sample may be extremely valuable in the diagnosis of anaemias, leukaemias and other diseases affecting the bone marrow.

Along each side of the sternum are seven indentations where the costal cartilages are attached. The sternum is divided into three parts: (*a*) the manubrium above; (*b*) the mesosternum or **body** in the middle; and (*c*) the xiphoid process below.

The **manubrium** is roughly triangular and it articulates on either side with the clavicle, at the sternoclavicular joint, and with the first and second costal cartilages. The adjoining body of the sternum also articulates with the second costal cartilage. The cartilages of the 3rd, 4th, 5th and 6th ribs are attached at intervals, while the seventh is attached at its junction with the xiphoid process.

The **xiphoid process** forms the lower extremity of the sternum. It sometimes remains cartilaginous but is usually ossified, in its upper part, in adults. It is a small, more or less triangular plate to which are attached the fibres of the linea alba and the rectus abdominis muscle. A part of the diaphragm is attached to its posterior surface.

The sternum is cut vertically down through its midline (*median sternotomy*) to give access to the heart for cardiac surgery. The sternum is also split in this way in operations on the thymus gland.

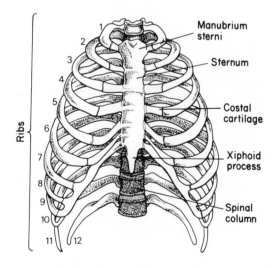

Fig. 5.40 Skeleton of the thorax.

The ribs

The ribs are arches of bone which are directed obliquely forwards and downwards from the spine, with which they articulate, to form the greater part of the thoracic cage. There are usually twelve pairs and they are classified into two groups.

(*1*) True ribs – the upper seven pairs, which are joined to the sternum by their costal cartilages.

(*2*) False ribs – the lower five pairs, which have no direct attachment to the sternum. The costal cartilages of the 8th, 9th and 10th fuse with the cartilage immediately above. The 11th and 12th (*floating ribs*) only partly encircle the thorax and are unattached in front (Fig. 5.40).

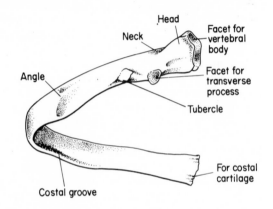

Fig. 5.41 A typical rib.

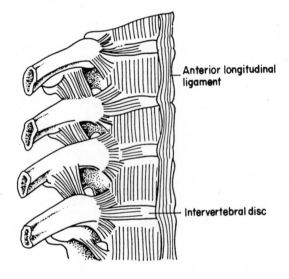

Fig. 5.42 Thoracic vertebrae with ligaments and articulations of ribs.

The first two ribs and the last three have special features but the intervening seven ribs are similar to each other and are regarded as typical ribs.

Each **typical rib** is a long bone with (*a*) a head, (*b*) a neck posteriorly and (*c*) a shaft. The **head** bears two facets for articulation with the numerically corresponding dorsal vertebra and the vertebra above. The **neck** is stout and lies in front of the transverse process of the corresponding vertebra. There is a well-marked tubercle on the outer surface of the rib where the neck joints the shaft. It has a facet for articulation with the transverse process of the corresponding vertebra.

The **shaft** of the typical rib is long, thin and flattened. It is not only curved and slightly twisted on itself, but is bent, having an angle about 5 or 6 cm lateral to the tubercle. The shaft has a smooth convex external surface, superior and inferior borders, and a smooth internal surface grooved inferiorly by the **costal groove** which is occupied by the intercostal artery and nerve (Fig. 5.43). The borders of the ribs give attachment to the intercostal muscles which pass to the ribs immediately above and below. The internal surface of the rib is lined by parietal pleura, which is firmly adherent to the periosteum of the bone.

The **atypical ribs** have distinctive features. The **first rib** is broad and is usually the shortest and flattest of the ribs. It is more curved than all the other ribs but unlike them it is not twisted. Its surfaces face upwards and downwards, i.e. it has *upper and lower surfaces*. Near the centre of the inner border of the upper surface is the scalene tubercle, a small projection to which the scalenus anterior muscle is attached. The subclavian vein crosses the rib in the groove in front of this tubercle and there is a groove for the subclavian artery behind the tubercle. The latter groove is also occupied by the lowest trunk of the brachial plexus; at this site the anaesthetist can infiltrate the plexus with a local anaesthetic.

The **second rib** is about twice as long as the first rib and much less curved. It bears a rough tubercle for the serratus anterior muscle. Its costal groove is short and poorly marked.

The **eleventh and twelfth ('floating') ribs** are short and thin. They have no tubercles and their heads bear only a single facet.

The ribs consist of highly vascular trabecular bone, containing red marrow, enveloped by a thin layer of compact bone.

Rib fractures are common. Sometimes the

broken end of the bone is driven into the lung, lacerating it and causing the leakage of air and blood into the pleural cavity, a condition known as haemopneumothorax. A severe crushing injury may fracture several ribs fore and aft so that a portion of the thoracic cage becomes loose ('stove-in chest') and moves paradoxically with breathing, bulging out with expiration and being sucked in with inspiration. The mediastinum swings with the abnormal movement and causes severe shock. Urgent treatment is necessary, a drain being inserted into the chest and connected to an underwater seal, and postive pressure ventilation being given through an endotracheal or tracheostomy tube.

A *cervical rib* is found in a few people. It articulates with the transverse process of the 7th cervical process. Its anterior end is either free, if the rib is short, or it articulates with the first thoracic rib. It may in part be a fibrous cord. Paraesthesiae ('pins and needles') in the ulnar aspect of the forearm and wasting of the small muscles of the hand may result from the pressure of a cervical rib (even if only a fibrous cord) on the overlying lower trunk of the brachial plexus. Occasionally, pressure on the overlying subclavian artery causes vascular problems (ischaemia or even gangrene) in the upper limb.

The costal cartilages

These are the flattened bars of hyaline cartilage which connect the upper seven ribs to the sternum and the 8th, 9th and 10 ribs to the cartilages immediately above them. The cartilages of the 11th and 12th ribs are tapered and end in the musculature of the abdominal wall.

The costal cartilages contribute to the mobility, elasticity and resilience of the thoracic cage. Were it not for them, the ribs and sternum would be fractured more frequently.

In old age, the costal cartilages tend to ossify and become less pliable. This makes cardiac massage more difficult (more force being necessary to achieve a useful cardiac output) and more likely to result in fractures of the ribs and sternum.

The intercostal spaces

Typically, each intercostal space (Fig. 5.43) contains three muscles and a neurovascular bundle (of intercostal nerve, artery and vein). Irritation of the nerves by diseases of the thoracic spine (e.g. spinal tuberculosis) may cause referred pain in the front of the chest, where the nerves terminate. Pus from a tuberculous vertebra tends to track along the neurovascular bundle and to erupt as a cold abscess at one of the three points of emergence of the cutaneous branches of the intercostal nerves, just lateral to the spine, in the mid-axillary line, or just lateral to the sternum.

The intercostal spaces, being occupied only by soft tissue, provide useful 'windows' for examination of the heart by ultrasound (echocardiography), which cannot be performed through the bone of the ribs.

In a conventional thoracotomy for excision of diseased lung tissue (e.g. lobectomy for bronchial carcinoma or bronchiectasis), access is gained by resecting a portion of the fifth or sixth rib, by a technique which preserves the neurovascular bundle. The elasticity of the thoracic cage permits the surgical intercostal space so created to be widely opened, using rib retractors.

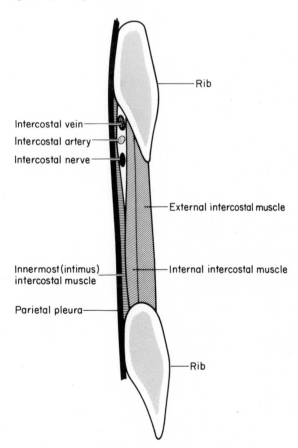

Fig. 5.43 Section through ribs and an intercostal space.

Questions

1. What are the constituents of bone and what factors influence its development?
2. What are the different varieties of bones? Give one example of each type and describe its structure.
3. What are the functions of the skeleton?
4. Briefly describe the skull and indicate its functions.
5. Describe the skeleton of the thorax.
6. Briefly describe the bones of the shoulder girdle.
7. Describe the bony framework of the pelvis and indicate some of the pelvic differences between the sexes.
8. Describe the femur. What is the relationship between age and the site of fracture of the femur?
9. Describe the mandible. What are its functions?

6
The Joints or Articulations

Joints or articulations are the meeting places of bones. They are concerned with growth, rigidity and movement. They may be divided into three groups:

(1) fixed or immovable joints (fibrous joints);
(2) slightly movable joints (cartilaginous joints);
(3) freely movable joints (synovial joints).

Fixed joints

These are exemplified by the sutures between the skull bones. These bones either have bevelled edges and overlap (in a squamous suture) or have irregular edges which interlock together in a jagged line (in a plane suture).

Freely movable joints

Movement at these joints is dependent upon the opposed surfaces being able to slide upon one another. Freely movable (synovial joints) have the following characteristics.

(a) The articulating ends of the bones are covered by hyaline cartilage.
(b) The joint is enclosed in a fibrous capsule, supported by ligaments.
(c) The capsule of the joint is lined by synovial membrane.
(d) The cavity of the joint contains viscous synovial fluid for its lubrication and for the nutrition of living cells in the articular cartilages, which contain no blood vessels.

Depending on their shapes, synovial joints have been given differing names, of which the following are examples.

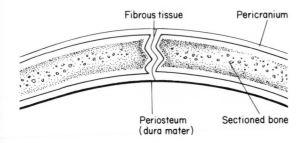

Fibrous tissue   Pericranium

Periosteum   Sectioned bone
(dura mater)

Fig. 6.1 A fixed joint or suture (skull).

Slightly movable joints

These are represented by the symphysis pubis, the manubrio-sternal joint, and the joints between the vertebral bodies. The movement at these joints depends upon the deformability of the tissue (cartilage) interposed between the bones.

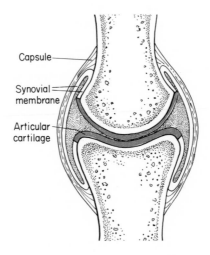

Capsule

Synovial membrane

Articular cartilage

Fig. 6.2 Typical synovial joint.

Plane joints (gliding joints)

In these, movement is mainly a gliding (translation) of one plane surface across another, as in the metatarsal joints.

Hinge joints

These roughly resemble door hinges in that their to-and-fro movement is largely restricted to one plane, a movement called flexion and extension. The elbow and interphalangeal joints are good examples.

Bicondylar joints

These are like hinge joints in that their principal movement is in one plane, but a limited amount of rotation is also possible. Two convex, knuckle-shaped condyles articulate with concave surfaces, as in the knee and temporomandibular joints.

Pivot joints

These permit only rotation, such as that of the atlas around the dens of the axis.

Ellipsoid joints

These are formed by an oval, convex surface of one bone fitting into an elliptical concavity of another. Primary movements are possible about two axes at right angles, as exemplified by the metacarpophalangeal joints where flexion-extension and abduction occurs. A combination of these movements results in circumduction, in which the movement of the finger outlines a cone with its apex at the joint and its base distally.

Sellar (saddle) joints

A saddle joint has apposing surfaces which are concavo-convex. Each surface has a direction in which it is maximally convex and at right angles to this the surface is maximally concave. In the joint, the convexity of the larger surface is apposed to the concavity of the other bone. Primary movements can occur in two planes at right angles and are accompanied by a degree of axial movement. The carpometacarpal joint of the thumb is the best known example.

Spheroidal (ball-and-socket) joints

Exemplified by the hip and shoulder joints, these are formed by the globular head of one bone fitting into the cup-like concavity of another. They are very freely movable in all directions, namely:

(*1*) Extension (straightening) and flexion (bending).

(*2*) Abduction, or movement of the part away from the mid-line of the body.

(*3*) Adduction, or movement towards the mid-line of the body.

(*4*) Rotation, or movement around the axis of the limb itself.

(*5*) Circumduction, which is seen particularly in the shoulder joint, as in the act of bowling a cricket ball (overarm). It involves a combination of all the other movements in some degree.

Joints of the shoulder girdle and upper limb

The sternoclavicular joint

This is a type of saddle joint and is formed between the medial end of the clavicle, the upper and lateral angle of the manubrium sterni, and the cartilage of the first rib. There is an articular disc between the articular surfaces of the sternum and clavicle. A fibrous capsule surrounds the joint but is thin above and below. The joint owes its strength to anterior and posterior thickening of the capsule, four other ligaments and especially the articular disc. Such is this strength that dislocation of the joint is rare and the clavicle breaks rather than dislocates from the sternum.

The **scapula** has two articulations:
(1) the acromioclavicular joint;
(2) the humeral (shoulder) joint.

The acromioclavicular joint

This joint is between the lateral end of the clavicle and the medial margin of the acromion. It is of approximately plane type, the apposing surfaces not being quite flat. The articular surfaces are covered with fibrocartilage and there is frequently also an articular disc of cartilage between the ends of the bones. The joint is surrounded by a fibrous capsule which is strengthened superiorly by the acromioclavicular ligament. Another ligament, the coracoclavicular ligament, is quite separate from the joint but helps to keep the two bones of the joint together. With dislocation of the joint, this ligament is torn and the scapula falls away from the clavicle.

The shoulder joint

This is a spheroidal (ball-and-socket) joint formed by the relatively large hemispherical head of the humerus and the small, shallow glenoid cavity of the scapula. The glenoid cavity is deepened by a ring of fibro-cartilage, the glenoid labrum, attached to its rim. The fibrous capsule of the shoulder joint is lax, allowing very free movement, and the stability of the joint depends almost entirely on the surrounding muscles. One of these is the long head of biceps which actually lies within the joint in a synovial sheath, together with which it passes through an opening in the capsule.

Movements

The shoulder joint is capable of an enormous variety of combinations of swinging and spinning movements. The movements usually described are:

(*1*) Flexion, or moving the arm forwards and medially across the chest, and extension, or moving it backwards and laterally.

(*2*) Abduction anterolaterally away from the trunk and adduction towards it.

(*3*) Circumduction and medial and lateral rotation.

A wide range of *accessory* movements is also possible at the shoulder joint because of the laxity of its capsule and the shallowness of the glenoid cavity. For example, the head of the humerus can be moved back and forth, and up and down, in relation to the glenoid cavity.

The price of great mobility in a joint is reduced stability and the shoulder joint is more often dislocated than any other. The lower and front part of the capsule is thin and unsupported by muscles and it is usually this part which is torn by the head of the escaping humerus as it is dislocated by violent abduction. The humeral head comes to lie below the level of the glenoid cavity and in a subcoracoid position. Occasionally, the circumflex nerve, which lies in relation to the surgical neck of the humerus, is torn during dislocation of the shoulder. The dislocated head of the humerus is drawn medially by the powerful adductors of the shoulder and the normal bulge of the deltoid muscle over the greater tuberosity is lost, so that the deltoid region of the upper arm appears flattened. In addition, the head of the humerus is internally rotated by the subscapularis muscle.

Movements at the shoulder joint are only a part of *movements of the shoulder girdle as a whole*.

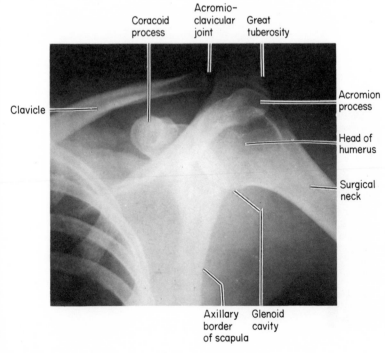

Fig. 6.3 X-ray of the shoulder (front view).

For example, abduction of the humerus initiated by the supraspinatus muscle and taken to 90° by the deltoid muscle, is continued smoothly to 180° (elevation) by upward rotation of the scapula by the trapezius and serratus muscles. Even if the shoulder joint is ankylosed (fused), considerable shoulder movement can be effected by elevation and depression, as in shrugging of the shoulders, forward (protraction) and backward (retraction) movement of the scapula around the chest wall, and rotation of the scapula forwards and upwards. This rotation involves movements at the sterno-clavicular and acromioclavicular joints.

The elbow joint

This is a hinge joint and also a compound joint because it includes two articulations in which three bones participate, namely the humerus above and the ulna and radius below. The trochlea of the humerus articulates with the trochlear notch of the ulna and the capitulum of the humerus articulates with the head of the radius. A fibrous capsule envelopes the joint and is continuous with the capsule of the superior radio-ulnar joint. The articular capsule is related anteriorly to the brachialis muscle and posteriorly to the tendon of the triceps muscle.

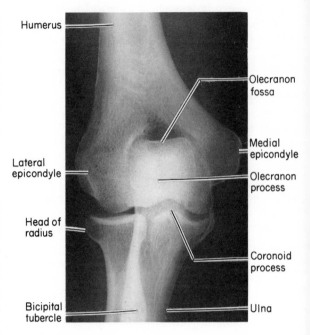

Fig. 6.4   X-ray of the elbow.

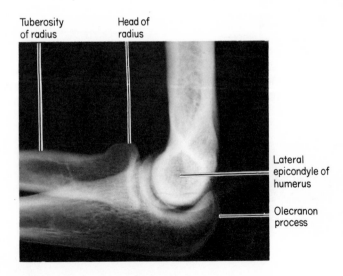

Fig. 6.5   X-ray of the elbow joint.

Movements
These are:

(*1*) flexion, in which the forearm swings up towards the humerus;

(*2*) extension, or straightening of the limb.
During flexion, the coronoid process of the ulna fits into the coronoid fossa of the humerus. In extension, the olecranon process occupies the olecranon fossa of the humerus.

A fall on the hand may, as a result of indirect violence, cause posterior dislocation of the elbow joint. This is occasionally accompanied by fracture of the coronoid process of the ulna.

The radio-ulnar joints

The radius and ulna articulate at their upper and lower extremities and their shafts are connected by the interosseous membrane which separates the forearm into anterior and posterior compartments. The proximal and distal radio-ulnar joints are uni-axial pivot joints. A strong fibrous band, the annular ligament of the proximal radio-ulnar joint, is attached to the anterior and posterior margins of the radial notch. It forms about four-fifths of the osseofibrous ring in which the circumference of the head of the radius rotates; the remaining fifth is the radial notch of the ulna. The annular ligament has a thin coating of cartilage on that part of its inner surface which comes into contact with the head of the radius. The distal radio-ulnar joint has a triangular articular disc which, together with the articular capsule, holds the ulna and radius together at their lower ends.

Movements
The movements between the radius and ulna, at the radioulnar joints, result in supination and prona-tion. When the forearm is in the position of anatomical description, with the palm of the hand facing forwards, the bones are parallel and the forearm is said to be supinated. The movement resulting in this position is called **supination**. Rota-tion of the radius so that its lower part crosses the front of the ulna, as in placing the palm of the hand on a flat surface, is called **pronation**. Supination is the more powerful movement and this has been taken into account in the design of nuts, bolts and screws, which are inserted and tightened by supina-tive movements (most people being right-handed).

The wrist joint or radiocarpal joint

This is an ellipsoid joint formed by the distal end of the radius and the lower surface of the articular disc above, and the proximal row of carpal bones below. The fibrous articular capsule of the joint is lined by synovial membrane and strengthened by ligaments.

The intercarpal joints

These are the joints between the eight carpal bones, which are connected together by an extensive system of ligaments.

Movements
Movements of the wrist joint are closely associated with those of the intercarpal joints, all forming part of the same mechanism and being acted upon by the same muscle groups. The movements include:

(*1*) Flexion, or bending the wrist forwards, and extension, or bending it backwards.

(*2*) Adduction, i.e. movement to the ulnar side and abduction, to the radial side. These movements are particularly important in the manipulation of precision tools.

(*3*) Circumduction, which is achieved by a sequence of flexion, adduction, extension and abduction or the same movements in reverse order.

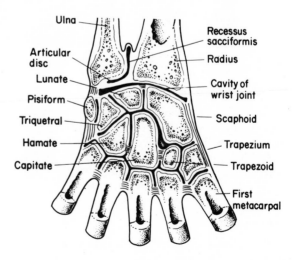

Fig. 6.6 Section showing wrist, carpal and carpophalangeal joints.

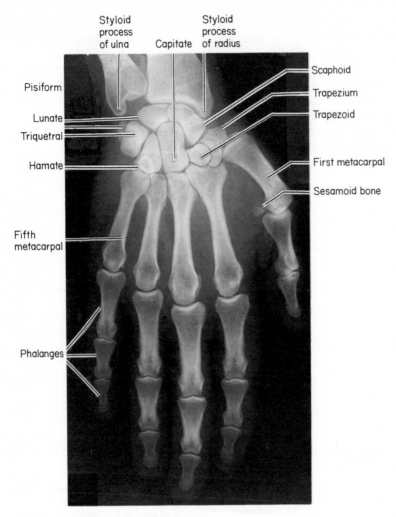

Fig. 6.7   X-ray of the wrist and hand.

The carpometacarpal joints

These are the joints between the distal row of carpal bones and the bases of the metacarpals. The carpometacarpal joint of the thumb is the most obviously saddle-shaped and, therefore, a sellar joint; its range of movement is large because of the way its articular surfaces are shaped.

The intermetacarpal joints

These are small synovial joints between the bases of the second to fifth metacarpals.

The **carpometacarpal** and **intermetacarpal joints** of the fingers permit only slight gliding movements of the articular surfaces upon one another. They are utilized most obviously in cupping the hand.

The **metacarpophalangeal joints** are usually classified as ellipsoid, and the **interphalangeal joints** are hinge joints. The active movements of the metacarpophalangeal joints are flexion and extension, adduction and abduction (as in spreading the fingers), circumduction and a small degree of rotation. The only active movements of the interphalangeal joints are flexion and extension.

The movement of applying the tip of the thumb to the tips of the slightly flexed fingers is called *opposition* and is dependent on movements at both the carpometacarpal and the intercarpal joints.

Joints of the pelvic girdle and lower limb

Joints of the innominate bone

Each innominate bone is concerned in the formation of three joints:
- (*a*) the sacro-iliac joint;
- (*b*) the pubic symphysis;
- (*c*) the hip joint.

The sacro-iliac joint

This is an important joint transmitting the weight of the body through the vertebral column, via the pelvis, to the lower limbs. It is formed by articulation of the ear-shaped area on the lateral mass of the sacrum with a similar area on the posterior part of the internal surface of the ilium. The articular surfaces exhibit ridges and depressions which fit into each other and limit the movement of the joint but contribute to its strength. Strong ligaments hold the bones in apposition. The sacro-iliac joints, which are synovial joints, are the first to be affected by inflammatory changes in the disease known as ankylosing spondylitis.

The pubic symphysis

This is the cartilaginous joint between the two pubic bones, where they meet anteriorly in the midline. The bones are held together by ligaments and by a disc of fibrocartilage. Very little movement normally occurs but this joint and the sacro-iliac joints have increased mobility in late pregnancy and during childbirth.

The hip joint

It has been seen that the essential feature of the shoulder joint is mobility. That of the hip joint is *stability* combined with a reasonable degree of movement.

The hip joint is a ball-and-socket joint formed between the head of the femur and the acetabulum of the innominate bone. The socket of the acetabulum, like that of the glenoid cavity of the scapula, is deepened by a surrounding ring of fibrocartilage, called in this case the acetabular labrum. The strong fibrous capsule of the joint is strengthened by three ligaments:
- (*a*) the iliofemoral ligament, the strongest ligament in the body, in front and above;
- (*b*) the pubofemoral ligament below; and
- (*c*) the ischiofemoral ligament behind.

The ligament of the head of the femur is a triangular, somewhat flattened band extending from a pit in the head of the femur to the notch in the rim of the acetabulum, where it blends with the transverse ligament of the acetabulum.

Movements

The active movements of the hip joint are:

(*1*) Flexion, or bending the thigh upwards towards the trunk; and extension, or movement of the thigh backwards.

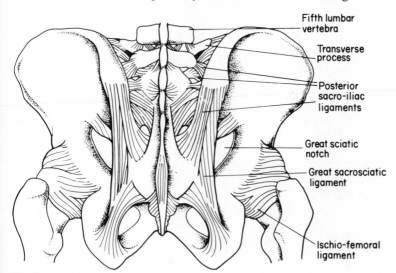

Fifth lumbar vertebra

Transverse process

Posterior sacro-iliac ligaments

Great sciatic notch

Great sacrosciatic ligament

Ischio-femoral ligament

Fig. 6.8 Ligaments of the pelvis viewed from behind.

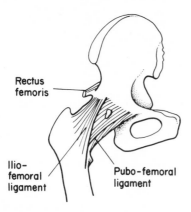

Fig. 6.9   The anterior ligaments of the hip joint.

(*2*) Abduction and adduction, or movement of the thigh away from and towards the mid-line.

(*3*) Circumduction, a combination of the foregoing movements.

(*4*) Rotation.

Simple flexion and extension of the hip are limited to about 90° forwards and only up to 10 to 20° backwards from the vertical. However, the amplitude of these movements is greatly increased by angulation of the vertebral column, by tilting and rotation of the pelvis, and by flexion of the knee and concomitant medial or lateral rotation at the hip joint.

These adjustments increase the extension of the thigh in walking, running, dancing and other activities. The range of abduction and adduction can be similarly increased. A patient with a fused hip (arthrodesis) can walk by swinging the affected limb along, utilizing pelvic tilt and spinal angulation.

Sometimes in the fetus there is defective formation of the acetabulum so that the head of the femur is not retained in its cavity. This condition is known as *congenital dislocation of the hip* and may not always be discovered until the child starts to stand. It requires special orthopaedic treatment.

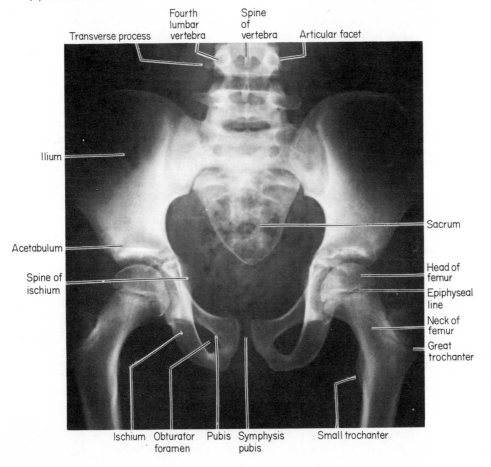

Fig. 6.10   X-ray of the pelvic girdle.

In late middle and old age, degenerative changes are common in the hip joint. The condition known as *osteoarthritis* limits mobility and causes pain, both of which may be overcome by total hip replacement with metallic prostheses (an artificial hip joint).

The knee joint

This is the largest joint in the body and is of great importance because of (a) movement in walking, etc., (b) stability, including maintenance of the erect posture and the transmission of the body weight to the feet.

The knee joint is formed by the articulation of the two condyles of the femur with the condyles of the tibia. The patella also takes part in its formation.

The capsule is strengthened by the important medial and lateral collateral ligaments. Violent abduction or adduction strains may tear these ligaments. Anteriorly, the capsule is strengthened by the ligamentum patellae or patellar ligament. The capsule does not extend above the patella; its absence at this site allows free communication between the cavity of the knee joint and the suprapatellar bursa. This bursa may become distended with fluid, when the knee joint is affected by inflammation or trauma, and cause swelling above the level of the patella.

The following structures are present within the joint:

(*1*) The **cruciate ligaments**. These are attached above to the intercondylar notch of the femur and below to the upper surface of the tibia, in such a way that they cross each other. They may be torn in severe abduction or adduction injuries.

(*2*) The **medial and lateral menisci** (semilunar cartilages). These are half-moon or C-shaped pads of cartilage which are attached to the upper end of the tibia and help to deepen the articular surfaces. They are liable to injury, especially in sports such as football, and may need surgical removal (meniscectomy). Postoperatively, the menisci regenerate but may be permanently damaged if the knee is subjected to continued and violent exercise during the period of regeneration.

The patella articulates with the anterior aspect of the femur between the two condyles.

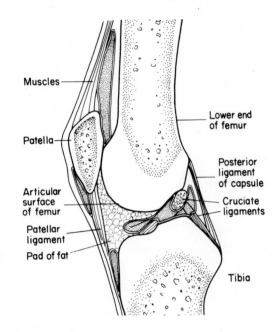

Fig. 6.11   Section through the knee joint.

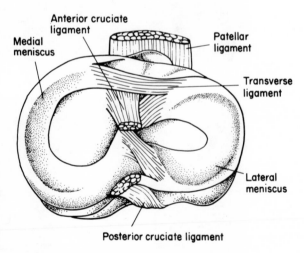

Fig. 6.12   The meniscus of the right knee joint and structures attached to the upper surface of the tibia.

The knee joint is of the compound variety, differing from a true hinge joint in that the movements of flexion and extension occur around a shifting axis and are accompanied by some rotation.

The stability of the knee depends on its powerful ligaments but to an even greater extent upon the strength of its neighbouring muscles. Hence the

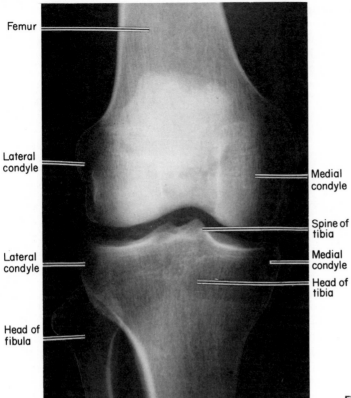

Femur

Lateral
condyle

Medial
condyle

Spine of
tibia

Lateral
condyle

Medial
condyle

Head of
tibia

Head of
fibula

Fig. 6.13   X-ray of the knee joint – anterior view.

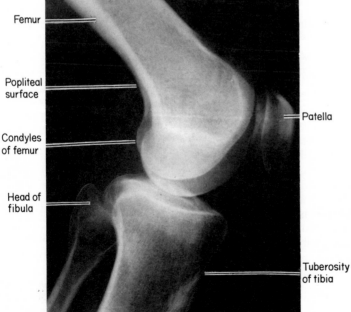

Femur

Popliteal
surface

Condyles
of femur

Head of
fibula

Patella

Tuberosity
of tibia

Fig. 6.14   X-ray of the knee joint – lateral view.

importance of quadriceps exercises during convalescence and rehabilitation after knee injuries and operations.

The tibiofibular joints

The **superior tibiofibular joint** is a plane synovial joint between the head of the fibula and the lateral condyle of the tibia. In contrast with the mobility of the radius on the ulna, very little movement takes place between the tibia and the fibula because the main function of the leg is weight-bearing, rather than movement. The inferior tibiofibular joint is a fibrous joint in which the lower ends of the bones of the leg are firmly bound together by ligaments. Firm union is necessary in order to give stability to the ankle joint below.

Stretching between the shafts of the bones is an interosseous membrane which divides the leg into anterior and posterior compartments like those of the forearm.

The ankle joint

The ankle or talocrural joint is a hinge joint formed between the tibia above and medially, the fibula laterally and the talus inferiorly. The fibrous capsule of the joint is strengthened by the medial and lateral ligaments (Fig. 6.15). Some of the fibres of the medial ligament are torn in abduction sprains

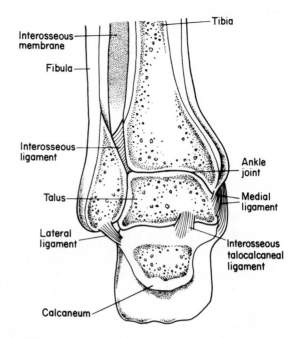

Fig. 6.15 Section through the ankle joint.

of the subtalar joints, the common sprains which are often referred to as sprained ankles. True sprains of the ankle usually result from forced plantar flexion, which tears the capsular ligament anteriorly.

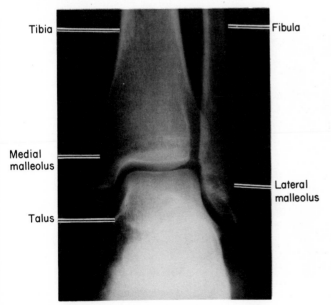

Fig. 6.16 X-ray of the ankle — anterior view.

The joints of the foot

These include the **intertarsal joints** (the subtalar joint between the talus and the calcaneus, and those between the other tarsal bones), the **tarsometatarsal joints**, the **intermetatarsal joints**, the **metatarsophalangeal joints** and the **interphalangeal joints**. The latter two sets of joints are similar to those of the hand and fingers.

Movements of the foot and toes
The active movements of the ankle joint, which is a hinge joint, are those of *dorsiflexion* (pointing the foot up towards the leg) and *plantar flexion* (point-ing the foot downwards). Movements at the inter-tarsal joints enable the foot to be moved so that the sole faces medially (*inversion*) or laterally (*eversion*). As in the corresponding joints of the hands, the metatarsophalangeal joints permit flexion, extension, adduction and abduction of the digits. Movement of the toes upwards towards the dorsum of the foot is called extension and movement down-wards towards the sole is called flexion. If the ankle and toes are both moved towards the leg, the ankle is flexed and the toes are extended. In the opposite movement, the ankle is extended and the toes are flexed.

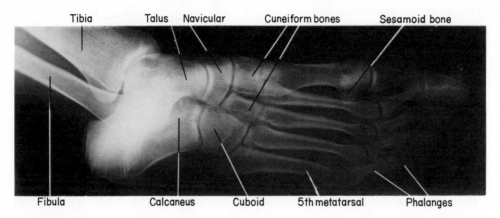

Fig. 6.17 X-ray of the foot from above

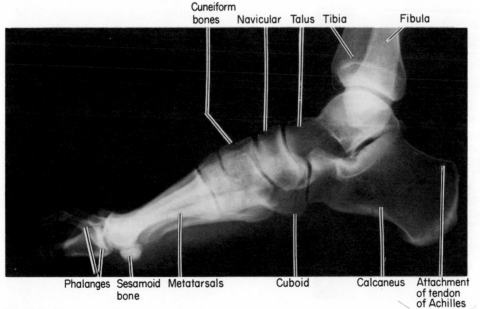

Fig. 6.18 X-ray of the foot, lateral view.

Joints of the trunk, etc.

Vertebral joints, see page 68.
Joints of the ribs, see page 71.

Practical notes
Joints are liable to inflammation which is called arthritis. They may be involved in fractures of the bones in their vicinity. Ligaments may be partially torn (sprains) or completely torn and the bones may be forced out of their natural positions — dislocations.

Questions

1. Describe the hip joint. What are its movements and how may they be affected by disease of the joint?
2. Describe the ankle joint and its movement.
3. Give an account of the types of joint found in the human body.
4. Give a description of the knee joint. Which of its components are particularly liable to injury?
5. Describe and compare the shoulder joint and the hip joint.

7

The Muscular System

One of the most important characteristics of humans is their ability to adjust to changes in their environment, and *movement* is an essential feature of this adjustment. The skeleton forms a supportive framework and consists of a series of bony levers capable, by virtue of the joints and muscles, of moving upon one another. Muscles which move the skeleton are called *skeletal muscles* (also known as voluntary or striated muscles p. 17). Skeletal muscle constitutes approximately 40 per cent of the total body weight.

Muscle tissue has the specialized function of *contraction* (i.e. the ability to shorten and thicken) and thus produce movement of the different parts of the body. Muscle tissue is also elastic, and can be stretched by a weight. When the weight is removed the muscle returns to its normal length.

The upright posture of humans means that they are constantly battling against the pull of gravity. The force which counteracts gravity and keeps the body upright is provided by the skeletal muscles. Muscles are continually in a state of partial contraction which enables the body to maintain stationary positions such as sitting and standing.

Skeletal muscle cells are highly active and the chemical reactions which take place during contraction result in the production of heat. Muscular activity therefore plays an important role in the homeostasis of normal body temperature (p. 212).

Structure of skeletal muscle

Skeletal muscle is composed of numerous elongated cells called **muscle fibres**. The fibres are arranged in bundles known as **fasciculi**. The spaces between the fibres within a fasciculus are filled by delicate connective tissue, the **endomysium**. Each fasciculus is surrounded by a stronger connective tissue sheath, the **perimysium**. The perimysium extends between the fasciculi and is continuous with the tough fibrous sheath, the **epimysium**, which envelops the whole muscle.

Muscle fibres

Muscle fibres are elongated cells ranging between 10 and 100 microns in diameter. The length of the fibres varies with the size and shape of a muscle, most fibres extending the entire length of the muscle, while a few end in connective tissue intersections within the body of the muscle.

Each fibre is surrounded by a cell membrane known as the **sarcolemma**, just beneath which lie many flattened nuclei. The cytoplasm of muscle fibres is called the **sarcoplasm** and contains numerous mitochondria lying between bundles of fine longitudinal threads or myofibrils.

Myofibrils are about one micron in diameter and are the contractile units of muscle fibres. They show the characteristic cross striations of skeletal muscle. Under the electron microscope it can be seen that each myofibril is composed of still smaller units, the thick and thin **myofilaments**.

Groups of *thick* myofilaments lie partially overlapping groups of *thin* myofilaments, giving rise to the alternating dark and light bands of the myofibrils (Fig. 7.1). The dark *A band* consists of overlapping thick and thin myofilaments, the lighter bands containing only thin myofilaments are called *I bands*. Halfway along the length of the I band the thin myofilaments are attached to a narrow zone known as the *Z line*. The *H zone* lies in the centre of the A band and consists of thick myofilaments only. The portion of a myofibril lying between two Z lines is called a **sarcomere**, each myofibril consisting of a series of sarcomeres.

The sarcolemma of a muscle fibre forms blind-ended sacs or **T-tubules** which penetrate the cell and lie in the spaces between the myofibrils. The T-tubules contain interstitial fluid and do not open into the interior of the muscle fibre. Within the sarcoplasm of the muscle fibre there is an extensive network of branching and anastomosing channels which forms the **sarcoplasmic reticulum** (this structure is comparable to the endoplasmic reticulum,

(a)

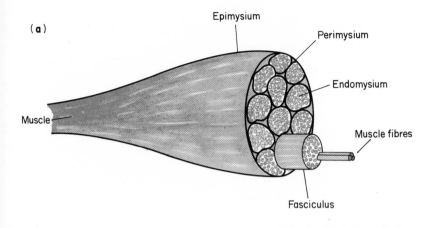

Epimysium

Perimysium

Endomysium

Muscle fibres

Muscle

Fasciculus

(b)

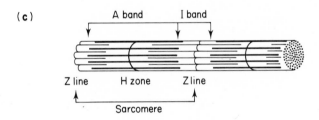

Muscle fibre

Myofibril

Nucleus

Endomysium

(c)

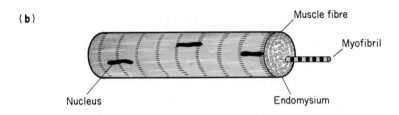

A band

I band

Z line

H zone

Z line

Sarcomere

Fig. 7.1 (a) Structure of skeletal muscle. **(b)** Single muscle fibre, showing characteristic striations. **(c)** Myofibril, illustrating a sarcomere.

p. 10). The channels of the sarcoplasmic reticulum lie in close contact around the ends of the T-tubules, and contain stores of calcium ions.

Contraction of muscle fibres

The thick myofilaments are composed of a protein called *myosin*. Each myosin filament has small regular projections known as **crossbridges**. The crossbridges lie in a radial fashion around the long axis of the myofilament. The rounded heads of the crossbridges lie in apposition to the thin myofilaments.

The thin myofilaments are composed of a complex protein called *actin*, arranged in a double-stranded coil. The actin filaments also contain two additional proteins called *troponin* and *tropomysin*.

In a resting muscle fibre the myosin crossbridges are prevented from combining with the actin filaments by the presence of troponin and tropomysin. When a nerve impulse reaches a muscle fibre it is conducted over the sarcolemma and into the T-tubules, then to the sarcoplasmic reticulum. The sarcoplasmic reticulum releases *calcium ions* into the sarcoplasm. The liberated calcium ions combine with troponin causing it to push tropomysin away from receptor sites on the actin filaments. The myosin crossbridges interact with the actin receptor sites and pull the actin myofilaments toward the centre (H zone) of each sarcomere. The bond between the myosin crossbridges and actin breaks down under the influence of enzymes and the crossbridges are then free to rejoin with other actin receptor sites. The actin filaments do not shorten but slide past the myosin filaments overlapping them so that the Z lines are drawn toward each other, thus shortening the sarcomere. As each sarcomere shortens the whole muscle fibre contracts (Fig. 7.2).

Relaxation of the muscle fibres occurs when the calcium ions are actively reabsorbed by the sarcoplasmic reticulum, thus allowing troponin and tropomysin to again inhibit the interaction of the actin and myosin filaments.

Innervation of skeletal muscle

Each skeletal muscle is supplied with one or more nerves containing both sensory and motor nerve fibres (p. 249). After entering the muscle the nerve divides into branches which pass through the connective tissues to supply the muscle fibres. A motor neuron and the muscle fibres it innervates is called a **motor unit**. The number of muscle fibres in each motor unit varies according to the function of a particular muscle. In the extrinsic muscles of the eyes a motor unit contains only a few muscle fibres which produce very precise movements, while in the muscles of the limbs a motor unit may contain 500 or more muscle fibres.

As the motor neuron approaches a muscle fibre it loses its myelin sheath and branches to form a complex of nerve terminals, the **motor end plate**. The motor end plate lies in an invagination of the sarcolemma called the **synaptic gutter**; the space surrounding it is the **synaptic cleft**. The sarcolemma at the base of the synaptic gutter is thrown into numerous folds. The area of contact between the motor end plate and the muscle fibre is known as the **neuromuscular junction**. (The neuromuscular junction is comparable with the synapse p. 20.)

Stored in the nerve terminals of the motor end plate are many small vesicles containing the neurotransmitter *acetylcholine*, which is synthesized by the cytoplasm of the nerve terminals.

When a nerve impulse reaches a motor end plate, calcium ions diffuse from the extracellular fluid into the nerve terminals causing the vesicles to rupture and release acetylcholine into the synaptic cleft. The acetylcholine, which acts at specific receptor sites in the muscle membrane, causes depolarization of the underlying sarcolemma and

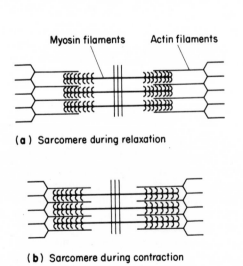

(a) Sarcomere during relaxation

(b) Sarcomere during contraction

Fig. 7.2   (a) Sarcomere during relaxation.
(b) Sarcomere during contraction.

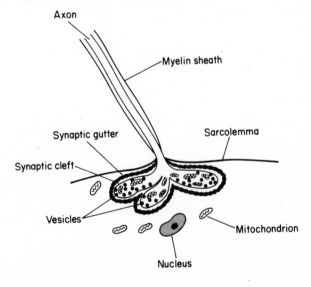

Fig. 7.3   Motor end plate.

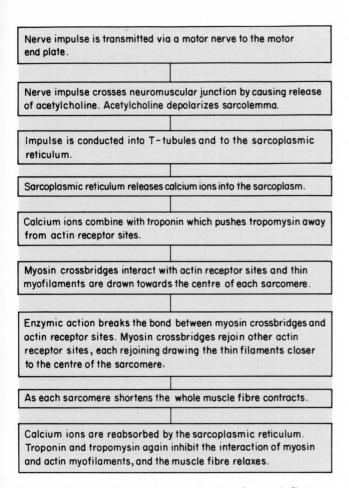

Fig. 7.4 Summary of events in the contraction of a muscle fibre.

initiates contraction of the muscle fibres. Almost immediately after depolarization of the muscle fibre the acetylcholine is destroyed by the enzyme *cholinesterase*, which is present in the synaptic cleft.

Other stimuli, such as an electric shock applied directly to the muscle or its nerves will cause the muscle to contract. This response is used in medicine to test nerves which may have been affected by disease or injury.

The transmission of nerve impulses across the neuromuscular junction can be blocked by a group of drugs known as *muscle relaxants*, (e.g. tubo-curarine, pancuronium). These drugs compete with acetylcholine for the receptor sites in the muscle membrane and block the action of acetylcholine resulting in paralysis of *all* skeletal muscles.

The all or none law of muscle contraction

Skeletal muscle fibres contract in response to nervous stimuli, and the weakest stimulus that can initiate contraction is known as the *threshold* or *liminal* stimulus. When muscle fibres receive a stimulus strong enough to elicit a response the fibres contract with the maximum force possible or they do not contract at all. They are unable to respond with partial contraction. This is known as the *all or none law*. This however is not true of a muscle as a whole. In a whole muscle contraction is graded, i.e. contraction may be weak or strong depending on the number of muscle fibres responding to a stimulus.

The strength of a muscle contraction may be affected by lack of oxygen and nutrients or by fatigue.

Energy for muscular contraction

A resting muscle uses very little energy and muscle metabolism is maintained by the uptake of nutrients from the bloodstream. During rest periods, muscle fibres store energy in the form of ATP (adenosine triphosphate, p. 184), creatine phosphate and glycogen.

When a muscle fibre is stimulated to contract, the myosin containing the myofibrils acts as an enzyme (ATPase) to cause the breakdown of ATP to ADP causing the release of energy. However, since the ATP stores are limited, ATP must be continually resynthesized within the muscle fibre. Creatine phosphate supplies the energy for the resynthesis of ATP.

These chemical reactions are only able to maintain energy for muscular contraction for a few seconds, and in *moderate exercise* oxygen consumption rises to oxidize glucose to supply the necessary energy. When continued maximum muscular contraction is required, as in *strenuous bouts of exercise*, the muscle fibres are unable to obtain sufficient oxygen and extra energy is derived from *anaerobic glycolysis* of stored glycogen (p. 186). This process results in the production of lactic acid which diffuses from the muscles into the bloodstream.

When energy is supplied by anaerobic metabolism an *oxygen debt* is incurred. Immediately after strenuous exercise this debt must be repaid. The accumulation of lactic acid stimulates an increased respiratory effort and a greater consumption of oxygen. The extra oxygen is utilized to reconvert lactic acid to pyruvic acid and glucose, and to replenish the muscle stores of ATP and creatine phosphate.

Blood supply of skeletal muscles

The arteries which supply blood to skeletal muscles branch to form small arteries and arterioles in the perimysium. These divide again to form a capillary network in the endomysium. When the muscles are relaxed the capillary network carries oxygen and nutrients to the muscle fibres. When the muscles are contracting, bloodflow through the nutritive capillary beds is impeded and the blood passes into **arteriovenous anastomoses** in the perimysium. During exercise muscle fibres alternately contract and relax, and the capillaries are able to supply the fibres with nutrients in the brief relaxation periods.

However, repeated or sustained muscle contraction maintains compression of the capillaries and blood flows through the arteriovenous anastomoses without passing through the capillaries. This results in a state of *muscle fatigue* due to depletion of nutrients and the accumulation of waste products (i.e. lactic acid and carbon dioxide). The strength and speed of contractions become progressively weaker. As fatigue increases the muscle fails to relax completely after each contraction, resulting in muscle spasm and pain.

In the condition known as *intermittent claudication* the blood supply to the muscles of the legs is reduced by atheroma of the arteries. The reduced blood flow causes a decrease in the oxygen supply to the muscle fibres and a rapid accumulation of waste products. The individual's exercise tolerance becomes increasingly restricted due to the resulting pain and spasm of the calf muscles.

Muscle tone

Even when a muscle appears to be at rest it is always partially contracted, and therefore ready for immediate action. At any given time a small number of motor units in a muscle are stimulated to contract and cause a tautness of the muscle, rather than full contraction and movement. The groups of motor units functioning in this way change periodically so that tone is maintained without fatigue. This state of partial contraction is called *muscle tone* and is important in maintaining body posture. Muscle tone is absent during sleep or when consciousness is lost.

Information about the state and position of the muscles is transmitted to the central nervous system by **muscle spindles**, which are specialized sensory receptors capable of detecting the degree of stretch in a muscle. A second type of sensory receptor, the **tendon apparatus**, is found in muscle tendons which detect the amount of tension applied to the tendon and therefore to the muscle. Changes in muscle tone are adjusted according to the information received from the sensory receptors.

Muscles with less than the normal degree of tone are said to be *flaccid*. When muscle tone is greater than normal the muscles become *spastic* (the muscle fibres are overcontracted and the muscle remains rigid).

A contraction which increases muscle tension but does not change muscle length is called an *isometric* contraction. An *isotonic* contraction is a contraction in which tension remains the same but the muscle shortens and produces movement. Body posture is maintained by isometric contraction. The majority of skeletal muscles can contract both isometrically and isotonically, and many physical activities such as walking or running involve both types of contraction.

Fast and slow muscle

Skeletal muscles display a wide variation in the speed and duration of contraction. Duration of contraction is adapted to the function of a particular muscle. For example the contraction of an eye muscle lasts for less than 1/100 second because ocular movements must be extremely rapid and precise. The contraction of leg muscles, such as the gastrocnemius, lasts for about 1/30 second; this moderate speed is needed for walking and running. Muscles which are mainly concerned with sustaining body posture, such as the soleus muscles, have a longer duration of contraction.

Fast muscle fibres contain an extensive sarcoplasmic reticulum which allows for the rapid release and reabsorption of calcium ions required for very rapid contractions. Fast muscle is paler in colour than slow muscle.

Slow muscle fibres are usually smaller and contain more mitochondria than fast muscle fibres. They also have a richer blood supply. The sarcoplasm of slow muscle fibres contains a large amount of myoglobin, a substance similar to the haemoglobin of red blood cells. *Myoglobin* can combine with oxygen and store it within the cell until the oxygen is needed by the mitochondria. It is the presence of myoglobin that gives slow muscle its characteristic dark red colour. Slow muscle fibres are more resistant to fatigue.

Muscle *hypertrophy* occurs as a result of forceful muscular exercise. The muscle enlargement is due to an increase in the diameter of individual muscle fibres rather than an increase in the number of muscle cells.

Muscle *atrophy* results if muscles are not used (e.g. due to prolonged bedrest or to the immobilization of a limb in a plaster cast), or when the motor nerve supply to a muscle is destroyed (e.g. by trauma or disease). The muscle fibres degenerate and with prolonged disuse become replaced by fibrous tissue.

Skeletal muscles

Bones may be regarded as a series of rigid levers, freely movable about fixed points, the joints, which act as fulcrums. The integration of the muscular and nervous systems, in association with the bones of the skeleton, forms an effective apparatus capable of a wide range of complex activities.

Each muscle is an individual organ, having its own blood supply, lymphatics and nerves. Muscles consist mainly of muscle fibres supported and strengthened by connective tissue. They are attached to each other and to surrounding tissues by the connective tissue components of the fasciae, tendons and aponeuroses.

Fasciae

The whole body is enclosed in an envelope of fibrous connective tissue called **fascia**. The fascia consists of two layers.

The **superficial fascia** lies just beneath the surface of the skin and is continuous with the dermis. It is a loose layer composed of fibrous areolar tissue impregnated with adipose tissue. The distribution of the adipose tissue is different in the two sexes, being more abundant in females. It acts as a thermal insulator, protecting the body from loss of heat. The superficial fascia varies in thickness, being thickest over the lower anterior abdominal wall and thinnest over the face, the neck and the dorsal aspects of the hands and feet.

The **deep fascia** is composed of dense fibrous tissue and lies immediately beneath the superficial layer. It encloses the muscles, blends with ligaments and provides attachment to bony surfaces. Extensions of the deep fascia separate muscles into functioning groups, enclose viscera, blood vessels and nerves, and help to maintain the position of these structures. In some parts of the body the deep fascia is particularly thick and strong, e.g. the *pelvic fascia* lining the pelvis; the *fascia lata* surrounding the muscles of the thigh; the *palmar fascia* in the hand and the *plantar fascia* in the sole of the foot (see Appendix).

Attachments of muscle

At the extremities of the muscles the connective tissue of the endomysium, perimysium and epimysium unite to form strong, fibrous, non-elastic cords called **tendons**. Tendons attach a muscle to the periosteum of a bone, and vary in length from a fraction of an inch to more than a foot. Sometimes they form a broad, flat expansion called an **aponeurosis**.

Most skeletal muscles are attached to bones, although a few, such as the muscles of facial expression, are attached to the soft tissues of the face, and others are attached to cartilage or ligaments.

Some tendons, for example those of the wrist and ankle, which pass under ligamentous bands or through bony tunnels, are enclosed in sheaths of synovial membrane called **tendon sheaths**. These facilitate smooth, frictionless movement of the tendons.

In situations where a muscle or tendon comes in contact with or moves over a bony prominence, or where the skin moves directly over bone, the pressure is relieved by a small synovial sac called a bursa. **Bursae** are usually located near joints, for example the olecranon bursa over the olecranon process of the ulna; the prepatellar bursa where the skin moves over the front of the patella.

Bursae may become inflamed, usually as a result of repeated minor injury (e.g. prepatellar bursitis or 'housemaid's knee').

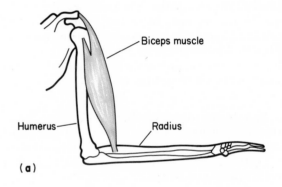

(a)

Actions of muscles

The main mass of a muscle is termed the **belly**, and lies along the shaft of a bone, never over a joint. Muscles are firmly attached at each end to different bones and it is the tendons which cross a joint. When a muscle contracts, a pull is exerted on both bones but one is stabilized by isometric contractions of other muscles and the contraction pulls the other bone toward it. For example, the belly of the biceps muscle lies parallel to the shaft of the humerus in the upper arm. The two tendons at the upper end of the biceps are attached to the scapula, while the tendon at the lower end crosses the elbow joint and is attached to the radius. Contraction of the biceps muscle draws the lower arm toward the upper (Fig. 7.5.).

The fixed point of muscle attachment is called the *origin*, while the movable point of attachment is

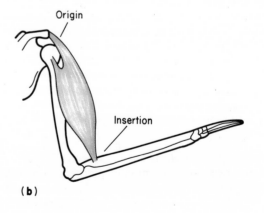

(b)

Fig. 7.5   (a), (b) Muscles are attached to two bones across a joint. Contraction of the muscle pulls on the bone into which the muscle is inserted, producing movement.

called the *insertion*. In the example above the muscle attachments into the scapula form the origin of the biceps muscle and the attachment into the radius forms the insertion. Muscles which stabilize the bone giving origin to the muscle are known as **fixation muscles**.

Muscles which bend a limb at a joint are called *flexors*. Muscles which straighten a limb at a joint are called *extensors*. Muscles which move a limb away from the midline of the body are known as *abductors*, while those that move the limb toward the midline are called *adductors*. Some muscles cause *rotation* of a limb. In movements of the wrist joint, *supinators* turn the hand palm upward and *pronators* turn the palm downward. In movements of the ankle joint *dorsiflexors* turn the foot upward and *plantarflexors* extend the foot toward the ground. Muscles which raise a part of the body are called *levators* and those which lower a part are known as *depressors*.

Movements are complex and the performance of any given movement, e.g. flexion of the elbow joint, requires the co-ordination of several muscles. Muscles which initiate and maintain a movement are called *prime movers* or *agonists*, while those which oppose a movement or reverse it are known as *antagonists*. Thus when the biceps muscle contracts to raise the lower arm towards the shoulder it is the prime mover. The triceps muscle, which can oppose this action and straighten the elbow, is the antagonist. However, actions which require a joint to be held rigid will cause simultaneous contraction of both prime movers and antagonists.

Synergists are muscles which assist a prime mover by stabilizing a joint crossed by the tendon of the prime mover, thus allowing it to produce a more effective movement.

Muscles may be named according to one or more of the following:
  (*1*) Function, e.g. flexors, extensors, abductors, etc.
  (*2*) Attachments, e.g. sterno-mastoid.
  (*3*) Shape, e.g. deltoid (like Greek letter D or △).
  (*4*) Position or direction, e.g. pectoralis major (large breast muscle), rectus abdominus (straight abdominal muscle), the oblique and straight muscles of the eye.
  (*5*) Formation e.g. biceps = two heads; triceps = three heads; quadriceps = four heads.

Principal groups of muscles

Muscles of the head and neck

The muscles of the head, face and neck are too numerous to describe in detail. They can, however, be divided into several main groups.

(1) Muscles of the scalp (occipito-frontalis) and the ear (p. 262) and of the eye (p. 266).

(2) The **muscles of facial expression** which are supplied by the VIIth cranial nerve.

The **orbicularis oculi** are circular muscles surrounding each eye which when contracted act as sphincters and close the eyes tightly. The **orbicularis oris** is situated in the lips between the skin and the mucous membrane of the mouth and has a similar type of action. The **buccinator** is the principal muscle of the cheek and forms the lateral wall of the mouth. It is used in chewing and sucking.

(3) The **muscles of mastication** which move the lower jaw are supplied by branches of the Vth cranial nerve.

The **temporal** muscle arises from the temporal fossa of the skull. Its fibres converge into a strong tendon which is inserted into the coronoid process of the mandible. The **masseter** muscle is quadrilateral in shape and arises above from the zygomatic arch of the temporal bone. It is inserted into the outer surface of the lower jaw anterior to the angle.

(4) Muscles of the neck, attaching the head to the trunk.

Superficial muscles
The **platysma** extends from the lower jaw in a thin flat sheet to the deep fascia on the front of the chest.

The **sternomastoid** is an important muscle extending upwards from the manubrium of the sternum and medial end of the clavicle to the mastoid process of the temporal bone. When operating singly each muscle rotates the head towards the opposite side; when acting together they flex the neck.

The **trapezius** is a large triangular muscle situated at the back of the neck. It arises from the occipital bone of the skull and from the spines of all the cervical and thoracic vertabrae. Its upper fibres are inserted into the lateral third of the clavicle, the middle fibres into the acromion process and the

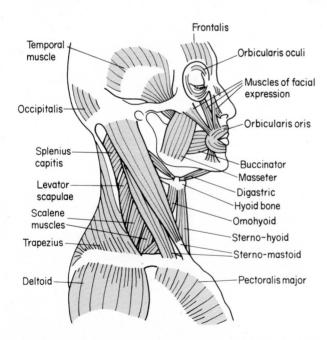

Fig. 7.6   Muscles of the face and neck.

lower fibres into the spine of the scapula. It may also be considered as one of the muscles of the shoulder girdle attaching the scapula to the trunk.

There are also a number of muscles in the front of the neck extending *(i)* from the lower jaw to the hyoid bone and *(ii)* from the hyoid bone and thyroid cartilage to the sternum. These are closely related to the trachea and thyroid gland.

Deep muscles
Examples are the **scalene** muscles extending from the cervical vertebrae to the first and second ribs.

(5) Muscles of:
(a) the pharynx – the constrictor muscles which take part in the act of swallowing; also the muscles of the tongue and floor of the mouth;

(b) the larynx – external muscles which move the larynx as a whole and internal muscles which affect the tension of the vocal cords and are used in voice production.

Muscles of the shoulder girdle and upper limb

(1) Muscles attaching the scapula to the trunk
Examples are: deep – **rhomboids**
superficial – **trapezius,**
serratus anterior

(2) Muscles attaching the humerus
(*a*) To the scapula, e.g. **supraspinatus, infraspinatus, subscapularis, deltoid.**
(*b*) To the chest wall, e.g. **pectoralis major and minor, latissimus dorsi.**

(3) Muscles of the arm
The most important muscles of the arm are:
(*a*) The **biceps**, which has its origin by two heads, long and short, from the scapula. The long head arises from the top of the glenoid cavity; the short head from the tip of the coracoid process. It is inserted into the bicipital tubercle below the head of the radius. Passing as it does over two joints, it can produce movements at both. Its main actions are to act as a powerful supinator of the forearm and to flex the elbow joint. It also helps in the forward movement of the shoulder joint.
(*b*) The **brachialis**, which arises from the front of the shaft of the humerus and is inserted into the ulna.

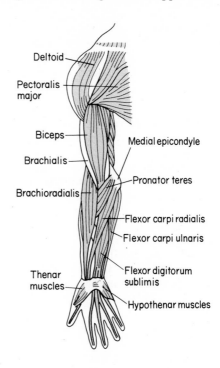

Fig. 7.7   Muscles of the right upper limb – anterior aspect.

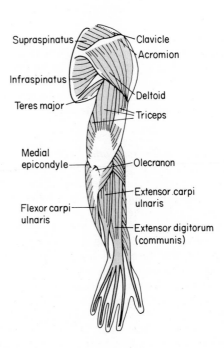

Fig. 7.8   Muscles of the right upper limb – posterior aspect.

(*c*) The **triceps**, situated at the back of the arm, passes from the scapula to the olecranon process of the ulna. It therefore extends the elbow joint, and helps to support the shoulder joint and draw the arm backwards.

(4) Muscles of the forearm

The muscles of the anterior aspect may be divided into three groups:

(*a*) The main group consists of superficial and deep muscles. The superficial muscles are attached above to the medial epicondyle of the humerus and pass to the fingers; these include the **flexor digitorum sublimus**, which not only flexes the fingers but also the wrist and elbow joints. The deep muscles include the **flexor digitorum profundus** arising from the ulna, which does not move the elbow joint but flexes the wrist and fingers.

(*b*) Muscles which flex the elbow and wrist only, passing from the humerus above to the wrist bones below (**flexor carpi radialis** and **flexor carpi ulnaris**).

(*c*) Muscles whose main action is concerned with pronation and supination (**pronator teres, pronator quadratus** and **brachio-radialis** or supinator longus).

The muscles of the posterior aspect may be divided into:

(*a*) The extensor muscles of the wrist and fingers (**extensor digitorum communis**) which, arising from the lateral epicondyle of the humerus, also extend the elbow joint.

(*b*) The extensors of the wrist (**extensor carpi radialis** and **extensor carpi ulnaris**).

(5) Muscles of the hand and fingers

The tendons of the flexor muscles in front and the extensor muscles behind are inserted into the bases of the terminal phalanges of the digits. Slips are also given to the other phalanges.

The thumb has separate muscles which, however, correspond to the main flexor and extensor groups in the forearm. Special *short* muscles situated in the palm of the hand move the thumb and form the **thenar eminence** at the base of the thumb. The prominence on the ulnar side of the hand at the base of the little finger is called the **hypothenar eminence**. Arising between the metacarpal bones and inserted into the phalanges are the **lumbrical** and **interosseous** muscles.

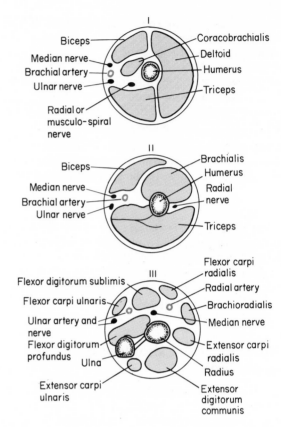

Fig. 7.9 Diagram illustrating the position of the most important structures in the upper limb. (Top) In Upper third of arm. (Middle) In lower third of arm. (Bottom) Middle of forearm.

Muscles of the trunk

The muscles of the trunk may be divided into *(i)* those of the thorax, and *(ii)* those of the abdomen.

Muscles of the thoracic wall

The superficial muscles include:

(*a*) The **pectoralis major**, a large fan-shaped muscle on the upper anterior part of the chest, which also forms the anterior part of the axilla; and the **pectoralis minor**, a smaller triangular muscle lying deep to the pectoralis major. These muscles arise from the anterior aspect of the sternum, ribs and costal cartilages and are inserted into the upper end of the humerus and coracoid process of the scapula respectively.

(*b*) The **serratus anterior** arising from the ribs passes backwards to the vertebral border of the scapula.

(*c*) The **intercostal muscles** (eleven pairs) which pass from the lower border of one rib to the upper border of the rib below. They are formed by two distinct layers, the fibres of which pass in opposite directions. The **internal intercostals** are directed downward and backward; the more superficial **external intercostals** pass downward and forward. They are important muscles of respiration.

The diaphragm

The diaphragm is the large dome-shaped partition separating the thoracic cavity from the abdominal cavity. It is formed partly by muscle (around the circumference) and partly by a flattened tendon called the central tendon of the diaphragm. It is attached to the circumference of the thoracic cavity:

(*1*) in front to the lower end of the sternum (sternal part);

(*2*) on either side to the lower six ribs (costal part); and

(*3*) posteriorly to the first two lumbar vertebrae by two slips called the crura (legs) of the diaphragm (lumbar part).

The heart and pericardium are related to the central portion of its upper surface. On either side it is covered by pleura and is related to the bases of the lungs. Its lower concave surface, largely covered by peritoneum, is related on the right side and centrally to the upper surface of the liver; on the left side it is in contact with the fundus of the stomach and the spleen.

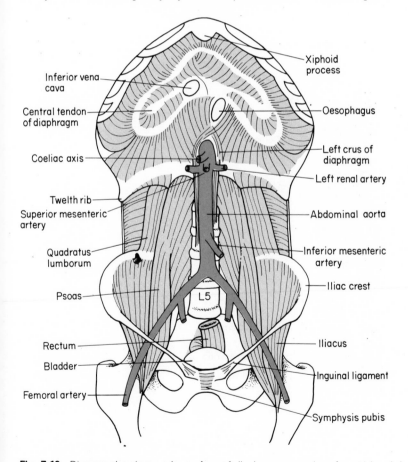

Fig. 7.10 Diagram showing: under surface of diaphragm; muscles of posterior abdominal wall; abdominal aorta dividing into common iliac arteries which give off internal and external iliac arteries.

In the posterior part of the diaphragm, close to its origin from the lumbar vertebrae, are a number of **openings (hiatuses)** for the passage of the following structures:

(*1*) the aorta, in the mid-line;

(*2*) the oesophagus, slightly to the left; and

(*3*) the inferior vena cava, slightly to the right.

The diaphragm is a very important muscle of respiration and is supplied by the **phrenic nerve** from the cervical plexus. During inspiration the muscle of the diaphragm contracts so that the diaphragm becomes flattened towards the abdomen, thus helping to enlarge the thoracic cavity. During expiration the diaphragm relaxes and resumes its dome-shaped appearance.

Damage to the phrenic nerve may cause paralysis of the diaphragm. Sometimes abdominal contents herniate through the openings in the diaphragm, the commonest being 'hiatal hernia' in which a portion of the stomach enters the chest through the oesophageal opening.

Muscles of the abdomen

(1) Anterior abdominal wall

The muscles of the anterior abdominal wall are arranged in sheets which protect the delicate abdominal organs. Contraction of the abdominal muscles aid in the act of defaecation.

(*a*) The **rectus abdominis** is the straight muscle of the anterior abdominal wall. It runs parallel with its fellow of the opposite side and is separated from it by a thin band of fibrous tissue, the **linea alba** (white line), which extends from the xiphoid process of the sternum to the symphysis pubis. The muscle arises from the pubic bone and is inserted into the xiphoid process and the adjacent costal cartilages. It is enclosed in a dense sheath of fibrous

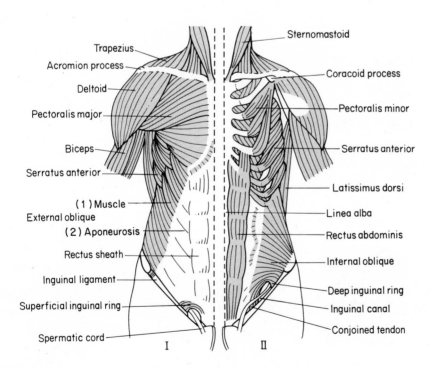

Fig. 7.11   Muscles of the front of the thorax and anterior abdominal wall. (Left) Superficial. (Right) Deep.

tissue, formed by the aponeuroses of the two oblique muscles, called the **rectus sheath**. The fibres of the rectus muscle are interrupted by three bands of fibrous tissue which cross it transversely – the tendinous intersections.

(*b*) Two **oblique muscles** help to form the side and anterior walls of the abdominal cavity. The fibres of the **external oblique** muscle pass downwards and forwards, arising from the lower eight ribs. It is inserted in a fan-shaped manner into the rectus sheath, the iliac crest and the pubic bone. The aponeurosis of the lower border, between the anterior superior iliac spine and the pubic spine, is thickened to form the **inguinal ligament** (Poupart's ligament).

The **internal oblique** muscle arises from the iliac crest and the inguinal ligament and is inserted into the rectus sheath and the lower ribs. Its fibres pass upwards and medially and cross those of the external oblique.

The **transversus** muscle arises from the lower ribs, the iliac crest and the inguinal ligament and is inserted into the linea alba and the pubic bone.

The point at which the aponeuroses of the anterior abdominal wall unite is known as the conjoined tendon. In the inguinal area the aponeuroses of these muscles are pierced by the inguinal canals.

(2) Posterior abdominal wall

The **psoas** muscle arises from the lumbar vertebrae. The **iliacus** muscle arises from the inner surface of the ilium. They are inserted together into the lesser trochanter of the femur.

The **quadratus lumborum** extends from the iliac crest upwards to the last (12th) rib. The posterior surface of the kidney is closely related to the psoas and quadratus lumborum muscles.

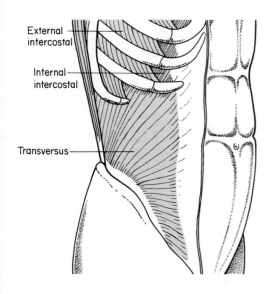

Fig. 7.12 The transversus muscle.

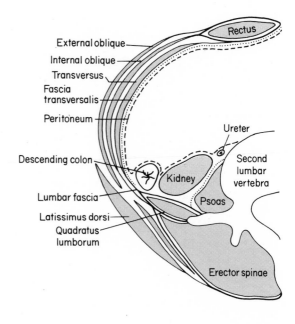

Fig. 7.13 Section of the abdominal wall through the lower part of the second lumbar vertebra.

Muscles of the back

The muscles of the back play an important part in the maintenance of an upright body posture. There are several groups of muscles on either side of the spine which extend for varying distances between the occiput above and the sacrum below. The upper ones extend the neck. The lower ones, including the **erector spinae**, straighten the spine. In the lumbar region they form a large aponeurosis, the lumbar fascia.

Muscles of the pelvis

The muscles of the pelvis collectively form the floor of the pelvic cavity. The most important of these muscles is the **levator ani**, which aids in defaecation. It is a muscular sheet extending across the outlet of the pelvis through which passes the rectum, urethra, and in the female, the vagina (p. 281).

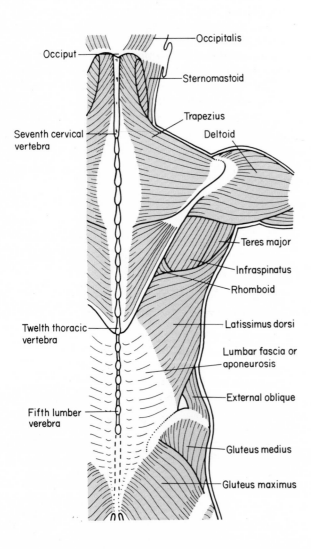

Fig. 7.14   Muscles of the back.

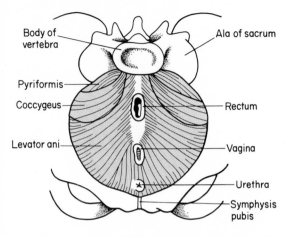

Fig. 7.15 Diagram of the muscles of the female pelvic floor.

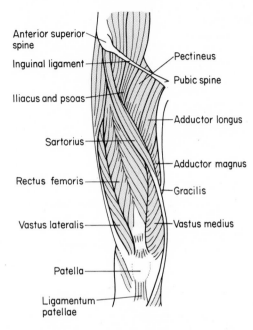

Fig. 7.16 Muscles of the front of the thigh.

Muscles of the lower extremity

The muscles of the lower extremities are among the largest and most powerful in the body.

Muscles of the buttock

The rounded eminence of the buttock is formed by the gluteal muscles. The **gluteus maximus** is the most superficial and the largest of this group. It arises from the ilium and is inserted into the gluteal tuberosity of the femur and the iliotibial tract (the fascia which envelops all the muscles of the thigh). The **gluteus medius** and the **gluteus minimus** arise from the ilium and are inserted into the greater trochanter of the femur. The main action of these muscles is to extend and abduct the thigh.

Muscles of the thigh

There are a number of muscles in the thigh which can be described in three groups.

(1) The anterior group The main muscle of this group is the **quadriceps extensor**, which is made up by the **rectus femoris, vastus medialis, vastus lateralis** and **vastus intermedius,** and terminates in a single tendon (the ligamentum patellae), in which the patella is developed as a sesamoid bone. It is inserted into the tuberosity of the tibia and its action is to extend or straighten the knee joint.

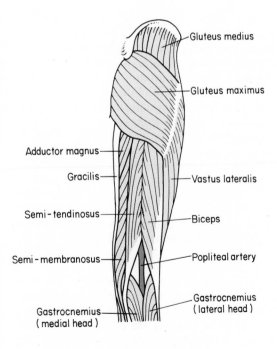

Fig. 7.17 Muscles of the back of the thigh.

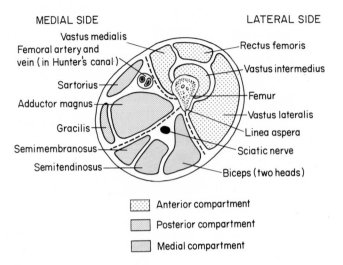

MEDIAL SIDE   LATERAL SIDE

Vastus medialis
Femoral artery and
vein (in Hunter's canal)
Sartorius
Adductor magnus
Gracilis
Semimembranosus
Semitendinosus

Rectus femoris
Vastus intermedius
Femur
Vastus lateralis
Linea aspera
Sciatic nerve
Biceps (two heads)

▦ Anterior compartment

▦ Posterior compartment

▦ Medial compartment

Fig. 7.18   Transverse section of the middle of the thigh.

The **sartorius** muscle arises from the anterior superior spine of the ilium and passes obliquely to be inserted into the medial side of the tibial tuberosity.

(2) The posterior group These are also called the *hamstring group*, which is formed by three muscles, the **biceps femoris, semimembranosus** and **semitendinosus**. They arise from the tuberosity of the ischium and are inserted into the upper ends of the tibia and fibula. The biceps femoris passes to the lateral side of the leg and also has a short head arising from the linea aspera of the femur. The other two muscles pass to the medial side of the leg. The biceps, therefore, forms the lateral boundary of the popliteal space, and the semimembranosus and semitendinosus form the medial boundary.

The hamstring muscles, having their origin above the hip joint and their insertion below the knee joint, are capable of producing movement at both (i.e. they straighten or extend the hip and flex the knee).

(3) The medial group This consists of three muscles, the **adductors longus, brevis** and **magnus**, which arise from the pubic bone and are inserted mainly into the linea aspera of the femur. Their function is described by their name (i.e. they adduct the thigh towards the mid-line).

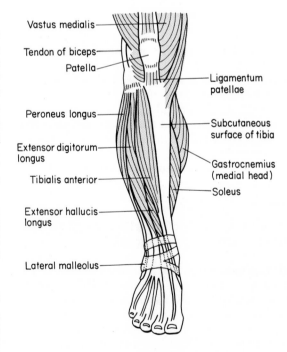

Vastus medialis
Tendon of biceps
Patella
Peroneus longus
Extensor digitorum longus
Tibialis anterior
Extensor hallucis longus
Lateral malleolus

Ligamentum patellae
Subcutaneous surface of tibia
Gastrocnemius (medial head)
Soleus

Fig. 7.19   Muscles of the front of the leg.

Muscles of the leg

The muscles of the leg function to provide movement of the foot. They can be described in three groups.

(1) Anterior group This is composed of those muscles which lie in front of the interosseous membrane between the tibia and fibula. It includes the **tibialis anterior**, passing from the tibia to the tarsal bones to dorsiflex the ankle, and the extensor muscles of the toes (**extensor digitorum longus**).

(2) Posterior group This consists of superficial and deep layers. The superficial muscles, i.e. the **gastrocnemius** and the **soleus**, form the back of the calf. The upper end of the gastrocnemius arises from the condyles of the femur, the soleus arises from the posterior aspects of the tibia and fibula. Both are inserted into the calcaneum by the tendo Achilles and plantar-flex the ankle joint.

The deep muscles include the **tibialis posterior** arising from the tibia and fibula, which also plantar-flexes the ankle joint as it passes to its insertion into the tarsal bones, and the flexor muscles of the toes (**flexor digitorum longus**).

In contrast to the muscles of the forearm and hand, the flexor muscles of the toes are situated on the posterior aspect of the leg and the extensors of the toes on the anterior surface. (It must be remembered that the palm of the hand corresponds to the sole of the foot.)

(3) Fibular group These are also called the **peroneal** muscles. They arise from the lateral surface of the fibula and are inserted into the tarsal and metatarsal bones of the foot. Their action is to evert the foot outwards.

There are a number of small muscles in the foot similar to those in the hands, e.g. short muscles in the sole which are attached to the big toe and the **interosseous** and **lumbrical** muscles for the toes.

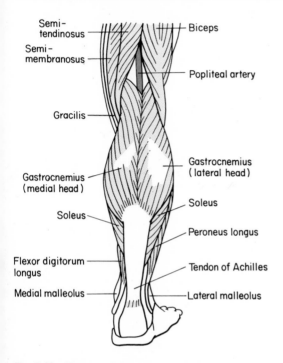

Semi-tendinosus — Biceps — Semi-membranosus — Popliteal artery — Gracilis — Gastrocnemius (medial head) — Gastrocnemius (lateral head) — Soleus — Soleus — Peroneus longus — Flexor digitorum longus — Tendon of Achilles — Medial malleolus — Lateral malleolus

Fig. 7.20 Muscles of the back of the leg.

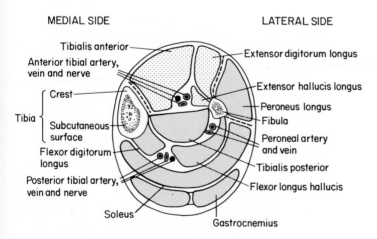

MEDIAL SIDE   LATERAL SIDE

Tibialis anterior — Extensor digitorum longus — Anterior tibial artery, vein and nerve — Extensor hallucis longus — Crest — Peroneus longus — Tibia — Fibula — Subcutaneous surface — Peroneal artery and vein — Flexor digitorum longus — Tibialis posterior — Posterior tibial artery, vein and nerve — Flexor longus hallucis — Soleus — Gastrocnemius

Fig. 7.21 Transverse section of the leg (middle of calf).

Questions

1. Describe the microscopic structure of a muscle fibre.
2. Describe the structure and function of the neuromuscular junction.
3. Discuss the chemical events which result in contraction of muscle fibres.
4. Describe the structure and function of whole muscle.
5. Describe the structure and function of fascia.

8
The Respiratory System

The essential features of respiration are the transference (*i*) of oxygen (O_2) from the atmosphere to the tissues, and (*ii*) of carbon dioxide (CO_2) from the tissues to the outer air. In addition to this interchange of gases, some water vapour is lost from the body.

There are two phases in the interchange of gases.

(1) *External respiration*, or the absorption of oxygen from the air into the blood and the excretion of carbon dioxide from the blood into the air. This takes place in the lungs.
(2) *Internal or tissue respiration*, in which the oxygen is transferred from the blood to the tissues of the body which at the same time give up carbon dioxide.

These exchanges take place through the walls of the capillaries. It follows that external respiration is mainly a function of the respiratory system and the act of breathing (ventilation); while tissue respiration depends on the efficiency of the circulation. The importance of the circulatory system in maintaining the respiratory process is evident in Fig. 8.1.

In summary, the function of the respiratory system is to facilitate the exchange of oxygen and carbon dioxide between the cells of an organism and its surroundings. In humans and other mammals this involves *(a)* breathing and *(b)* the transport of gases in the bloodstream between the two sites of gas exchange, the lungs and the tissues.

Carbon dioxide dissolves in body fluids to form carbonic acid, which dissociates into hydrogen ions (H^+) and bicarbonate ions (HCO_3^-). The speedy elimination of carbon dioxide by breathing helps to keep the pH of the blood at 7.40.

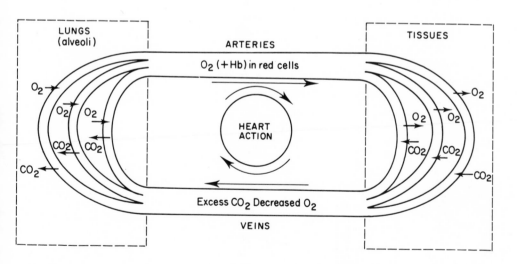

O_2 and CO_2 represent molecules of oxygen and carbon dioxide respectively

Fig. 8.1   Gaseous exchange in lung and other tissues.

The air passages

The respiratory system consists of the lungs and air passages. In its course to the lungs the air passes through the nose, mouth, pharynx, larynx and upper trachea (extrathoracic parts) and the lower trachea, bronchi, bronchioles and alveoli (intrathoracic parts or tracheobronchial tree). Blood vessels and nerves (mainly branches of the vagi or Xth cranial nerves) accompany the bronchi.

Nasal cavities

(1) The anterior nares or nostrils form the entrances to the nasal cavities. Small hairs inside the anterior nares act as a coarse filter for dust in the inspired air.

(2) The nasal cavities are separated into right and left portions by the nasal septum (formed above by the perpendicular plate of the ethmoid, behind by the vomer and in front by the cartilage of the septum). Each cavity is lined by mucous membrane covered with ciliated columnar epithelium and is plentifully supplied with blood. The surface area of the nasal mucous membrane is increased by the presence of three (upper, middle and lower) turbinate bones (conchae) which project medially from the lateral wall of each cavity. The impor-

tance of this increased surface area is that the air entering the respiratory tract may be warmed and moistened before reaching the lungs. The floor of the nasal cavities is formed by the upper surface of the hard palate, and the roof by portions of the frontal, ethmoid and sphenoid bones. The maxilla forms the main part of the lateral wall.

(3) The posterior nares are situated at the back of the nasal cavities and constitute the entrances to the nasopharynx.

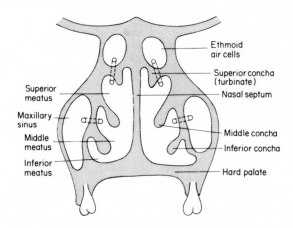

Fig. 8.3 Diagram of the nasal cavities, showing the conchae (turbinate bones), and the ethmoid and maxillary sinuses. (Note that the maxillary sinus opens into the middle meatus.)

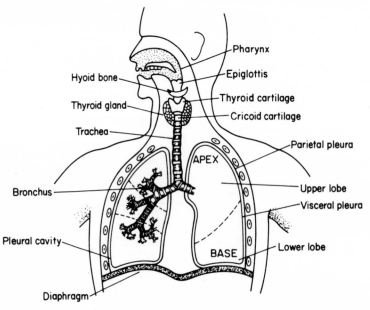

Fig. 8.2 The respiratory system.

In addition to the anterior and posterior nares, each nasal cavity has openings connecting with

(*a*) the maxillary antrum;

(*b*) the frontal, ethmoidal and sphenoidal air sinuses (Fig. 8.3); and

(*c*) the nasolacrimal duct which conveys the tears from the conjunctival sac to the nose (see also p. 265).

The pharynx

The pharynx has three parts which, from above downwards, are the nasal part (nasopharynx), the oral part (oropharynx) and the laryngeal part.

The **nasopharynx** lies at the base of the skull, immediately behind the nasal cavities, which open into it via the posterior nares. On its lateral walls are the openings of the Eustachian tubes, which connect it to the middle ears. In its posterior wall is the lymphoid tissue which is known, when enlarged, as 'adenoids'. The nasopharynx is continuous below with the oropharynx.

The **oropharynx** is continuous in front with the buccal cavity ('mouth') and below with the laryngeal part of the pharynx. The **tonsils** lie in its lateral walls.

The **laryngeal part** of the pharynx is continuous with the oesophagus. Near its upper end, on each side, is a small recess, the **pyriform fossa**, in which foreign bodies may lodge.

The larynx

Besides acting as a part of the air passages the larynx is modified in structure to enable it to perform the special function of voice production (p. 113). It is situated in the mid-line of the neck between the pharynx above and the trachea below. It is placed in front of the oesophagus and corresponds with the levels of the fourth, fifth and sixth cervical vertebrae.

Structure of the larynx

The larynx consists of a framework of the following hyaline cartilages:

(a) The **thyroid cartilage**, which is the largest and consists of two side wings united in the mid-line in front to form an angular projection, sometimes called the Adam's apple.

(b) The **cricoid cartilage**, situated below the thyroid cartilage, forms the lowest part of the larynx which it connects to the trachea. It is circular in shape with an expansion at the back, giving it a resemblance to a signet ring.

The lobes and upper poles of the thyroid gland lie on each side of the cricoid and thyroid cartilages.

(c) The **arytenoid cartilages** are two small structures situated on the upper surface of the expanded signet portion of the cricoid cartilage. They are shaped like pyramids and give attachment to the posterior ends of the true vocal cords. In front the vocal cords are attached to the posterior surface of the thyroid cartilage. The tension of the cords is varied by muscles which rotate the arytenoid cartilages and, in this way, the pitch of the voice is altered.

(d) The **epiglottis**. This is a leaf-shaped plate of yellow elastic fibrocartilage situated in an upright position between the base of the tongue and the upper opening of the larynx. Its lower stalk-like portion is attached by a ligament to the thyroid cartilage. The main function of the epiglottis is to prevent food from entering the larynx during the act of swallowing.

(e) The **hyoid** is a horseshoe-shaped bone lying between the mandible above and the larynx below. It is situated at the base of the tongue and gives

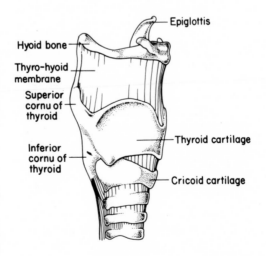

Fig. 8.4 The cartilages of the larynx (right lateral view).

attachment to this and various other muscles. It does not actually take any part in the formation of the true larynx.

The larynx is lined by mucous membrane which, except over the vocal cords, is covered with ciliated columnar epithelium.

The vocal cords

The true **vocal cords** are fibro-elastic bands extending from the posterior aspect of the thyroid cartilage in front to the arytenoid cartilages behind. The false vocal cords are two loose folds of mucous membrane situated above the true cords which do not appear to play any special part in voice production (see Fig. 12.6, p. 159).

The trachea

The trachea or windpipe is 12 cm ($4\frac{1}{2}$ in) long and about 2.5 cm (1 in) in diameter. Its upper half is situated in the mid-line of the neck, its lower half in the superior mediastinum of the thorax. It lies in front of the oesophagus and ends opposite the fourth thoracic (dorsal) vertebra, where it divides into the two main bronchi (see Fig. 8.2).

The trachea consists of a number of C-shaped rings of cartilage connected by fibrous tissue and having the opening of the C posteriorly. It is lined by mucous membrane and therefore the posterior wall, which lies in front of the oesophagus, consists of mucous membrane unsupported by cartilage. The function of the rings of cartilage is to keep the windpipe permanently open so that its walls do not collapse like those of the oesophagus. It is also lined with ciliated columnar epithelium and cells which secrete mucus. Its upper four rings are crossed by the isthmus of the thyroid gland.

Tracheostomy is an operation in which an opening is made through the upper rings of the trachea below the cricoid cartilage and a tube is inserted. It is performed in cases of laryngeal obstruction and other conditions in which there is severe respiratory embarrassment.

The bronchi

The trachea ends by dividing or bifurcating into the two main bronchi (right and left) opposite the level of the fourth thoracic vertebra. Each bronchus passes to the corresponding lung. From each main bronchus numerous smaller bronchi are given off, like the branches of a tree, and the smallest bronchial tubes are called **bronchioles**.

The structure of the bronchi is similar to that of the trachea, consisting of incomplete hoops of hyaline cartilage lined with mucous membrane covered with ciliated columnar epithelium. In addition, they have some smooth (involuntary) muscle in their walls. The structure of the bronchioles is similar to that of the bronchi, but they contain no cartilaginous hoops. Instead, there are more muscle fibres.

A moving carpet of mucus covers the epithelium, the mucus being wafted upwards by the cilia to the larynx and pharynx, from which it is swallowed or expectorated. The mucus is secreted by goblet cells interspersed between the ciliated cells and by submucosal mucus-secreting glands.

Spasm of the muscle fibres in the walls of the bronchi and bronchioles occurs in asthma. The contraction of these muscles causes narrowing of the bronchial tubes and obstruction to the passage of air through them. This spasm is relaxed by adrenalin (p. 201), aminophylline and similar drugs.

The alveoli or air sacs

Each bronchiole terminates in an irregular sac made up of a number of air pockets. These pockets are lined with a delicate layer of flattened epithelial cells and are surrounded by numerous capillaries, through the walls of which the interchange of gases takes place. The blood in the capillaries is conveyed to the lungs by the pulmonary artery and from the lungs to the left atrium by the pulmonary veins.

The alveolar wall consists of three layers, namely surface epithelium, a supporting connective tissue layer of fine reticular collagenous and elastic fibres, and blood vessels (mainly capillaries). The connective tissue layer is absent in most of the alveolar wall and only a shared basement membrane lies between the surface epithelial cells and the capillary endothelial cells. There is thus only the thinnest wall (alveolar-capillary membrane) between the air in the alveoli and the blood in the capillaries. Some of the epithelial cells (Type II) secrete *surfactant*, a detergent-like substance which reduces surface tension in the alveoli and helps to prevent their collapse. Alveolar macrophages (dust cells) in the walls and air spaces of the alveoli engulf debris and any fine dust particles which reach the alveoli.

The lungs

The lungs are a pair of conical-shaped organs, each enveloped in a serous membrane, the pleura. They occupy the greater part of the thoracic cavity. The lungs are separated from each other by the mediastinum which contains the heart and great vessels, the oesophagus and, in its upper part, the trachea. Each lung is divided by deep fissures into lobes. The right lung has three lobes, upper, middle and lower; and the left two, upper and lower. The lung is described as having a mediastinal and a costal surface, an apex and a base.

The outer or costal surface is in contact with the wall of the pleural cavity which consists of the ribs and intercostal muscles and is lined by pleura.

The medial surface of the lung is applied to the mediastinum. Its chief feature is the presence of the **hilum** where the main bronchus and pulmonary artery enter and the pulmonary veins leave the lung. Also at the hilum are lymph nodes which may be enlarged by disease, for example tuberculosis or cancer.

The apex rises into the root of the neck for about one inch above the clavicle. The base is concave and is related to the upper surface of the diaphragm.

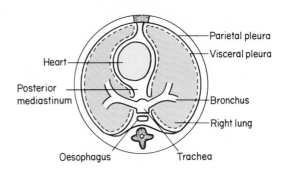

Fig. 8.5 Horizontal section of the chest, showing normal positions of the heart and lungs.

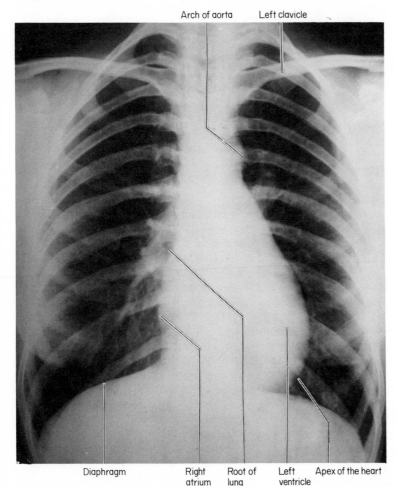

Fig. 8.6 X-ray of the chest.

Structure of the lungs

On examination the lung feels spongy and, if a portion is dropped into water, it will float because of the air which it contains.

Each lobe of the lung is made up of a large number of small lobules consisting of the alveoli and their bronchioles, which join with each other to form the larger bronchi. In addition, there is a framework of fibrous or interstitial tissue in which run the blood vessels and lymphatics. The pulmonary artery which supplies the lung with venous blood arises from the right ventricle of the heart, while the pulmonary veins pour their oxygenated blood into the left atrium.

Groups of lymph glands which drain the lung are situated at the hilum of the lung and at the bifurcation and along the sides of the trachea.

The pleura

The pleura is a serous membrane which, like the pericardium, consists of two layers, the visceral and the parietal. The **visceral layer** forms the outer covering of the lung which it encloses completely except at the hilum, where it is reflected over the structures entering the lung and becomes continuous with the parietal layer. The **parietal layer** lines the interior of the chest wall and upper surface of the diaphragm. The two layers are smooth and shiny and are moistened by a small amount of serous fluid resembling lymph which acts as a lubricant so that the two surfaces can glide smoothly over each other during the act of breathing. In disease states, the two layers of the pleura may become separated by fluid (pleural effusion) or by air (pneumothorax).

Respiratory movements

The renewal of the air in the lungs is secured by the respiratory movements of *inspiration* (breathing in) and *expiration* (breathing out). The thorax may be regarded as a completely closed box which alters its size and shape with each ventilation. With inspiration, the cavity of the thorax is enlarged and the lungs, being elastic, expand to fill up the increased space. This expansion of the lungs causes air to be sucked in through the upper air passages and trachea. Inspiration is an *active* muscular process.

With expiration, the capacity of the thorax returns to its former size and air is expelled from

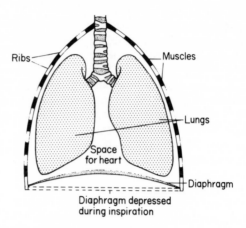

Fig. 8.7   The chest in section.

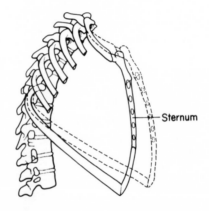

Fig. 8.8   Movements of the ribs and sternum during respiration.

the lungs. Expiration is largely *passive* in a person at rest.

The increase in the size of the thoracic cavity during inspiration is brought about by two factors.

(*1*) Upward movement of the ribs.

(*2*) Downward movement of the diaphragm.

Upward movement of the ribs results mainly from contraction of the **external intercostal muscles**. In forced, deep inspirations the muscles of the neck and shoulder girdle may be brought into operation, viz. trapezius, sternomastoid and pectoralis major muscles.

When at rest the diaphragm is dome-shaped, having its concavity towards the abdomen. When the muscle of the diaphragm contracts during inspiration it becomes flattened and, therefore, depressed towards the abdominal cavity.

During quiet expiration the chest returns to its resting size mainly on account of the elasticity of the lungs and the chest wall and by the upward pressure of the abdominal contents on the diaphragm as it relaxes. The **internal intercostal muscles** provide the small active element of quiet expiration. In forced expiration, such as occurs in coughing and during exercise, the abdominal and other accessory muscles are employed.

Normal ventilation, therefore, is a combination of two sets of movements, thoracic and diaphragmatic, sometimes known as thoracic and diaphragmatic or abdominal breathing. In men quiet ventilation is mainly carried out by movements of the diaphragm, while in women the thoracic type of ventilation usually predominates.

Another important function of the respiratory movements is to aid in the return of venous blood to the heart (p. 141).

Special resiratory movements

(1) Sighs and yawns are types of prolonged inspiration.
(2) Cough is a forcible expiration usually preceded by a prolonged inspiration. The sound of a cough is produced by forcing air through the narrow opening between the vocal cords.
(3) Hiccough is a noisy inspiration caused by muscular spasm of the diaphragm at irregular intervals. The noise is produced by the sudden sucking of air through the vocal cords.
(4) Changes in the breathing pattern can signify emotions.

Movement of air in the respiratory tract

It has been seen that the expansion of the chest by the movements of the thorax and diaphragm causes air to enter the lungs with each inspiration. Further, by an added effort, a forced inspiration will result in still greater expansion and an additional amount of air will enter the lungs.

In the same way, a normal expiration can be supplemented by a forced expiration. Even after a forced expiration, however, some air still remains in the alveoli of the lungs.

The amount of air passing in and out of the lungs with ordinary quiet breathing is called **tidal air** and measures about 500 ml. The additional volume taken in by forced inspiration is called the **inspiratory reserve**. That expelled by forced expiration after an ordinary breath is referred to as the **expiratory reserve**, while that remaining in the alveoli is the **residual air**.

The term **vital capacity** may be defined as the volume of air that can be expelled by the deepest possible expiration after the deepest possible inspiration.

It will be seen from Table 8.1 that the vital capacity (VC) is the sum of the tidal volume, the inspiratory reserve and the expiratory reserve – also that the total lung capacity (TLC) is the sum of the vital capacity and the residual volume (RV), i.e.

$$TLC = VC + RV$$

Table 8.1   An example of respiratory volumes

	Volume in ml	*Comment*
Tidal volume (TV)	500	Quiet breathing
Inspiratory reserve (IR)	2500	Forced inspiration
Expiratory reserve (ER)	1000	Forced expiration
Vital capacity (VC)	4000	
Residual volume (RV)	1000	Always left in lung
Total lung capacity (TLC)	5000	

The term hyperpnoea is sometimes used to express an increased depth of respiratory movement. Apnoea means a temporary cessation of breathing. Difficult or laboured breathing is called dyspnoea.

The stages of respiration are:

(*1*) *Ventilation* of the lungs so that air moves freely in and out.

(*2*) *Interchange of gases* between the blood and the air in the alveoli. The term 'diffusing capacity' or 'gas transfer factor' refers to the ability of each square millimetre of the alveolar-capillary membrane to transfer gas by the process of diffusion. It is measured in ml/min/mm.Hg.

In addition there is:

(*1*) loss of water vapour; and

(*2*) supply of air to the larynx for the purpose of voice production.

Atmospheric air is a mixture of gases and, as a result of oxygen being absorbed and carbon dioxide being excreted by the lungs, it follows that the amount of oxygen in expired air is diminished, while the amount of carbon dioxide is increased. With quiet breathing, the oxygen uptake and carbon dioxide excretion is about 250−300 ml per minute. The percentage of nitrogen remains constant. The amount of carbon dioxide in expired air is 100 times greater than in atmospheric air (Table 8.2).

The following is a summary of the differences between expired air and inspired air:

(1) expired air contains less oxygen and more carbon dioxide;

(2) expired air is nearer to body temperature;

(3) expired air is saturated with water vapour. The minute droplets of water may pick up bacteria during their passage through the respiratory tract and become a source of infection (droplet infection) to others.

The regulation of respiration

The normal rate of ventilation in adults is 14 to 18 breaths per minute. In children the rate is more rapid and in infants approaches 40 per minute. The rate is increased in certain diseases, for example pneumonia, and may also be abnormally slowed, especially in some cases of poisoning, for example with morphine.

Respiration is controlled by nervous impulses and by the chemical composition of the blood.

Nervous control

Although for a short time the rate and depth of respiration can be controlled by the will, ordinarily it is an automatic act under the unconscious control of the nervous system. Situated in the medulla oblongata of the brain is a collection of nerve cells called the **respiratory rhythm generator** or, in older terminology, the respiratory centre. Afferent inputs to the rhythm generator arise from various sources. Those which arise from the **stretch receptors** in the lungs and which inhibit inspiration travel in **vagal nerve fibres. Glossopharyngeal nerve** fibres carry impulses from the peripheral **chemoreceptors** to stimulate breathing. From the medulla efferent nerve fibres pass down the spinal cord to the diaphragm and the respiratory muscles.

A **pneumotaxic centre** in the upper pons inhibits the inspiratory cells of the rhythm generator and provides a *rate-controlling mechanism*.

Chemical control

It has been seen that carbon dioxide passes from the tissues into the blood and thence to the lungs, where it is excreted. The central chemoreceptors are particularly sensitive to the amount of carbon dioxide (carbonic acid) in the blood. If the amount rises as a result of more being formed in the tissues, such as during muscular exercise, they are stimu-

Table 8.2   Gas content of inspired and expired air

	Inspired or atmospheric air (%)	*Expired air (%)*
Oxygen	20	16
Carbon dioxide	0.04	4
Nitrogen	79	79

lated and send impulses to the rhythm generator. This in turn sends impulses to the respiratory muscles to produce deeper and quicker breathing so that carbon dioxide can be excreted more rapidly by the lungs and the amount in the blood reduced to its normal level. In other words, the function of the rhythm generator is to send out impulses to the respiratory muscles which maintain the rate and depth of breathing so that the level or concentration of carbon dioxide in the blood remains constant.

Chemoreceptors are sensory areas which detect changes in carbon dioxide, hydrogen ion or oxygen in the blood. There are both central and peripheral chemoreceptors.

The **central chemoreceptors** are on the surface of the medulla, where they are bathed by the cerebrospinal fluid (CSF) in the course of its circulation. Carbon dioxide diffuses through the walls of capillaries from the blood into the CSF and stimulates the receptors, which 'notify' the respiratory rhythm generator. Changes in the blood hydrogen ion (H^+) concentration or partial pressure of oxygen (PO_2) do not influence the medullary chemoreceptors.

The main **peripheral chemoreceptors** are the carotid bodies and the aortic bodies, situated close to arteries. When stimulated by a decrease in the partial pressure of oxygen in arterial blood (P_aO_2), they provoke a reflex increase in breathing. The peripheral chemoreceptors are also stimulated by increases in blood H^+ concentration and increases in the partial pressure of carbon dioxide in arterial blood (P_aCO_2).

Under normal conditions, the principal chemical factor in the regulation of respiration is the partial pressure of carbon dioxide in the arterial blood (P_aCO_2), acting mainly through the central chemoreceptors.

Two important practical applications of this fact deserve particular mention.

(*1*) Carbon dioxide may be administered mixed with oxygen during and after a general anaesthetic. It will be clear from the facts just mentioned that the effect of inhaling carbon dioxide will be to increase the amount in the alveolar air and, thus, in the blood. The raised level of carbon dioxide in the blood acts as a stimulus to the respiratory centre which sends out impulses causing increased ventilation of the lungs, hence a more rapid excretion of the anaesthetic and at the same time a greater intake of oxygen. Carbon dioxide is sometimes used to produce increased ventilation of the lungs for other purposes.

(*2*) When administering oxygen, care must be taken to ascertain the percentage of oxygen required and the appropriate type of apparatus for administering it. In some respiratory disorders (e.g. chronic bronchitis) the amount of carbon dioxide in the blood may be persistently raised and the respiratory centre no longer sensitive to it. Deficiency of oxygen (hypoxia) may then be supplying the main stimulus to the respiratory centre. If too high a concentration of oxygen is given, this hypoxic stimulus is removed. Consequently breathing becomes shallower and there is further retention of carbon dioxide, which can result in coma and ultimately death.

Voice production

It is convenient to consider this subject in connection with the respiratory system. The voice sounds are produced in the larynx. They are modified in character by the resonance afforded by the nasal cavities and the accessory air sinuses and finally, by means of the tongue, lips and jaw movements, the actual sounds of speech are produced.

The human voice has the following characteristics:

(*1*) loudness
(*2*) pitch
(*3*) quality

Fig. 8.9 The glottis, showing the vocal cords closed (left) on phonation and open (right) during inspiration.

(*1*) *Loudness* The variable loudness of the voice is dependent upon the force of the air currents expelled from the lungs through the vocal cords and their consequent vibration. The vibration of the vocal cords is called phonation.

(*2*) *Pitch* By pitch is meant the variation in note, i.e. a high note or a low note. This is dependent on two factors, the length and tension of the vocal cords. In children the vocal cords are relatively short and, therefore, the pitch of the voice is high. Alterations in pitch can be produced by voluntary action by using certain of the muscles of the larynx which increase or decrease the tension of the vocal cords. Variation in pitch by the alteration of tension can easily be demonstrated by twanging a stretched string, wire or elastic, viz. the greater the tension of the cord or string, the higher the note produced.

(*3*) *Quality* The quality of a note is due to the resonance produced in the mouth, nose and accessory nasal sinuses in the skull. The difference in quality of sound is easily demonstrated by speaking through the mouth and 'speaking through the nose'. The soft palate plays an important part in this act, and if imperfectly formed (cleft palate) or paralysed (e.g. in diphtheria) a typical nasal voice develops. Full use of the possible variations in quality is made in singing.

Speech

The sounds of the spoken word or articulation are modifications of the primary laryngeal sound which are brought about by movements of the lips, tongue and jaw working either independently or together.

There are two types of broken sound:

(1) Vowels: sounds produced with the mouth open and with the vocal cords vibrating continuously without interruption.

(2) Consonants: there is a sharp interruption or curtailment of the vocal cord vibration. Some consonant sounds are produced mainly by the movement of the tongue against the teeth, for example t and d, are called dentals. The sounds p and b are dependent upon closure of the lips and are called labials. Throaty sounds, g and k, are gutterals.

Whispering In the act of whispering the sound is produced entirely by movement of the air in the mouth. The vocal cords are relaxed (or open) and do not vibrate, i.e. there is no phonation. The formation of word sounds in whispering is carried out by the movements of the mouth and tongue. Inflammation of the vocal cords interferes with their contraction and vibration and consequently the voice becomes hoarse or is entirely lost, so that whispering only is possible. If it is necessary to rest the vocal cords the patient must be instructed to speak only in whispers.

Questions

1. What is the purpose of respiration?
2. Give an account of the anatomy of the lungs. What is the mechanism by which air enters and leaves the lungs?
3. Write notes on the bronchi and alveoli.
4. How is respiration regulated?
5. Describe the nose and pharynx.

9

The Blood
(Haemopoietic System)

The blood is a transport system which plays a very important part in the maintenance of life. It flows throughout the body and, when in the capillaries, it is in intimate relationship with the tissues, taking oxygen and other nutritive substances to them and at the same time removing their waste products.

The total volume of blood in the body is about 6 litres (10 pints) and constitutes about one-twentieth of the total body weight.

The composition of blood

When blood is examined under a microscope it is seen to consist of cells or corpuscles floating in a yellowish fluid, the blood plasma. Normal blood contains approximately:

red corpuscles	$5. \times 10^{12}$ per litre (l)
white cells	8×10^{9} per litre
platelets	250×10^{9} per litre

Red corpuscles or erythrocytes (RBC)

These are the most numerous cells in the blood and contain a substance called haemoglobin, which gives them their red colour. They are correctly called corpuscles rather than cells, because they have no nucleus, but they are most often called red cells. They are discs, the two outer surfaces of which are concave. In short, they are bi-concave, non-nucleated discs. They are very small, having a diameter of 7.5 thousandths of a millimetre (7.5 μm) or 1/3200 inch and are flexible or deformable. They are, therefore, able to pass through the capillaries. Their function is to carry oxygen and carbon dioxide.

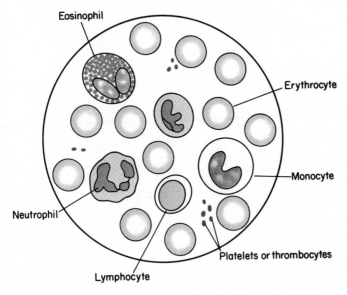

Fig. 9.1 Cells of the blood.

Haemoglobin

This is a complex protein which gives the red colour to the erythrocytes. It consists of a protein, globin, combined with an iron-containing pigment called haem. Iron, therefore, is essential for the formation of haemoglobin.

A daily intake of 10–25 mg of iron is provided by the average Western diet but not all of it is 'available' (or normally needed) for absorption. Meat and green vegetables are rich sources of iron. It is absorbed from the duodenum and upper jejunum. Men lose only about 1 mg of iron each day (in red cells and epithelial cells shed into the gastrointestinal tract), and absorb a similar amount but a woman loses about 20 mg during each normal menstrual period. The need for iron increases greatly during pregnancy when a woman's blood volume and red cell mass increase and iron is transferred across the placenta to the fetus.

Although most of the iron in the body is used for haemoglobin synthesis, it is also vital for the synthesis of myoglobin (a compound, resembling haemoglobin, in skeletal muscle) and many tissue enzymes. Iron is transported to the tissues bound to *transferrin*, a protein found in plasma and serum. Serum transferrin is measured by its capacity for binding iron, a measurement known as the total iron binding capacity (TIBC). The serum iron (normal range 12–26 μmol/ℓ) and TIBC (normal range 45–70 μmol/ℓ) are useful measurements in elucidating some kinds of anaemia. Normally transferrin is about 35% saturated with iron.

Iron is stored in the tissues as *ferritin* and *haemosiderin*. Large deposits of these substances accumulate in the tissues, and interfere with their functions, in conditions of chronic iron overload known as haemosiderosis and haemochromatosis. The former may occur in patients with thallassaemia after numerous blood transfusions.

The normal level of haemoglobin in the blood is 13.0–18.0 g/dℓ for men and 11.5–16.5 g/dℓ for women. The normal mean corpuscular haemoglobin (MCH) is 27–32 pg and the normal mean corpuscular haemoglobin concentration (MCHC) is 31–35 g/dℓ.

Haemoglobin has a very strong affinity for oxygen, and when they come into contact the oxygen is taken up, forming oxyhaemoglobin. This process normally takes place in the lungs where the venous blood, deficient in oxygen, is able to absorb and bind oxygen from the air in the alveoli and leave the lungs by the pulmonary veins as arterial blood.

Oxygenated arterial blood has a bright red colour, while that in the veins, having lost its oxygen, has a bluish-purple hue.

Haemoglobin also has a strong affinity for the poisonous gas, carbon monoxide, with which it forms carboxyhaemoglobin. Haemoglobin bound in this way cannot carry oxygen and the victim may die of anoxia. Carboxyhaemoglobin causes the individual to develop a cherry-red hue in contrast to the cyanosis of anoxia.

Development

The erythrocytes are formed in the red bone marrow. In the infant the cavities of all the bones contain red marrow. In adults much of this is replaced by yellow marrow and erythrocytes are mainly produced in the red bone marrow of the sternum, ribs, vertebrae, cranial bones and the proximal epiphyses of the femur and humerus. They commence as large cells which have nuclei and are called **proerythroblasts**. Their next stage is a smaller cell, the **normoblast**, which also contains a nucleus and is red in colour, because haemoglobin is present in its substance. The normoblast loses its nucleus and passes into the circulation as a mature **erythrocyte**. In order that these changes may take place:

(a) vitamin B$_{12}$ (the erythrocyte-maturing factor, cyanocobalamin) and folic acid are necessary for the development of the proerythroblast into the normoblast; and

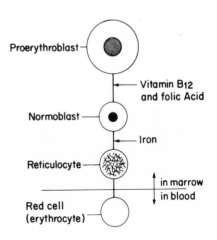

Fig. 9.2 Diagram illustrating the development of the red cell (erythrocyte) from the proerythroblast.

(*b*) iron is necessary for the synthesis of haemoglobin to fill the normoblasts before they
can become mature red corpuscles. Lack of
iron in the diet or loss of iron in bleeding
causes iron-deficiency anaemia.

The life of a red cell is limited to about four
months. When worn out, it is destroyed by the
spleen, which, however, saves and stores the iron
for use in the preparation of fresh haemoglobin.

Cyanocobalamin, vitamin B$_{12}$

This is a substance, occurring mainly in animal
proteins and bacteria but absent from vegetables,
which is necessary for the development of the
proerythroblast into the mature erythrocyte. Its
absence leads to the development of Addisonian or
pernicious anaemia, the study of which has helped
in the understanding of red cell formation.

Although it was over 100 years ago that Addison
first described this type of anaemia, it was not until
Minot and Murphy in 1926 showed that it could be
cured by giving uncooked liver that its cause began
to be understood. Ten years later Castle showed
that a special blood-forming factor was necessary
for the maturing of the proerythroblast into the
normoblast. He further suggested that before this
could be used another factor was necessary and
that this was secreted by the mucous membrane of
the stomach. He, therefore, called vitamin B$_{12}$ the
extrinsic factor, because it was present in the diet,
and called the substance secreted by the stomach
the *intrinsic factor*.

It is now known that vitamin B$_{12}$ cannot be
absorbed from the lowest part of the small intestine
in the absence of intrinsic factor with which it
probably temporarily combines.

In 1948 it was found that from over a ton of liver
about 1 gram of minute red crystals could be
obtained which had the same effect as the whole
liver. Later it was discovered that this substance
could be prepared commercially by other methods
and it was identified as *vitamin B$_{12}$* and shown to
contain traces of the metal cobalt. Regular
injections of vitamin B$_{12}$ are now the recognized
treatment of pernicious anaemia.

It is clear that in pernicious anaemia the absence
of intrinsic factor from the gastric juice leads to
failure of vitamin B$_{12}$ to be absorbed and stored in
the liver, whence it is conveyed to the bone marrow.
Its absence results in defective formation of the red
cells in the bone marrow and the proerythoblasts

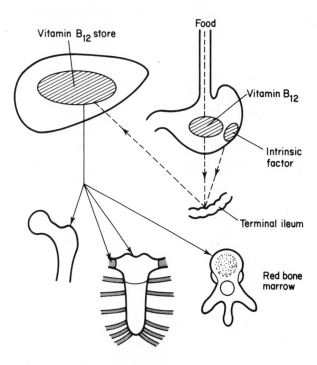

Fig. 9.3 The absorption of vitamin B$_{12}$, its storage in the liver,
and its distribution to red bone marrow.

fail to develop into normoblasts, so that anaemia
results. The number of red cells is considerably
reduced, often to between 1×10^9 and 2×10^9 per
litre but they are abnormally large (macrocytes)
and their haemoglobin content is high.

Folic acid is also necessary for normal red cell
production. It is found in green vegetables, liver,
kidney and yeast. Its deficiency, like that of cyanocobalamin, results in a macrocytic anaemia.

White blood cells (WBC)

These are colourless cells containing nuclei and are
a little larger in size than the red cells (10 μm,
1/2500 inch) but much less numerous (1 WBC to
500 RBC). Their number normally varies between
4.0×10^9 and 11.0×10^9 per litre, but 8.0×10^9 is a
normal average.

Two main varieties are found in the blood:
(*1*) polymorphonuclear leucocytes (75 per cent);
(*2*) lymphocytes (25 per cent).

The polymorphonuclear or granular leucocyte

This is so called because its nucleus is irregular and variable in shape and its cytoplasm contains granules. It forms about 75% of the white cells. The distinguishing feature of these cells is their power of independent movement; in this respect they resemble the amoeba.

They have two important functions:
(*1*) to protect the body against the invasion of bacteria;
(*2*) to remove dead or injured tissue.

If bacteria enter the tissues, the granular leucocytes immediately attempt to surround the organisms and engulf them into their own bodies. Sometimes the toxins of the bacteria are powerful enough to kill the leukocytes and it is the accumulation of leucocytes destroyed in this manner, together with liquefied dead tissue, which forms pus.

After the bacteria have been killed it is the leucocytes which remove the tissues which have been damaged or destroyed by the action of toxins and, thus, they play their part in the first stages of the process of healing.

It is because of this power of engulfing bacteria and debris and digesting them by means of various enzymes that the granular leucocytes are sometimes called phagocytes and the process they carry out, phagocytosis. This is similar to the way in which the amoeba takes in food (see Fig. 2.6, p. 9).

Development
The polymorphonuclear leucocytes or granulocytes are derived from special cells in the bone marrow called **myeloblasts**.

Varieties
If special dyes are used it is possible to show several varieties according to the actual stains which the granules in each take. Some stain evenly and are called **neutrophil** cells; others take a red, acid stain and are called **eosinophil** cells; a few pick up an alkaline dye and are referred to as **basophil** cells. The number of eosinophils is sometimes increased in allergic states such as asthma, certain skin diseases and infections with intestinal parasites.

The lymphocytes

These are non-granular cells and constitute about 25% of the total number of white cells. They are distinguished from the polymorphonuclear cells by having a large round nucleus. Two types, large and small, are recognized. They play a vital role in the defence of the body against infection but do not show the power of phagocytosis shown by the polymorphonuclear cells.

Lymhocytes are concerned in the immunity mechanisms of the body (see Chap. 3). Hence the use of anti-lymphocytic serum (ALS) to prevent rejection of the foreign tissues introduced in an organ transplant operation.

Development
Lymphocytes are derived from the bone marrow, the lymph glands, the thymus and the spleen and other masses of lymphoid tissue in the body. Their total number is relatively increased in infancy.

Leucocytosis and leucopenia
In most bacterial infections, the body responds by increasing the number of granular leucocytes circulating in the blood. This is called a *leucocytosis*, and the number of cells may rise to $20.0-30.0 \times 10^9/\ell$ especially if an abscess (pus) is present.

In some conditions, such as bacterial pneumonia, if the body fails to produce a leucocytosis the outlook becomes serious. Virus infections do not often cause leucocytosis but frequently cause a *leucopoenia*. This is the term used when the leucocytes are diminished in number. It may also occur when the bone marrow is depressed by the action of certain drugs, toxins or radiation.

In leukaemia there is a proliferation of abnormal leucocytes and a deficiency of normal white blood cells.

The blood platelets (thrombocytes)

These are minute spherical structures present in the blood, normally numbering from 150 to $400 \times 10^9/\ell$. They are produced by large cells (megakaryocytes) present in the bone marrow. They are smaller than the red corpuscles and their main function is concerned with the clotting of blood. In certain diseases their number may be seriously decreased (thrombocytopoenia) and the patient shows a tendency to bleeding into the skin (purpura). Often, bleeding from the mucous membranes also occurs.

The blood plasma

The yellowish, slightly alkaline fluid in which the blood corpuscles float is called the plasma. It has the following composition:

> Proteins 60–80 g/ℓ; albumin 35–50 g/ℓ; globulin 17–39 g/ℓ; and fibrinogen 3 g/ℓ
> Salts including chlorides, sulphates, phosphates of sodium, potassium and calcium
> Urea 2.5–6.5 mmol/ℓ (15–40 mg/100 ml)
> Creatinine 60–120 μmol/ℓ (0.7–1.4 mg/100 ml)
> Glucose (fasting) 3.0–5.3 mmol/ℓ (54–95 mg/ 100 mℓ)
> Water (90%)

Other substances such as prothrombin, vitamins, enzymes and antibodies are also present.

It will be noticed that plasma contains three types of protein:

(*a*) albumin and globulin
(*b*) fibrinogen

By various means the fibrinogan can be removed from the plasma. The fluid then remaining is called serum, i.e.

$$plasma = serum + fibrinogen$$

or

$$serum = plasma - fibrinogen$$

It will be noticed that, since it consists of plasma from which fibrinogen only has been removed, serum still contains the valuable antibodies to disease which are special proteins called immunoglobulins.

The total amount of protein may be reduced in certain diseases. This occurs particularly in the nephrotic syndrome when large quantities of albumin are lost in the urine. In this instance the serum albumin is particularly lowered. This loss of protein from the blood is an important factor in the production of oedema in this condition because it results in a lowering of the osmotic pressure of the blood.

Blood has a slightly salty taste, which a patient may mention if there has been any haemorrhage into the mouth.

The coagulation (clotting) of blood

Since humans are liable to injury and the shedding of blood, a mechanism is provided within the body whereby there is a spontaneous tendency for the loss of blood to be limited.

The actual mechanism is a complicated one, but the general principles are simple and important. Shortly after being shed, the blood becomes sticky and sets into a jelly-like mass. After a few hours this mass contracts and from it is squeezed a yellowish fluid, the serum.

The essential change in the coagulation of blood is the conversion of the protein **fibrinogen** into a substance called **fibrin**. This forms fine threads which entangle the blood cells and then contract. It is this contraction of fibrin which expresses the serum and binds the clot into a firm mass. The formation of this firm mass in a wound acts as a plug and effectively seals off the opened blood vessels, thereby preventing further loss of blood.

The coagulation of blood involves a 'cascade' of reactions (Fig. 9.4) resulting in the conversion of **prothrombin** (factor II) into **thrombin** and the action of this on fibrinogen (factor I), converting it into fibrin. The latter traps red and white cells and platelets in its meshes, forming a clot from which serum is expressed by contraction of the fibrin threads.

Among the factors necessary for coagulation under varying circumstances are tissue thromboplastin, platelet factors, calcium (factor IV), anti-haemophilic factor (factor VIII) and Christmas factor (factor IX). The international nomenclature lists factors I to XIII. Prothrombin (factor II) is synthesized in the liver and vitamin K is necessary for its synthesis.

Certain conditions hasten the clotting of blood, while others retard it. These often have a practical importance.

Factors hastening coagulation or clotting

(*1*) Injury to the tissues. A clean cut with a sharp knife or razor bleeds much more freely than a crushed wound in which there is considerable bruising and damage to the surrounding tissues.

(*2*) Contact with a foreign body. The application of a surgical dressing such as gauze aids very considerably in the speedy formation of a clot and arrest of haemorrhage.

(*3*) Temperature slightly higher than that of the body. Use of this fact is sometimes made during surgical operations when bleeding surfaces are packed with swabs soaked in hot saline at 49°C (120°F).

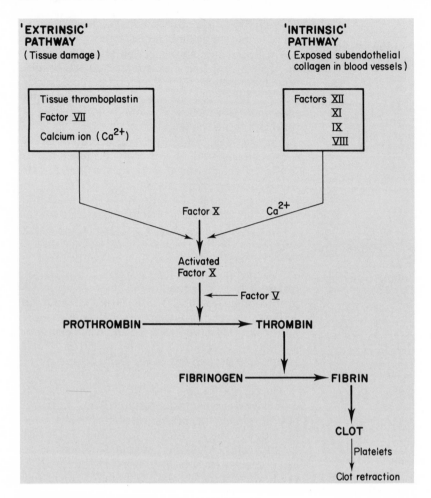

Fig. 9.4   The blood coagulation cascade.

Factors retarding coagulation or clotting

(*1*) The addition of sodium or potassium citrate. This acts by removing the activity of calcium salts. It is used to prevent blood clotting in blood transfusions. The blood from the donor is taken directly into a solution of potassium citrate (3.8%) and may be kept for very long periods without risk of clotting.

(*2*) Contact with oil, grease or paraffin wax. This means that ointments and greasy materials should not be applied to wounds until bleeding has ceased. An old-fashioned method of blood collection for transfusion was to take the blood into a bottle which had been previously coated with paraffin wax.

(*3*) Local cold.

Certain other substances are used to alter the coagulability of the blood. A substance which can be obtained from the liver, called heparin, prevents coagulation, and drugs such as phenindione (Dindevan) produce a similar effect. The leech secretes a substance (hirudin) that delays clotting, which explains why bleeding continues for so long after a leech bite. The poison of certain snakes (especially Russell viper venom) hastens clotting and has been used for this purpose.

Adrenalin is sometimes applied to bleeding surfaces in order to arrest haemorrhage. This has no effect on the mechanism of clotting but acts by causing the blood vessels to contract, thereby diminishing the actual escape of blood.

Summary of the functions of the blood

(1) To convey oxygen to the tissues by means of the haemoglobin in the red cells.
(2) To remove waste products from the tissues and convey them to the appropriate organs for excretion, e.g:
  (a) carbon dioxide is carried to the lungs;
  (b) urea is carried from the liver to the kidneys for excretion; and
  (c) water is carried to the kidneys, where any excess is removed.
(3) To carry nutrients to all parts of the body.
(4) To carry the hormones or chemical messengers of the body, i.e. the internal secretions of the endocrine glands (e.g. insulin from the pancreas).
(5) To aid in the defence of the body by the phagocytic action of the polymorphonuclear leucocytes.
(6) To participate in the circulation of lymphocytes required for the immune response (see Chap. 3).
(7) To carry antibodies (immunoglobulins) to sites of infection.

Blood groups

In connection with the important procedure of blood transfusion, individuals may be divided into four main groups and it is essential that the correct type of blood is employed in each transfusion. Table 9.1 shows the groups and the names given to them.

Table 9.1 The ABO blood groups

Name	Percentage of persons
AB	5
A	40
B	10
O	45

The division of persons into these four groups depends on the following facts:

(*1*) Human blood serum contains substances called **agglutinins**; that is, substances which have the power of causing the red cells of persons belonging to another group to clump together or agglutinate if they are mixed with this serum.

(*2*) Human red cells possess substances called **agglutinogens**; i.e. substances which have the power of stimulating the production of agglutinins.

These substances are designated in the following way:

The agglutinins are called α (alpha) and β (beta).
The corresponding agglutinogens are called A and B.
Human serum may contain α or β or both (α + β) or no agglutinogens.
Human red cells may contain A or B or both (A + B) or no agglutinogens (O).

It follows that if a person possesses red cells belonging to Group AB his serum cannot contain α or β agglutinins, otherwise his own red cells would agglutinate and life would be impossible. In the case of an individual of Group A, his red cells will contain the agglutinogen A and his serum the agglutinin β which will not affect them. If both the agglutinins α and β are present in the serum the red cells can contain no agglutinogen and the individual will belong to Group O.

Table 9.2 ABO blood group factors

Group	Agglutinogens (cells)	Agglutinins (serum)
AB	A and B	None
A	A	β
B	B	α
O	None	α and β

The main point in blood transfusion is that *the cells of the donor must be compatible with the serum of the recipient* who receives them. (For practical purposes, the effect of the serum of the donor on the cells of the recipient may be ignored as it becomes so well diluted in the act of transfusion that, at the worst, only minor ill-effects can be produced.) Table 9.3 sets out the principles on which blood transfusions are based.

Table 9.3 Principles of blood group cross-matching

Recipient's serum		Donor's cells (agglutinogens)			
Group	Agglutinins	AB	A	B	O
AB	nil	−	−	−	−
A	β	+	−	+	−
B	α	+	+	−	−
O	α and β	+	+	+	−

+ = agglutination of donor's cells by recipient's serum
− = no agglutination of donor's cells by recipient's serum

It will be seen that individuals of Group AB have serum containing no agglutinins and so they can receive cells from any other group by transfusion. They are therefore sometimes called *universal recipients*.

On the other hand, those of Group O have red cells which contain no agglutinins so that their cells can produce no reactions in members of the other groups. Persons having Group O blood are called *universal donors*.

Groups A and B can only receive blood from their own groups or Group O. Group O patients can only receive Group O blood, since A and B agglutinogens in donor blood would be agglutinated by the recipient's α and β agglutinins.

In practice, the group to which an individual belongs is ascertained by testing that person's cells against the serum of a known group. Both the recipient and the donor having been appropriately labelled, it is customary to take the added precaution of testing the serum of the recipient against the corpuscles of the donor to confirm that the mixture is compatible (*cross-matching*).

If incompatible blood is transfused, very serious symptoms may follow. The foreign red cells are first of all agglutinated into clumps which may block capillaries in various parts of the body. Later, these clumps are broken down (haemolysis) and free haemoglobin is liberated into the bloodstream. This is partly converted into bilirubin by the liver and will produce jaundice. Part, however, is excreted by the kidneys and may block the renal tubules, especially if the urine is acid, leading to suppression of urine and kidney failure which may prove fatal. Distress and fever may occur early in the transfusion and should be an indication to dis-

continue the procedure before too much damage has been done.

Another interesting feature is that blood groups are inherited, the important factors being the A and B agglutinogens. This is sometimes of importance in proving that a child could not be the offspring of an alleged father.

The Rhesus (Rh) factor

This substance was first discovered in experiments on the rhesus monkey and was consequently given the name of 'the rhesus factor', abbreviated to Rh.

Further experiment showed that this agglutinogen was also present in the red cells of 85% of human beings, who can therefore be divided into two groups:

Rh positive = 85%
Rh negative = 15%

Under certain circumstances antibodies may develop in the blood of an individual which are capable of causing the agglutination and destruction of Rh positive cells. These antibodies are called anti-Rh.

If two parents are both Rh positive their offspring will be Rh positive. If one of the parents is Rh positive their offspring will probably be Rh positive but, if the mother is Rh negative and her child is Rh positive, she may form antibodies (agglutinins) to the rhesus factor in the child's blood.

Father Rh +ve plus mother Rh −ve child Rh +ve
Mother Rh −ve plus child Rh +ve mother Rh −ve with anti-Rh

In any future pregnancy this anti-Rh antibody may affect the Rh ±ve red cells of the fetus, causing their agglutination and destruction when it passes through the placenta and enters the fetal circulation. This condition is known as erythroblastosis fetalis and results in severe and often fatal jaundice (icterus gravis neonatorum). It is also responsible for hydrops fetalis and some cases of repeated stillbirths or miscarriages.

Father	Mother	Child	
Rh +ve	Rh +ve	Rh +ve	= normal
Rh −ve	Rh +ve	Rh +ve	= normal
Rh −ve	Rh −ve	Rh −ve	= normal
Rh +ve	Rh −ve	Rh +ve	= 4% risk of abnormality in first child

A similar state of affairs can be produced in the maternal blood by transfusion. If a woman who is Rh negative is transfused with blood of the correct ABO group but which is Rh positive she may develop anti-Rh antibodies in her blood. If she subsequently has an Rh positive child (from an Rh positive father), the child may be affected by erythroblastosis fetalis.

In other words, the commonest cause of haemolytic disease in the newborn is the possession or formation by Rh-negative women of antibodies against the red cells of an Rh-positive fetus.

It follows that Rh negative females should only be transfused with Rh negative blood if they are of childbearing age or younger.

These are the general principles which apply to the Rh factor but the subject is complicated by the existence of a number of subgroups.

Questions

1. What are the normal numbers of red cells, white cells and platelets in the blood? What is a deficiency of each of these called?
2. What is haemoglobin and what is its function?
3. What are the dietary sources of iron? In which part of the gastrointestinal tract is it absorbed and how is it transported in the plasma?
4. What are the normal stages of development of an erythrocyte? What substances are required for this process?
5. What are the functions of polymorphonuclear leucocytes?
6. Where are blood platelets produced? How does a serious deficiency of platelets manifest itself?
7. What are the constituents of plasma?
8. Describe the essentials of coagulation of the blood.
9. What are the functions of the blood?
10. Write notes on the ABO and Rhesus blood groups.

10
The Circulatory System

The circulatory system is a *transport system*, carrying oxygen, nutrients, hormones and other substances to the tissues and conveying carbon dioxide to the lungs and other waste products to the kidneys. The blood is the vehicle or carrier and the blood vessels are the channels along which it travels. The motive power is supplied primarily by the heart, which is a muscular pump. However, the venous return of blood to the heart is assisted by gravity (for those parts which are higher than the heart), skeletal muscle activity squeezing the veins, and the inspiratory phase of breathing, which sucks blood towards the thorax. In addition to the circulation of the blood there is a lymphatic circulation (see Chap. 11).

Blood is pumped by the heart (left ventricle) into the **aorta**, from which it is distributed by **arteries** to all parts of the body. The arteries branch and narrow down to **arterioles** and these in turn lead to microscopic **capillaries** which ramify throughout the tissues. The diameter of a capillary is about the same as that of a red blood cell. The walls of the capillaries consist of only one layer of cells, which are joined together but between which there are small spaces through which white blood cells can pass by means of their power of amoeboid movement. It is in the capillaries, and in them only, that the blood equilibrates with the interstitial fluid in the tissues; nutrients, water, oxygen, carbon dioxide, and other waste products (arising from the metabolic activities of cells) being exchanged.

The capillaries drain into **venules** and these unite to form **veins**, which convey the deoxygenated blood back to the heart (right atrium) to complete the **systemic circulation**, sometimes referred to as the *greater circulation*. The *lesser circulation* (lesser in size but not in importance) is the **pulmonary circulation** in which venous (de-oxygenated) blood is pumped from the right side of the heart (right ventricle) into the **pulmonary artery** and its branches. Again, after traversing arterioles it reaches capillaries; this time the capillaries of the lungs. *Gaseous exchange* takes place between the blood in these capillaries and the air in the alveoli, across what is called the **alveolar-capillary membrane**. The blood thereby loses its excess of carbon dioxide and becomes enriched with oxygen. However, the alveolar-capillary membrane may be thickened by a disease process such as pulmonary oedema or fibrosis. Then, oxygen cannot diffuse across it so readily and cannot fully oxygenate the blood so that the patient may show *cyanosis*, or blueness. Carbon dioxide is a more diffusible gas than oxygen and is not retained in the blood in these cases of what is often called alveolar-capillary block.

Leaving the capillaries of the lungs, the oxygenated blood returns to the heart via venules which join to form the **pulmonary veins**. Four large pulmonary veins, two from each lung, enter the receiving chamber (left atrium) of the left side of the heart. In contradistinction to other arteries of the body, the pulmonary artery contains deoxygenated or 'venous' blood and the pulmonary veins, unlike any other veins, contain oxygenated or 'arterial' blood.

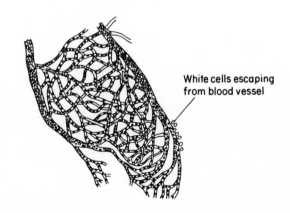

White cells escaping from blood vessel

Fig. 10.1 Network of capillaries between an arteriole and a venule.

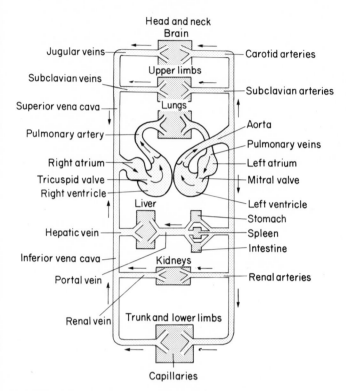

Fig. 10.2 Diagram of the circulation.

The heart

The heart is a hollow, muscular organ which lies obliquely in the thorax, in the middle mediastinum, between the lungs and immediately above the diaphragm. It is situated behind the sternum and adjoining left costal cartilages and ribs, two-thirds of it lying to the left of the median plane. In adults, its average weight is 300 g in males and 250 g in females. Being irregularly conical in shape, it is described as having a base, facing posteriorly and to the right, and an apex pointing anteriorly downwards and to the left. Clinically, the apex can be palpated in many individuals and is normally within the left mid-clavicular line, a line drawn on the chest wall from the mid-point of the left clavicle and running parallel with the median line, or midline of the sternum. Displacement of the apex beat leftwards may be due to cardiac enlargement (cardiomegaly) or to a shift of the heart caused by lung disease. In dextrocardia, the apex of the heart is located in the right side of the chest.

The heart is partitioned into right and left parts which, although anatomically joined together, are physiologically separate. Clinically they are often referred to as the 'right heart' and the 'left heart.' Each consists of two chambers, an upper thin-walled atrium and a lower thick-walled ventricle. The atria act as *receiving chambers* and the ventricles as *pumping chambers*. The right heart pumps blood into the pulmonary circulation and the left heart serves the systemic circulation. The greater circulation being the larger, and demanding a higher pressure, the left ventricle is thicker walled and more powerful than the right ventricle. The opening between each atrium and ventricle is guarded by a valve which permits blood to flow from the atrium to the ventricle but not in the opposite direction; in short it prevents back-flow.

The heart, therefore, actually consists of *two* pumps, one for each of the circulations, anatomically combined in one organ. The left ventricle normally contracts slightly in advance of the right and its ejection phase (when it is ejecting blood) is

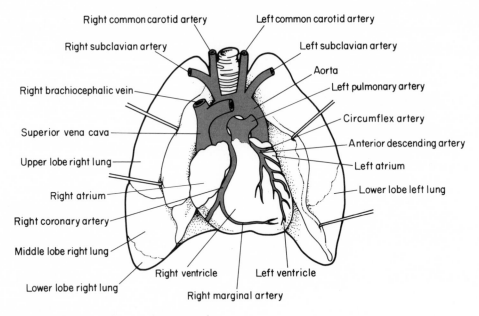

Fig. 10.3   The lungs, heart and great vessels.

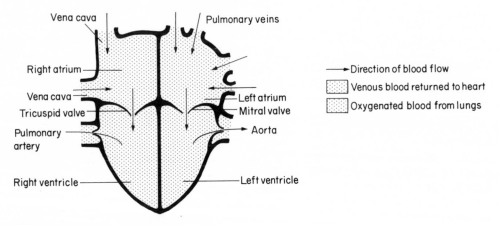

Fig. 10.4   The heart (diagrammatic).

shorter. The heart sounds are therefore double, the first and second heart sounds each having two major components which can be appreciated by auscultation (listening) with a stethoscope. At first, in clinical training, a medical student usually hears two sounds, resembling the words 'lubb dup,' with each heart beat. These are the first and second heart sounds. Soon, the student comes to appreciate that each of these is itself double, the sounds actually being more like 'lurr-ubb durr-up' than simply lubb dup. The first heart sound results from the closure of the atrioventricular valves, the left one

(the mitral valve) closing before the right-sided (tricuspid) valve. Similarly the second heart sound is due to closure of the exit valves of the heart in the roots of the aortic and pulmonary arteries, aortic valve closure normally preceding pulmonary valve closure.

The atria, the receiving chambers of the heart, are separated from one another by the interatrial septum. A congenital defect in this is fairly common and is one of the forms of 'a hole in the heart.' Blood is shunted through the defect from one side of the heart to the other, usually from left

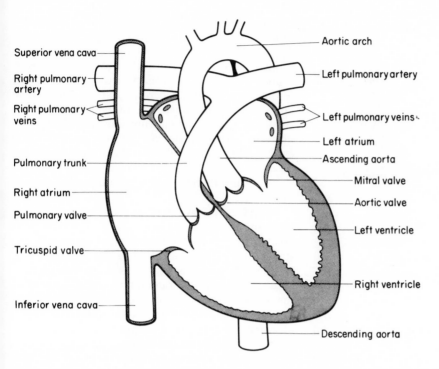

Superior vena cava

Right pulmonary artery

Right pulmonary veins

Pulmonary trunk

Right atrium

Pulmonary valve

Tricuspid valve

Inferior vena cava

Aortic arch

Left pulmonary artery

Left pulmonary veins

Left atrium

Ascending aorta

Mitral valve

Aortic valve

Left ventricle

Right ventricle

Descending aorta

Fig. 10.5 The heart and great vessels. The chordae tendineae and papillary muscles cannot be adequately represented in a single plane and have therefore been omitted. The reader should work out the direction of blood flow in the various chambers and vessels and check the result by reference to Fig 10.4.

to right because the pressure is usually greater in the left side of the heart than in the right side. A left-to-right interatrial shunt is then said to be present. This has no effect on the patient's appearance but central cyanosis (blueness of the skin and mucous membranes) may result if deoxygenated blood is shunted in the opposite direction, namely right-to-left.

Two large veins, called the venae cava, enter the right atrium. Blood from the veins of the head, neck and upper limbs enters the right atrium by the superior vena cava and from the rest of the body and lower limbs by the inferior vena cava. The coronary sinus, which drains venous blood from the heart muscle, also opens into the right atrium.

From the right atrium blood passes through the **tricuspid valve**, so called because it has three cusps or leaflets, into the right ventricle. The valve is anchored by fine cords, the **chordae tendinae** ('heart strings') to two or three small muscles which arise from the walls of the ventricle and are known as **papillary muscles**.

Blood leaves the right ventricle through the **pulmonary valve**, which has three half-moon-shaped

or semilunar cusps, and enters the **pulmonary artery**. It flows through the right and left pulmonary arteries and their many branches in the two lungs to reach the pulmonary capillaries, where gaseous exchange takes place. The oxygenated blood is returned to the heart by the pulmonary veins. Usually, four large **pulmonary veins** enter the left atrium, two from each lung.

The left atrium empties its blood through the **mitral valve** into the left ventricle. The mitral valve has two cusps and derives its name from its resemblance to a bishop's mitre. Like the tricuspid valve, the mitral valve is anchored to the left ventricular wall by chordae tendinae and papillary muscles, of which there are two (anterior and posterior). Rupture of the chordae tendinae results in sudden severe incompetence of the mitral valve, with regurgitation of blood from the left ventricle to the left atrium, a condition called acute mitral regurgitation. The inner surfaces of the ventricles are ridged or trabeculated, the left more so than the right. The outflow tract regions of both ventricles are, however, smooth. The walls of the left atrium and ventricle are thicker than those of the cor-

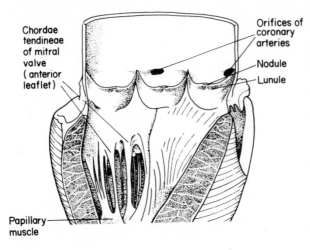

Chordae
tendineae
of mitral
valve
(anterior
leaflet)

Orifices of
coronary
arteries

Nodule

Lunule

Papillary
muscle

Fig. 10.6   The cusps of the aortic valve.

responding right-sided chambers, in keeping with the greater pressures in the left side of the heart and in the greater circulation.

Blood leaves the left ventricle by the large, main artery of the body called the **aorta**. This, like the main pulmonary artery, is about 3 cm in diameter. The opening from the left ventricle into the aorta is guarded by the **aortic valve** which, like the pulmonary valve, consists of three semilunar cusps. The aortic valve is generally stouter than the pulmonary valve, being a high-pressure valve. Just above the attached margin of each cusp, there is a dilatation of the wall of the aorta and this is called an **aortic sinus**. There are similar sinuses in the root of the pulmonary artery but they are less prominent. The right and left **coronary arteries** arise from two of the aortic sinuses and the third or right posterior aortic sinus is known as the non-coronary sinus.

Like any other organ, the heart requires oxygen and nutrients and therefore has its own blood supply. This consists of what clinicians call the coronary arterial tree, which has two trunks and many branches. The two trunks are the right and left coronary arteries, which are the first branches of the aorta.

The **right coronary artery** arises from the anterior aortic sinus and passes forwards between the pulmonary artery and the right atrium until it comes to lie on the front of the heart in the atrioventricular groove, between the right atrium and right ventricle. It curves around the right margin of the heart and continues posteriorly to its termination a little to the left of the posterior interventricular groove. It terminates by anastomosing or joining with terminal branches of the circumflex branch of the left coronary artery. Near to its termination, the right coronary artery gives off a large branch, the **posterior interventricular (descending) artery**, which runs downwards in the posterior interventricular groove towards the cardiac apex. The right coronary artery supplies blood to the right atrium and right ventricle and to parts of the interventricular septum and left side of the heart.

The **left coronary artery** is usually larger than the right and it supplies most of the left ventricle, left atrium and interventricular septum and a narrow strip of the right ventricle. It arises from the left posterior aortic sinus and passes first behind and then to the left of the pulmonary artery. It reaches the atrioventricular groove on the left side of the heart and divides into an **anterior interventricular (descending) artery** and a **circumflex artery**. The former descends in the anterior interventricular groove to the apex and usually turns around the apex into the posterior groove for a few centimetres to anastomose with the posterior interventricular branch of the right coronary artery. The circumflex artery lies in the atrioventricular groove anteriorly and turns around the left border of the heart into the posterior part of the groove.

The above-mentioned are the main branches of the coronary arterial tree. They and their sub-branches may be seriously affected by atheroma, a disease process which narrows the lumen of the affected vessel and restricts the flow of blood through it. The symptom which may result from this stenosis (narrowing) of one or more coronary arteries is angina pectoris, which is cardiac pain usually induced by exertion or emotion. If the artery becomes completely occluded ('coronary thrombosis'), a portion of the cardiac muscle may be critically deprived of blood and may undergo necrosis, a process known as myocardial infarction.

The state of the coronary arteries may be assessed in the living subject by *coronary arteriography*, in which the coronary arteries are injected with a radio-opaque medium and X-rayed.

The *venous drainage* of the heart is mainly via veins which accompany the branches of the coronary arteries. Most of them drain into the **coronary sinus**, which lies in the posterior part of

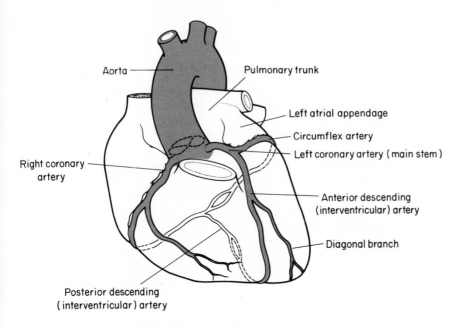

Fig. 10.7 The coronary arteries. For clarity, the smaller branches are not shown.

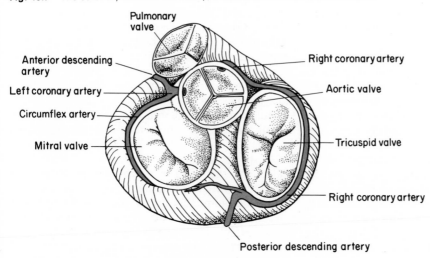

Fig. 10.8 Base of the ventricles seen from above, showing the relative positions of the valves and the courses of the coronary arteries. (Note that the atria have been removed.)

the artrioventricular groove and opens into the right atrium. The **anterior cardiac veins**, which drain the anterior part of the right ventricle, open independently into the right atrium. Some venous blood also drains directly into the chambers of the heart via small veins in their walls known as **Thebesian veins**.

The heart lies enclosed within a fibroserous sac called the **pericardium**. The outer fibrous layer of the sac is continuous above with the external coats of the great vessels and with the pretracheal fascia. Below, it is fused with the central tendon of the diaphragm. In addition, there are attachments to the posterior surface of the sternum. Thus the fibrous pericardium is anchored within the thorax and the general position of the heart is maintained. The fibrous pericardium may also be of importance in preventing over-distension of the heart.

Within the fibrous pericardium, is the serous pericardium which is a delicate double-layered sac; the **parietal layer** lines the fibrous sac and is reflected around the roots of the great vessels to become continuous with the **visceral layer** or **epicardium** on the surface of the heart. A film of fluid lubricates the two layers of serous pericardium so that they glide over one another freely with each heartbeat. An abnormal collection of fluid in the pericardial sac is called a pericardial effusion and may result from inflammation (pericarditis) or from some other process such as myxoedema (thyroid deficiency).

The myocardium or heart muscle has some resemblance to skeletal muscle in being cross-striated but is like smooth muscle in not being under voluntary control. Like smooth muscle, cardiac muscle is under inherent, autonomic and hormonal control. Between the cardiac muscle cells is delicate connective tissue which is extremely rich in blood capillaries, reflecting the *high metabolic demand* imposed by its continuous rhythmic contractility.

The atria have only to pass on their contained blood through the comparatively wide atrioventricular valve openings into the ventricles. Their work is therefore light and in consequence they have relatively thin walls. The right ventricle has more work to do, in pumping the blood through the lungs, and its wall is therefore thicker than that of the right atrium. The left ventricle, having to pump blood to all parts of the body, must be even stronger and its wall is therefore about three times as thick as that of the right ventricle. Connective tissue separates atrial and ventricular myocardium except at a central point where the specialized fibres of the atrioventricular bundle of His provide continuity.

The **endocardium** or innermost lining of the heart consists of a single layer of flattened epithelial cells lying on a bed of fine areolar connective tissue rich in elastic fibres; this supportive layer of connective tissue accommodates movements of the myocardium which might otherwise damage the endothelium. The endothelial lining of the heart is continuous with that of the blood vessels entering and leaving the heart. The endocardium lines the whole of the cavities of the heart and, at the valvular openings is folded back upon itself to form the cusps of the **valves** which, therefore, consist of two layers of endocardium sandwiching a little fibrous tissue to give them extra strength.

The functioning of the heart

The lay person sometimes wonders how the heart manages to pump day and night for seventy years or more without resting. The answer, of course, is that the heart does rest, between every beat. The resting period is called *diastole* and, at a heart rate of around 65 beats per minute, its duration is twice as long as that of *systole*, which is the period of muscular contraction. Passive filling of the ventricles occurs during diastole, when the aortic and pulmonary valves are closed. About 70 per cent of ventricular filling occurs passively but as the ventricles become distended, their rate of filling declines and complete filling of the ventricles depends upon atrial contraction (atrial systole) in late diastole. It will be noted that when the unqualified terms systole and diastole are used they are taken to apply to the ventricles. *Atrial systole* occurs towards the end of ventricular diastole and is followed by closure of the atrioventricular valves. Loss of co-ordinated atrial contraction, and therefore of the atrial transport mechanism, may precipitate cardiac failure in a patient with a diseased heart. This loss may result from heart block or from atrial fibrillation but in the latter case the tachycardia may be the more important factor because it reduces the time available for ventricular filling, tachycardia shortening diastole more than systole.

At the beginning of systole, the inlet and outlet valves of the ventricles are closed. When the semilunar valves open, blood is ejected into the aorta and pulmonary artery. The volume of blood ejected during a single heartbeat is called the *stroke volume* (SV). The *cardiac output* (CO) is the total volume ejected over one minute and is therefore the product of the stroke volume and the heart rate (HR), i.e.

$$CO = SV \times HR$$

The cardiac output is normally $4.0 - 8.0$!/min in the resting state and increases appropriately with exercise. It is common knowledge that the heart rate increases with exercise and emotion. The resting heart rate in normal adults is often quoted as being 72 beats per minute but in some people, especially athletes, it may be much lower.

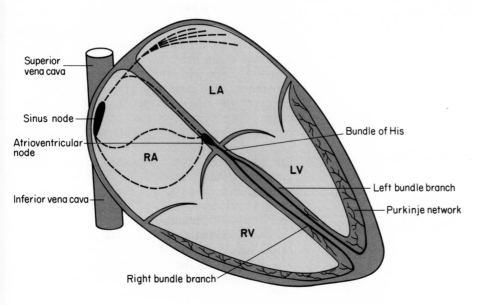

Fig. 10.9   The conducting system of the heart. RA = right atrium; RV = right ventricle; LA = left atrium; LV = left ventricle. The interrupted lines indicate specialized pathways between the nodes and between the sinus node and the left atrium. Their roles are disputed, however, and the impulses from the sinus node may spread by *direct excitation* of ordinary atrial muscle fibres.

Electrical activation of the heart

The stimulus for contraction of the atria and ventricles is electrical. Various parts of the myocardium are capable of discharging spontaneously but normally the *sinoatrial node* (SA node or sinus node) discharges most rapidly and is the normal *cardiac pacemaker*. The SA node lies in the right atrium, close to the entrance of the superior vena cava. The impulse generated in the SA node is propagated through the atria and reaches the *atrioventricular node* (AV node) which lies beneath the endocardium of the septal wall of the right atrium just behind the attachment of the tricuspid valve. Conduction of the impulse is delayed in the AV node whilst excitation and contraction of the atria is completed. The impulse is then conducted into the various parts of the ventricles in an orderly sequential manner so that papillary muscle contraction is able to control the atrioventricular valves whilst a wave of excitation and contraction spreads from the apices of the ventricles to their outflow tracts, resulting in ejection of blood.

Having passed through the AV node, the impulse passes along the *atrioventricular bundle* or *bundle of His*, which is the only direct muscular communication between the atria and ventricles, atrial and ventricular muscle fibres otherwise being separated by a fibrous tissue ring. High in the interventricular septum, the bundle of His divides into *right and left bundle branches* which are distributed respectively to the right and left ventricle. The bundle branches are insulated from the surrounding myocardium by connective tissue sheaths and the impulse reaches the ventricular muscle via the *Purkinje system* of fibres in which the bundle branches terminate.

The whole system, from the SA node to the Purkinje fibres, is known as the *conducting system* of the heart. It is composed of modified cardiac muscle cells and the reader may think in terms of working myocardium and specialized conducting pathways. The sequence of electrical discharge (depolarization) and recovery (repolarization) and the phases of the cardiac cycle may be appreciated in the *electrocardiogram* or ECG (see Fig. 10.10) recorded from the body surface. The P wave is produced by atrial depolarization and the QRS complex by ventricular depolarization. The ST segment and T wave represent ventricular repolarization. The evidence of atrial repolarization is submerged in the QRS complex.

The function of the conducting system of the heart is to ensure a co-ordinated sequence of events

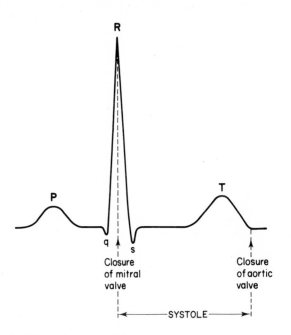

Fig. 10.10   The electrocardiogram (of one heart beat).

in the cardiac cycle. This involves

(*1*) atrioventricular conduction delay so that ventricular systole properly follows atrial systole;

(*2*) Mechanisms for a precise pattern of ventricular contraction.

Disease of the conducting system may result in complete atrioventricular block ('complete heart block'), so that impulses arising in the SA node cannot be conducted into the ventricles. Cells in one of the ventricles will then take over the function of pacemaker but, as their rate of rhythmic depolarization and repolarization is slow, the heart rate will be slow (bradycardia), often 40 beats or less per minute. More seriously, the ventricular pacemaker may be unreliable, resulting in periods of asystole when no ventricular contraction occurs and the patient loses consciousness in an Adams-Stokes attack. Such an attack may also result from ventricular fibrillation, to which these patients are also liable. The treatment is by direct electrical pacing of the right ventricle, bypassing the AV node. The lay person sometimes asks how a patient can ever die with a normally functioning implanted pacemaker continuing to stimulate the heart. The answer is that pacemakers only control the *rate* of the heart and cannot stimulate dead muscle.

Nervous control of the heart-beat

The electrical stimulus for contraction of cardiac muscle is generated spontaneously within the heart and this continues to occur if the heart is stripped of its nerve supply. In other words the heart-beat is spontaneous and continues in the denervated heart. However, the autonomic nervous system exercises control over the *rate* of the heart-beat. Impulses passing down the *parasympathetic fibres* of the **vagus nerve** are *inhibitory* to the heart and slow it. The heart also has a *sympathetic nerve supply*, however, from the cervical and upper sympathetic ganglia, and sympathetic nerve impulses *quicken* the heart. If the effects are balanced, the 'normal' heart rate is maintained. If there is sympathetic overactivity, possibly due to acute anxiety, tachycardia occurs. On the other hand, increased parasympathetic tone results in bradycardia. Strong vagal stimulation may even abolish spontaneous nodal discharge temporarily so that fainting (syncope) occurs. The pulse of a person who is recovering from a simple faint is slow.

The autonomic nerve fibres exert their effects upon the cardiac cells via neurotransmitters or chemical messengers. Certain proteins at the cell surface are receptors for these neurotransmitters. The *parasympathetic fibres* of the vagus nerve act by releasing *acetylcholine* which is taken up by acetylcholine receptors at the surface of the cardiac cell. These are also called *muscarinic receptors* because the actions of acetylcholine on them is similar to that of muscarine, an alkaloid which is present in toadstools in poisonous amounts. The chemical mediators for the action of the sympathetic nervous system on cells are *adrenalin* and *noradrenalin* and the receptors are known as *adrenergic receptors*, of which there are two types, alpha (α) and beta (β), each with two subtypes. The adrenergic receptors of the heart are β_1 receptors. Drugs which block these receptors are used to treat several cardiovascular diseases, particularly coronary heart disease in which they reduce the rate and force of cardiac contraction and shield the heart from the effects of stress, which causes the release of catecholamines (adrenalin and noradrenalin) from sympathetic nerve endings and from the suprarenal glands.

Cardiac investigations

The electrical events of the heart are recorded on the *electrocardiogram* which may be altered in a variety of ways by disease. Diagnostic changes occur in some conditions, such as acute myocardial infarction, and the electrocardiogram is invaluable in the detection of disturbances of cardiac rhythm, or arrhythmias. *Echocardiography* is the examination of the heart by means of ultrasound which is directed through the heart from a transducer placed on the skin and which is echoed back by the structures which it encounters. By this means, pictures are obtained of the four cardiac chambers and of the valves of the heart. The movements of the valves and of the walls of the heart may be studied and measurements may be made of the size of the cardiac chambers and of the thickness of their walls. Atrial and ventricular septal defects and other congenital abnormalities may be diagnosed. Pericardial effusions and tumours or thrombi inside the heart may be detected. A variety of further investigations is available for studying the alterations in anatomy and physiology in diseases of the heart.

The blood vessels

The **arteries** which convey blood from the heart to the capillaries are thick-walled tubes consisting of three coats – outer, middle and inner. The outer coat is called the **tunica adventitia** and is composed of fibrous tissue which gives protection and

strength to the vessel. The middle coat, or **tunica media**, consists of smooth (unstriated) muscle fibres with some yellow elastic fibres. The muscle fibres are arranged in a circular manner and by their contraction and relaxation the calibre of the vessel can be altered. This function is under the control of the autonomic nervous system and plays an important part in determining the amount of blood supplied to an organ and in the maintenance of the blood pressure. The inner coat, or **tunica intima**, consists of a layer of flattened endothelial cells which line the artery and are separated from the middle muscular coat by a layer of elastic fibres. The amount of elastic tissue is greatest in the large arteries, especially the aorta, while in the smaller arteries and arterioles the muscular tissue predominates. These structural characteristics are in keeping with the fact that the blood supply to an organ or part is mainly controlled by variation in the calibre of the small arteries and arterioles, while the size of the aorta and of large vessels remains constant.

The **veins** also possess three coats, corresponding to those found in the arteries, but they are all much thinner so that the walls of a vein collapse and fall together when the vessel is opened. An artery, however, remains open when divided. Many veins of the limbs and abdomen have **valves** in their interior, so arranged that they allow blood to flow only towards the heart and prevent flow in the

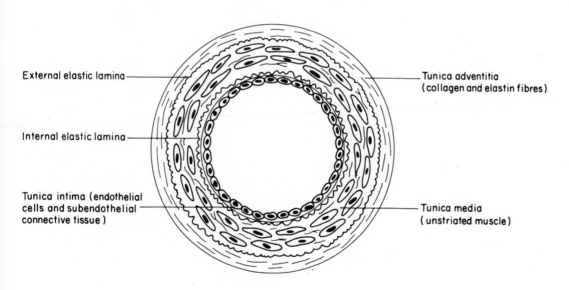

Fig. 10.11   Cross section of an artery.

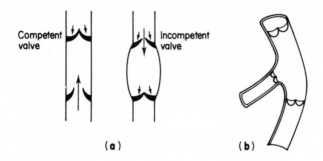

Fig. 10.12　Valves of veins and vein cut to show valves.

opposite direction. These valves consist of pouch-like folds with their openings directed towards the heart. They play a very important part in the return of the blood to the heart. In the condition known as varicose veins, the valves of the saphenous veins are defective and the vessels are therefore over-distended with blood and varicose. The high back-pressure and stagnation of blood in these veins impairs the nutrition of the skin of the lower leg which is liable to break down into a varicose ulcer if subjected to quite minor trauma.

The circulatory system is specially adapted to meet the following requirements.

(*1*) To maintain a constant blood supply to the brain and vital centres at all times. A sudden temporary interruption of the blood supply causes fainting whilst prolonged lack of oxygen causes permanent brain damage or death.

(*2*) To adjust the blood flow to other organs according to their immediate requirements, e.g. (*a*) the blood supply to the muscles is increased during exercise, (*b*) that to the abdominal organs during digestion and (*c*) that to the surface of the body is varied in order to regulate the body temperature.

The arteries

The **aorta** transports the oxygenated blood needed by all the tissues and is the largest artery of the body. It arises from the left ventricle and passes through the thorax and into the abdomen, where it terminates opposite the lower border of the fourth lumbar vertebra.

The aorta is described in three parts, namely
   the ascending aorta,
   the arch of the aorta, and
   the descending aorta.

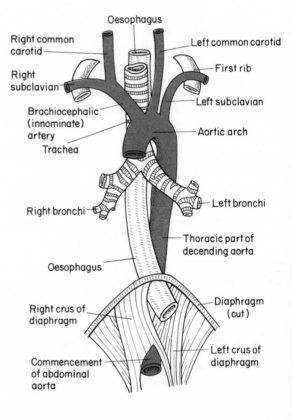

Fig. 10.13　Diagram of the thoracic aorta.

The **ascending aorta** arises from the left ventricle and passes upwards, forwards and to the right for about 5 cm (2 in) to reach the level of the upper border of the second right costal cartilage. The vessel continues as the **aortic arch** which begins behind the manubrium sterni and runs in a curved course backwards and to the left, finally passing downwards on the left side of the body of the fourth thoracic vertebra. At the lower border of this bone, it is continuous with the **descending aorta**. The latter vessel continues downwards in front of the remaining thoracic vertebrae and passes into the abdomen through a hole (the aortic hiatus) in the diaphragm at the level of the lower border of the twelfth thoracic vertebra. The portion of the descending aorta above the diaphragm is called the thoracic part and the portion below is called the abdominal part or simply the abdominal aorta (see Fig. 10.13).

The abdominal aorta descends in front of the lumbar vertebrae and terminates at the level of the 4th by dividing into the two (right and left) **common iliac arteries**. On the surface of the body this division is represented by a point just below and to the left of the umbilicus and is approximately at the level of a line joining the highest points of the iliac crest (see Fig. 7.10, p. 97).

As a result of disease, the wall of the aorta may be weakened and bulge outwards, forming an aneurysm. The thinning of the wall may result in rupture and fatal haemorrhage. The same outcome may result from the condition known as aortic dissection in which blood driven through a tear in the tunica intima destroys the tunica media and strips the intima from the tunica adventitia.

In the plan of the arterial system (Fig. 10.14, p. 136) it will be noted that certain branches of the aorta are paired and pass respectively to each side of the body, while others, which arise from the front of the aorta, are single.

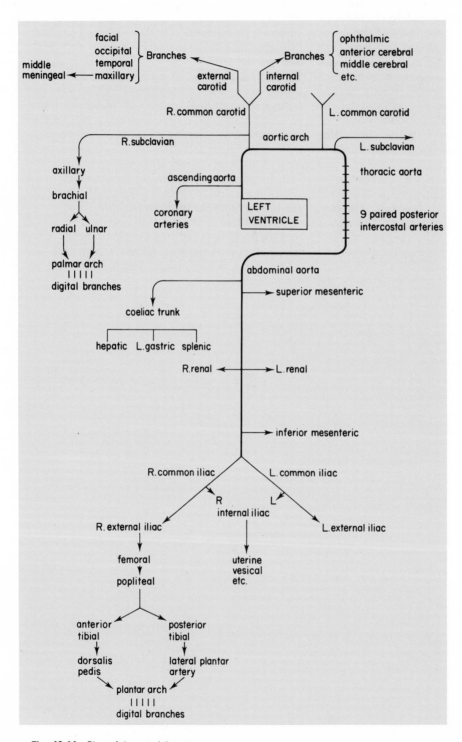

Fig. 10.14   Plan of the arterial system.

The coronary arteries

The right and left coronary arteries are the first branches of the aorta after it leaves the left ventricle. Each runs in the groove between its corresponding atrium and ventricle and gives branches to the heart muscle.

The brachiocephalic (innominate) artery

This arises behind the manubrium sterni from the convexity of the aortic arch, of which it is the largest branch. It passes upwards, backwards and to the right and, after 5 cm (2 in), divides into the right subclavian and right common carotid arteries (Fig. 10.13).

The common carotid artery

It must be remembered that the right common carotid arises from the brachiocephalic artery while the left common carotid springs directly from the arch of the aorta. Thereafter, their course and distribution are identical. The common carotid artery passes upwards in a sheath (the carotid sheath) along with the internal jugular vein and vagus nerve. It is surrounded by the muscles of the neck and, at the level of the upper border of the thyroid cartilage of the larynx, divides into two branches, the internal and external carotid arteries (Fig. 10.15).

The external carotid artery

This artery, as its name suggests, supplies the outer surface of the head and neck. It has four main branches.

(1) The **facial artery**, which passes up over the outer surface of the lower jaw just in front of the angle and supplies the lower part of the face.

(2) The **occipital artery**, which passes behind the ear and supplies the occipital part of the scalp.

(3) The **superficial temporal artery**, which passes upwards in front of the ear to supply the frontal, temporal and parietal portions of the scalp.

(4) The **maxillary artery**, which supplies the structures around the jaws and gives off the important **middle meningeal artery** to the interior of the skull.

The pulse can be felt both in the facial artery as it crosses the lower jaw and in the temporal artery where it is placed immediately in front of the external acoustic meatus.

The internal carotid artery

This commences at the bifurcation of the common carotid and extends upwards to enter the interior of the skull through the carotid canal in the petrous portion of the temporal bone. In the neck it lies deeply amid the muscles. It enters the middle cranial fossa of the skull and terminates in two branches, the **anterior and middle cerebral arteries**, which supply the brain. It also sends a branch, the **ophthalmic artery**, to supply the eye.

The anterior and middle cerebral arteries communicate with each other and also with the **basilar artery**, a continuation of the vertebral arteries, thereby forming the **circle of Willis**, which ensures an even distribution of blood to the brain.

The subclavian artery

The right subclavian artery arises from the brachiocephalic artery, the left directly from the arch of the aorta just beyond the origin of the left common carotid. It passes over the first rib, which it grooves, and behind the clavicle to enter the upper part of the axilla where it becomes the **axillary**

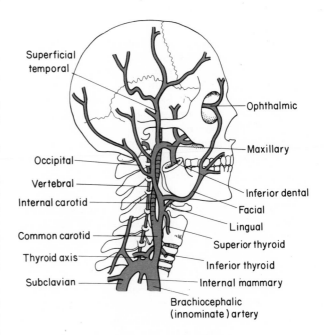

Fig. 10.15 Main arteries of the head and neck.

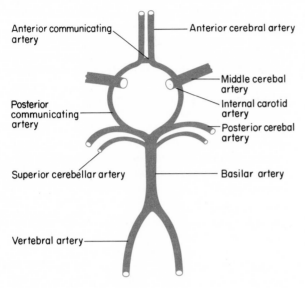

Fig. 10.16   The arterial circle of Willis at the base of the brain.

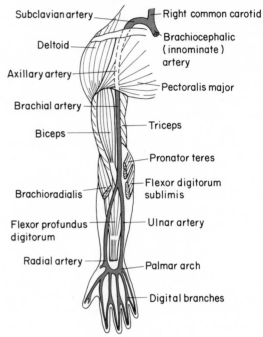

Fig. 10.17   Arteries of the upper limb.

artery. It is the main artery of supply to the upper limb, but has three major branches.

(1) The **vertebral artery**, which passes upwards through the foramina in the transverse processes of the cervical vertebrae (p. 65) to enter the skull by way of the foramen magnum. It is distributed to the posterior part of the brain and cerebellum.

(2) The **internal thoracic (mammary) artery**.

(3) The **thyrocervical trunk**.

The axillary artery

This is the direct continuation of the subclavian artery and, at the lower boundary of the axilla, becomes the brachial artery (Fig. 10.17).

The brachial artery

This runs down the arm from the lower border of the axilla to about 1 cm below the elbow joint, where it divides into the radial and ulnar arteries. In its course it runs close to the humerus on the medial side of the biceps muscle (Fig. 10.17).

The radial artery

This follows the lateral bone of the forearm and in the first part of its course is covered by muscles. In its lower part, just above the wrist, it is superficial and can be felt lying in front of the radius. It is in this position that the pulse is usually taken.

The ulnar artery

This runs down the medial side of the forearm close to the ulna.

The palmar arches

The ends of the radial and ulnar arteries pass in front of the wrist into the palm of the hand where they join in the form of two arches running transversely. One of the arches thus formed is superficial, the other deep, and from them branches pass to the fingers.

Branches of the abdominal aorta

The coeliac trunk (axis)

This is an artery which springs from the front of the abdominal aorta immediately after it has passed through the diaphragm. It is a short wide trunk which divides into three branches:

- The **left gastric artery**, which helps to supply the stomach.
- The **hepatic artery**, which supplies the liver, pancreas, gall bladder, stomach and duodenum.
- The **splenic artery**, a tortuous artery which passes to the left behind the stomach and along the upper border of the pancreas to the spleen.

The superior mesenteric artery

Arising from the front of the aorta just below the coeliac trunk is the superior mesenteric artery. This passes forwards to reach the fold of peritoneum called the mesentery, between the layers of which it descends as its name suggests. It supplies the whole of the small intestine and the first portion of the large intestine.

The renal arteries

Immediately below the superior mesenteric artery, the right and left renal arteries arise from the sides of the abdominal aorta at the level of the second lumbar vertebra. Each passes to the corresponding kidney. Because the inferior vena cava is situated to the right of the aorta, the right renal artery has to pass behind this large vein in order to reach the kidney.

The inferior mesenteric artery

This takes origin lower down from the front of the aorta and sends branches to the distal parts of the colon and to the rectum.

The common iliac arteries

These are the two branches into which the abdominal aorta divides opposite the lower border of the fourth lumbar vertebra. Each passes downwards and laterally for 4–5 cm (about 2 in) and then divides into the internal and external iliac arteries.

The internal iliac artery

This artery descends immediately into the cavity of the pelvis, the contents of which it supplies. In the female it gives origin to the important **uterine artery** which supplies the uterus.

The external iliac artery

This is the direct continuation of the common iliac artery and runs downwards along the brim of the pelvis, passing under the inguinal (Poupart's) liga-

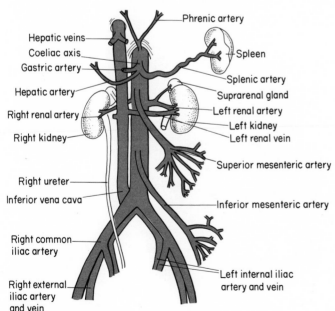

Fig. 10.18 Diagram showing abdominal aorta and inferior vena cava with important branches.

ment to become the femoral artery, which is the main artery of the lower limb.

The femoral artery

Commencing at the inguinal ligament, this is the continuation of the external iliac into the thigh. Its direction is indicated by the line drawn from a point midway between the anterior superior spine of the ilium and the symphysis pubis to a point just above the medial condyle of the femur (adductor tubercle). The upper half of the artery is comparatively superficial and lies in the space known as the femoral or Scarpa's triangle (p. 301). In the lower part it lies more deeply among muscles in a special tunnel called the adductor canal (Hunter's canal). It ends by passing backwards behind the femur to enter the popliteal space where it becomes the popliteal artery. The femoral artery gives branches to the surrounding muscles and to the knee joint.

The popliteal artery

This is the continuation of the femoral artery in the popliteal fossa. At the lower end of the fossa it gives off the **anterior tibial artery** which passes forwards between the tibia and fibula to supply the front of the leg and is continued downwards on the dorsum of the foot as the **dorsalis pedis artery**. The **posterior tibial artery** is the direct downward continuation of the popliteal artery in the back of the leg, where it lies deeply among the muscles. It passes behind and below the medial malleolus to divide into the medial and lateral **plantar arteries** in the sole of the foot. The lateral plantar artery unites with the dorsalis pedis artery to form the plantar arch, similar to the palmar arch of the hand, from which digital branches pass to the toes.

The pulses of the posterior tibial and dorsalis pedis arteries are palpated in the physical examination of a patient suspected of having peripheral vascular disease.

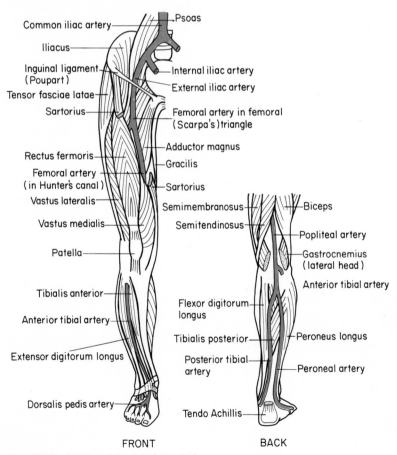

Fig. 10.19   Arteries of the right lower limb.

Pressure points

It is sometimes possible to arrest severe haemorrhage by compressing the artery which supplies the bleeding part. This can only be done in the case of those arteries which are near the surface and which can be compressed against some firm underlying structure, for example:

Arteries of the head and neck

Facial: against the lower jaw.
Temporal: in front of the external acoustic meatus.
Occipital: against the occipital bone about 6.5 cm (2½ in) behind the ear.
Common carotid: against the cervical vertebrae at the side of the larynx.

Arteries of the upper limb

Subclavian: against the first rib in the hollow above the clavicle.
Brachial: against the medial aspect of the humerus in the middle of the arm.
Radial: at the lower end of the radius just above the wrist on its anterior surface.
Ulnar: against the anterior surface of the ulna.

Arteries of the lower limb

Femoral: against the pubic bone under the inguinal ligament.
Posterior tibial: against the posterior surface of the medial malleolus.
Dorsalis pedis: against the upper surface of the navicular bone.

The veins

The veins commence at the termination of the capillaries, and convey blood back to the heart. The small venules unite with one another to form larger and larger vessels until two main trunks are formed which enter the right atrium of the heart. The upper of these trunks is the **superior vena cava** conveying the blood from the head and neck and the upper limbs. The lower trunk is the **inferior vena cava** which receives through its various branches blood from the rest of the body, including the contents of the abdominal cavity and the lower limbs.

The pumping action of the heart is the prime force in moving the blood around the circulatory system. However, the *venous return* of blood to the heart is aided by

(1) the thoracic pump
(2) the skeletal muscle pump
(3) the heartbeat.

The term 'thoracic pump' refers to the fall of intrathoracic pressure with each inspiration. The negative intrapleural pressure is transmitted to the great veins, reducing the pressure in them and thereby aiding venous return during inspiration.

The deep veins in the limbs are surrounded by skeletal muscles which compress them when they contract; hence the term skeletal muscle pump. The venous compression propels the blood towards the heart, reverse flow being prevented by the valves in the veins. If these valves are incompetent, possibly as a result of previous thrombophlebitis, venous stasis and ankle oedema may occur.

The heartbeat contributes to venous return by sucking blood into the right atrium (1) when the tricuspid valve ring is pulled downwards during ventricular systole and (2) during early diastole when rapid ventricular filling is taking place.

Venous pressure

This varies in the different parts of the venous system. Clinically it is useful to measure the jugular venous pressure and the central venous pressure. The former can be assessed by inspection of the neck of the seated patient who is reclining at 45† from the horizontal. In this position, venous pulsation is not usually seen in the necks of normal people. If there is pulsation in the internal jugular vein, the height to which it rises above the manubriosternal angle is measured. The jugular venous pressure is elevated in patients with congestive cardiac failure, hypervolaemia (fluid overload), constrictive pericarditis or cardiac tamponade.

A fine catheter inserted into the superior vena cava is used for measurements of the *central venous pressure* (CVP), which is read off a manometer. A high CVP is recorded in those conditions mentioned for elevation of the jugular venous pulse. A low CVP results from hypovolaemia due to dehydration or haemorrhage. The nurse must make a note of the *reference point* or 'zero level' (e.g. the sternal angle) from which manometer readings are being made, so that serial readings are all made from the same zero point.

A central venous pressure line may be used not only for making measurements but for giving certain drugs and for parenteral feeding, using concentrated solutions which might cause chemical

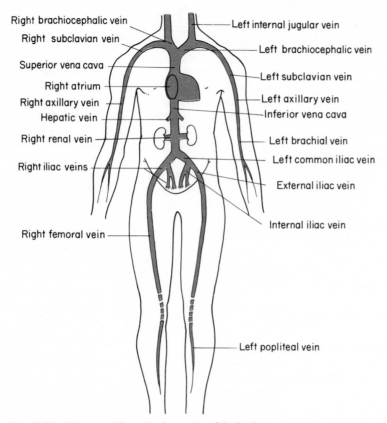

Right brachiocephalic vein
Right subclavian vein
Superior vena cava
Right atrium
Right axillary vein
Hepatic vein
Right renal vein
Right iliac veins
Right femoral vein

Left internal jugular vein
Left brachiocephalic vein
Left subclavian vein
Left axillary vein
Inferior vena cava
Left brachial vein
Left common iliac vein
External iliac vein
Internal iliac vein

Left popliteal vein

Fig. 10.20 Diagram of the main deep veins of the body.

irritation of the walls of smaller veins and lead to thrombophlebitis if infused into them.

The veins are divided into two main groups:
- (*a*) the **superficial veins** which are situated on the surface of the body just under the skin, some of which can be easily seen, and
- (*b*) the **deep veins** many of which accompany the main arteries of the body.

For convenience, the veins which go to form the inferior vena cava will be described first.

Veins going to form the inferior vena cava

Veins of the lower limb: (*a*) superficial
(*b*) deep

Veins of the abdomen: (*a*) outside the peritoneal cavity
(*b*) inside the peritoneal cavity (the portal circulation)

Veins of the lower limb

(*a*) There are two main superficial veins in the lower limb.

The **long saphenous vein** which commences on the medial aspect of the dorsum of the foot and passes up in front of the medial malleolus, up the medial side of the leg, behind the medial aspect of the knee, and up to just below the medial end of the inguinal ligament, where it passes deeper to enter the femoral vein. The point where it pierces the deep fascia of the limb is called the **saphenous opening**. The **short saphenous vein** commences on the lateral side of the foot and passes upwards along the centre of the *back* of the calf and pierces the deep fascia of the limb over the popliteal fossa at the back of the knee joint to reach the **popliteal vein**.

(*b*) The main deep veins of the leg commence as the anterior and posterior tibial veins which accompany the corresponding arteries and unite to form

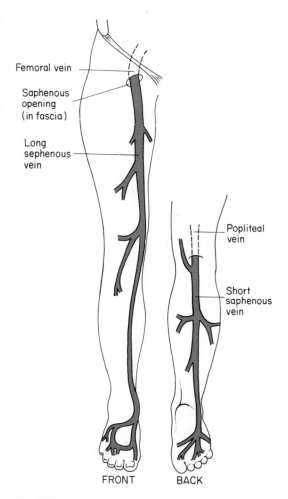

Femoral vein

Saphenous
opening
(in fascia)

Long
saphenous
vein

Popliteal
vein

Short
saphenous
vein

FRONT   BACK

Fig. 10.21 The superficial veins of the lower limb.

the **popliteal vein**. This passes upwards in company with the artery and enters Hunter's canal as the **femoral vein**. In the femoral (Scarpa's) triangle it lies in a sheath of fibrous tissue, called the femoral sheath, together with the femoral artery. In the femoral triangle it receives the long saphenous vein and passes under the inguinal ligament to become the **external iliac vein**.

Veins of the abdomen

On the posterior abdominal wall, the external iliac vein joins the internal iliac vein to form the common iliac vein and the right and left **common iliac veins** unite to form the inferior vena cava just below the bifurcation of the aorta. The **inferior vena cava** continues upwards in front of the bodies of the lumbar vertebrae, lying to the right of the abdominal aorta. In the abdomen it receives the right and left **renal veins** from the kidneys and, just before it pierces the diaphragm to enter the thorax, two or three groups of **hepatic veins** pour the blood from the liver into it. In the upper part of the abdomen it lies behind the liver, the posterior aspect of which it grooves. Shortly after entering the thorax it reaches the right atrium, where it terminates.

The portal circulation (Fig. 10.22)

The portal circulation inside the abdominal cavity concerns the blood which is supplied to and removed from the organs of digestion and its conveyance to the liver. It will be remembered that the blood to these organs comes from the following branches of the abdominal aorta:

> the coeliac trunk supplying the stomach, the spleen, the pancreas and the liver;
> the superior mesenteric artery to the small intestine and first part of the large intestine;
> the inferior mesenteric artery to the rest of the large intestine and the rectum.

These arteries all break up into capillaries in the organs which they supply and these capillaries unite in the usual way to form veins. The individual veins from the stomach, spleen and intestines in turn unite to form one large vein, the **portal vein**, which carries all their blood to the liver. Here, in the substance of the liver, the portal vein breaks up into smaller veins and finally into a *second set of capillaries*. These capillaries in the liver again unite to form the three major **hepatic veins** which leave the liver to join the inferior vena cava just before it pierces the diaphragm to enter the thorax. It will be noted that the portal vein, unlike other veins in the body, both commences and ends with capillaries. The importance of this is that the digested foodstuffs in the alimentary canal are absorbed into the capillaries which go to make up the portal vein and are carried by it to the liver. In order that these materials can come into contact with the individual liver cells for further chemical action or storage it is necessary for the blood to pass through a second set of capillaries, for it will be remembered that interchange of substances between the blood and tissues can only take place through the capillaries. The portal vein itself is quite short (8 cm, 3 in) and is formed just behind the neck of the pancreas whence it passes upwards, behind the duodenum, to reach the portal fissure of the liver.

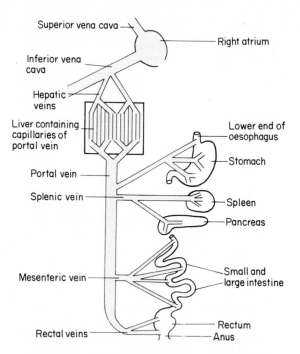

Fig. 10.22   Diagram illustrating the portal circulation.

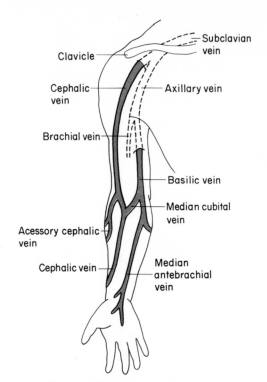

Fig. 10.23   Superficial veins of the right upper limb.

In order to avoid any confusion, it must be pointed out that the hepatic artery supplies arterial blood to the liver in order to provide an adequate amount of oxygen to the organ and takes no part in the formation of the portal circulation.

Obstruction of the portal vein or its branches (e.g. by thrombosis or by hepatic cirrhosis) causes a rise in blood pressure in the portal venous system, a condition known as portal hypertension. This results in enlargement of the spleen, oesophageal varices and haemorrhoids. It also contributes to the formation of ascites, an abnormal collection of fluid in the peritoneal cavity.

Veins going to form the superior vena cava

(*a*) Veins of the upper limb: } superficial
                                    deep
(*b*) Veins of the head and neck: { superficial
                                    deep

Veins of the upper limb

The superficial veins commence mainly on the back of the hand and the more important of them con-

verge on the antecubital space in front of the elbow joint. Passing up the front of the forearm is the **cephalic vein**. At the bend of the elbow this gives off the **median cubital** vein which goes to the medial side to join the **basilic vein**. The basilic vein runs up the medial side of the arm and the cephalic vein continues its course up the lateral side.

The veins at the bend of the elbow are of clinical importance on account of their relatively large size and the fact that they are readily accessible. They are frequently the veins selected for the removal of blood or for the injection of fluids or drugs into the bloodstream.

The cephalic and basilic veins pass upwards and pierce the deep fascia to accompany the axillary artery as the **axillary vein**. The deep veins of the arm accompany the radial, ulnar and brachial arteries and terminate in the axillary vein.

The axillary vein becomes the **subclavian**, which is finally joined by the **internal jugular vein** bringing blood from the head and neck to form the **brachiocephalic vein**. This unites with its fellow of the opposite side to become the superior vena cava. It will be noticed that although there is only one brachiocephalic artery, there are two (right and left) brachiocephalic veins.

Veins of the head and neck

The superficial veins of the scalp unite just behind the angle of the jaw to form the **external jugular vein**. This passes directly downwards, superficial to the sternomastoid muscle which it crosses obliquely, and enters the subclavian vein. Most of the veins of the face are, however, collected up into the main **facial vein** which crosses the lower jaw in front of the angle and then passes deeply to join the internal jugular vein. The facial vein actually commences as a small vein close to the medial angle of the eye (the **angular vein**). There is also a communication between the angular vein and the veins inside the skull which passes through the orbit. It is for this reason that boils and carbuncles on the face may be particularly dangerous, for infection may spread through this connection and cause inflammation within the skull (cavernous sinus thrombosis).

Apart from the veins of the brain, which pour their blood into the internal jugular vein, there are some other important venous channels within the skull. These are the **venous sinuses** which lie between layers of the dura mater (the outer covering of the brain).

The **superior sagittal sinus** commences in the frontal region of the skull and runs directly backwards in the mid-line to the occipital region, in a fold of dura mater called the **falx cerebri**. This fold of dura mater separates the right from the left cerebral hemisphere. In the infant (under 18 months) before the bones of the skull are completely fused, there is a diamond-shaped opening, the anterior fontanelle, between the frontal and two parietal bones which is only covered by membrane (see Fig. 5.31). The superior sagittal sinus passes directly beneath this opening in the mid-line, and it is possible to insert a needle into the sinus for purposes of removing blood. This method is sometimes used in infants as the limb veins are very small and difficult to find.

At about the centre of the occipital bone, the superior sagittal sinus continues directly into one of the transverse sinuses, usually the right, at the torcular Herophili or confluence of sinuses. The **straight sinus**, in the lower portion of the falx cerebri, continues into the other transverse sinus. The transverse sinuses pursue a curved course, in close relation to the inner aspects of the mastoid processes, to reach the base of the skull which they leave by the jugular foramena to become the internal jugular veins. The lower end of the transverse sinus is sometimes called the sigmoid sinus.

The close association of the transverse sinus with the mastoid process is of importance. The sinus may be inadvertently exposed or opened during operations on the mastoid process. Sometimes

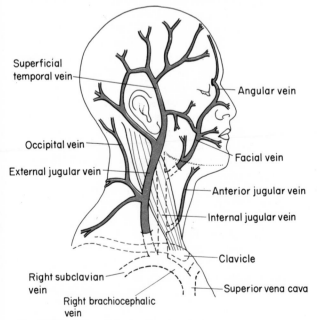

Labels: Superficial temporal vein, Angular vein, Occipital vein, Facial vein, External jugular vein, Anterior jugular vein, Internal jugular vein, Clavicle, Right subclavian vein, Superior vena cava, Right brachiocephalic vein

Fig. 10.24 Veins of the head and neck.

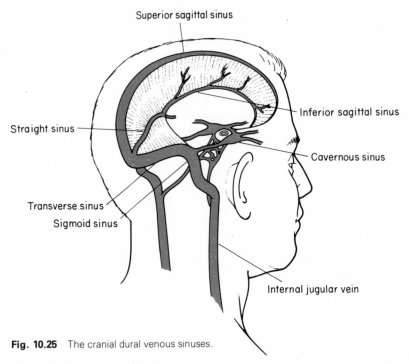

Superior sagittal sinus

Inferior sagittal sinus

Straight sinus

Cavernous sinus

Transverse sinus

Sigmoid sinus

Internal jugular vein

Fig. 10.25 The cranial dural venous sinuses.

infection may spread from the mastoid into the sinus, a serious condition which may result in general blood infection.

The **cavernous sinus** is situated at the side of the sella turcica. It communicates with its fellow of the opposite side, with the angular vein on the face, and posteriorly with the transverse sinuses.

The **internal jugular vein** commences at the jugular foramen in the base of the skull as the direct continuation of the sigmoid portion of the transverse sinus. It passes downwards, deep to the sternomastoid muscle, in close association with the carotid artery with which it is included in the carotid sheath. Finally it unites with the subclavian vein to form the brachiocephalic vein (Fig. 10.24).

The **brachiocephalic vein**. Each brachiocephalic vein is formed behind the medial end of the clavicle by the junction of the internal jugular and subclavian veins. The left brachiocephalic vein passes obliquely behind the manubrium sterni to join the right brachiocephalic vein and by their union they form the superior vena cava.

The **superior vena cava**, which returns the blood from the head and neck and upper extremities to the heart, is 8 cm (3 in) long and passes downwards behind the right margin of the sternum to end its course in the right atrium of the heart.

Cardiac catheterization

A catheter can be passed into the basilic vein in the arm and thence onwards into the superior vena cava and right atrium, right ventricle and pulmonary artery. Its exact position can be located by an X-ray image intensifier. Pressures in the right-sided cardiac chambers and pulmonary artery can be measured and specimens of blood can be obtained for determination of their oxygen content, the latter determinations enabling the size of a left-to-right intracardiac shunt to be determined (e.g. in cases of atrial septal defect).

Catheterization of the heart from a vein is called *right heart catheterization*. A full cardiac catheterization includes left heart catheterization, which is usually performed using a catheter introduced retrogradely (against the blood-flow) into an artery and on into the aorta and left ventricle.

Venous thrombosis

Blood sometimes thromboses in veins. If such a thrombus becomes dislodged (embolizes) it will travel in the bloodstream to the heart and thence via the pulmonary artery into a lung, cutting off some of its blood supply and resulting in a pulmonary infarct.

The blood pressure

Blood pressure may be defined as the force or pressure which the blood exerts on the walls of the artery in which it is contained.

When an artery is cut across, the blood spurts out from the end nearest the heart with considerable force and is obviously under high pressure within the artery. It will further be observed that, in addition to the continuous stream, there will be regular spurts of increased pressure corresponding with each heart-beat. The continuous pressure is partly dependent on the elasticity of the arteries and is called the *diastolic* pressure. The increased pressure occurring with each beat of the heart is the *systolic* blood pressure.

If a definite amount of fluid is pumped through a large tube with a certain amount of force it will flow out at the far end at a definite rate and pressure. If the same amount of fluid is pumped through a narrower tube with the same force the pressure with which it leaves the tube will be increased. The narrower the opening of the tube the greater will be the friction of the fluid on the walls of the tube.

An illustration of this can be seen in the use of the ordinary garden hose. If no nozzle is attached and water is turned on at moderate pressure a steady stream will emerge from the end of the hose and project for about a foot. If a narrow nozzle is then attached without altering the supply of water this will leave the nozzle in a fine jet projected for a number of yards with much friction against the walls of the nozzle as it leaves. Compression of any part of the hose without the nozzle will be easy, in other words, the pressure within is relatively low. The application of the nozzle, by narrowing the opening and causing resistance to the outflow, will cause an increase in the pressure within the whole of the rest of the hose which can no longer be so easily compressed.

The same mechanical factors are applicable in regard to the blood pressure within the arteries. The heart may pump with a constant force. The aorta is a relatively wide channel but the arteries gradually become smaller and by their resistance to the flow of blood maintain a high pressure in the whole of the arterial system. If the heart beats more strongly the pressure will be increased, but it will fall if the force of the heart-beat is reduced. Again, if the normal size of the arteries is reduced the resistance will be increased and the blood pressure

will rise. Lastly, if the volume of circulating blood is diminished (e.g. after haemorrhage) the blood pressure will fall.

Blood pressure is therefore maintained by the following factors:

(*1*) the force of the heart-beat;
(*2*) the resistance to blood flow in the arterioles i.e. the peripheral resistance; and
(*3*) the volume of circulating blood.

The measurement of blood pressure

Clinically, the blood pressure is measured by an instrument called the **sphygmomanometer**. This consists of a rubber bag which is placed around the arm. The interior of the bag is connected by a rubber tube to a mercury pressure-gauge. When the pressure in the bag equals the pressure in the artery, the latter is compressed flat and the flow through it temporarily arrested. The pulse, therefore, disappears. This may be appreciated either by the finger or by listening over the brachial artery at the bend of the elbow with a stethoscope. The height to which the column of mercury has been forced at this moment is therefore the systolic blood pressure. The height is measured in millimetres and the blood pressure is referred to as being so many millimetres of mercury (mm Hg).

As the column of mercury is allowed to fall there comes a point at which the stethoscopic sounds suddenly become muffled or disappear. The height of the mercury column at this point is taken as the diastolic blood pressure.

The blood pressure varies with rest and activity, emotion and age. A resting young adult would be expected to have a blood pressure of about 120/80 mm Hg, the numerator being the systolic pressure and the denominator the diastolic pressure. There is, however, a wide range of normality. A WHO guideline asserts that a normotensive individual has a systolic blood pressure of less than 140 and a diastolic pressure of less than 90 mm Hg at rest. Repeated readings above 160/95 mm Hg may indicate hypertension and intermediate readings are described as 'borderline.'

A high blood pressure, or *hypertension*, may damage the arteries and eventually lead to a stroke or a heart attack (myocardial infarction). A low blood pressure, or *hypotension*, occurs acutely in cases of shock due to blood loss, septicaemia, myocardial infarction or other causes. Renal perfusion may be threatened if the systolic pressure falls below 80 mm Hg and renal failure may ensue. Chronic hypotension may result from Addison's disease of the adrenal gland.

The difference between the diastolic and systolic arterial pressures is called the pulse pressure. This may also be affected by disease.

The pulse

Each time the left ventricle contracts it forces its contained blood into the aorta. The aorta being elastic expands in order to accommodate this additional amount of blood. At the same time, the blood which was present before the ventricular contraction is now pushed on into the next section which, in turn, also expands, and so on. Thus a wave of expansion is created which starts at the root of the aorta and spreads over the whole of the arterial system, gradually dying away as it reaches the capillaries. This wave of expansion constitutes the pulse and it travels rapidly over the arteries, much more rapidly, in fact, than the velocity of the bloodstream.

The pulse can be felt and often seen in the superficial arteries of the body, but it is customary to study it in the radial artery. The other arteries in which it can be felt with ease are the temporal in front of the ear, the facial as it passes over the lower jaw, the carotid artery in the neck and the dorsalis pedis on the dorsum of the foot.

Nervous control of the blood vessels

Most of the arteries of the body are directly under the control of the autonomic nervous system. Special centres, known as the **vasomotor centres**, exist in the hypothalamus and medulla oblongata to exercise this control. Two sets of nerves are present.

(1) Vasoconstrictor
(2) Vasodilator

These nerves may affect the whole of the circulatory system at a time, or their action may be limited to a localized organ or part.

Vasoconstrictor nerves   As their name suggests, these nerves narrow the lumen of a blood vessel and thereby diminish the amount of blood to the part or organ which it supplies. If all the vasoconstrictor nerves are sending out impulses, a general effect will be produced; the whole of the arterial system will be narrowed and therefore the blood pressure will be raised (see p. 147).

Vasodilator nerves   These act by dilating the blood vessels and allowing a greater blood supply to the organ. Thus, the blood vessels of the alimentary tract are dilated by this action during the process of digestion; also during muscular exercise the blood vessels to the muscles are dilated so that they are able to carry more blood.

In the serious condition known as 'shock' the blood vessels as a whole become dilated and, in consequence, the blood pressure falls. One of the methods of treatment is to give drugs which stimulate the α-adrenergic receptors causing the arteries to narrow, thereby raising the blood pressure.

Most sympathetic impulses stimulate vasoconstriction and parasympathetic impulses result in vasodilatation.

An example of the nervous control of blood vessels which everyone has either felt or observed is the phenomenon of *blushing*. This is a purely local modification of the circulation and variation in the amount of blood in the skin of the face, and surrounding parts if the blush is extensive. An emotion, pleasure, embarrassment, disgust, or offended modesty perchance, possesses the mind and with inconsiderate haste the skin grows red and a hot flush is felt. This is due to the conscious nervous system affecting the vasomotor centre over which the individual has no control. Vasodilator impulses are sent out and the small arteries in the skin of the affected part dilate, bringing an excess of blood to the surface.

Conversely, in extreme terror or rage the face may become very pale and cold ('white with rage'). In this case the vasoconstrictor nerves are limiting the flow of blood through the arteries.

Questions

1. Describe the general plan of the circulation.
2. Describe the structure of the heart.
3. Describe the conducting system of the heart. What may happen if this is blocked and what is the principle of treatment?
4. Which nerves affect the heartbeat and how?
5. What are the similarities and differences between arteries and veins? What is the principal defect in varicose veins of the legs?
6. Name the parts of the aorta and one of the branches from each part. Name two diseases which may affect the aortic wall.
7. Name the main branches of the abdominal aorta.
8. What is meant by the term 'venous return' and how is it achieved?
9. Describe how measurement of the venous pressure may help in the diagnosis of disease.
10. Describe the portal circulation. What is meant by the term portal hypertension and what are its effects?
11. Write brief notes on the dural venous sinuses.
12. What is meant by the term blood pressure and how is it measured in clinical practice? Write brief notes on hypertension and hypotension.

11

The Mononuclear Phagocytic and Lymphoid Systems

Phagocytes are cells which can engulf and digest or encapsulate foreign material, such as inert particles and microorganisms. The polymorphonuclear leukocytes of the blood can do this but are not included under the **mononuclear phagocytic (or macrophage) system**. The mononuclear phagocytic cells of which this system consists, and which are distributed throughout the body, are called **macrophages**. They are of two types, those *circulating* in the blood and known as **monocytes** and those which are derived from the peripheral blood monocytes but which are relatively *fixed* in the tissues and referred to as **tissue macrophages**. In the connective tissues, the latter cells are called histiocytes and in the thymus, lymph nodes and spleen they are known as sinus-lining macrophages. In the lungs, alveolar macrophages migrate from the connective tissue into the alveoli and may later be found in the sputum, having engulfed dust particles and bacteria which have been inhaled. The importance of macrophages in both innate and acquired immunity, providing first-line and second-line defence against invading microorganisms, is emphasized in Chapter 2. Macrophages are also involved in the phagocytosis and destruction of effete red blood cells in the spleen and liver. The name mononuclear phagocytic system is now used in preference to the older term, reticuloendothelial system.

The lymphoid system

The principal cells of the immune system, the **lymphocytes**, operate in close association with phagocytic cells and, like them, are disseminated throughout the body. They may be found singly or in diffuse aggregations or within the lymphoid organs. Separate central and peripheral lymphoid tissues are recognized. The bone marrow, where B-lymphocytes are produced, and the thymus, where T-lymphocytes are formed, are the **central lymphoid organs** and the lymph nodes, spleen and lymphoid nodules associated with the gut and bronchial tree are the **peripheral lymphoid organs**.

The tissues of the body are permeated by a vast network of capillaries containing blood. The walls of the capillaries consist of a single layer of cells and, except for the white cells which at certain times are able to make their way through these walls, the blood does not actually come into direct contact with the tissues. The tissues are, however, bathed in tissue fluid which may be regarded as an intermediary between the blood and the tissues; all interchange of nourishment and waste products between them takes place through the medium of the tissue fluid.

The lymphatic system is a subsidiary or second circulatory system which drains the tissue fluids.

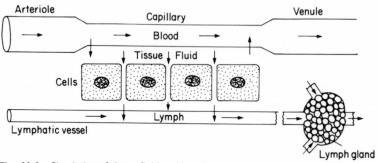

Fig. 11.1 Circulation of tissue fluid and lymph.

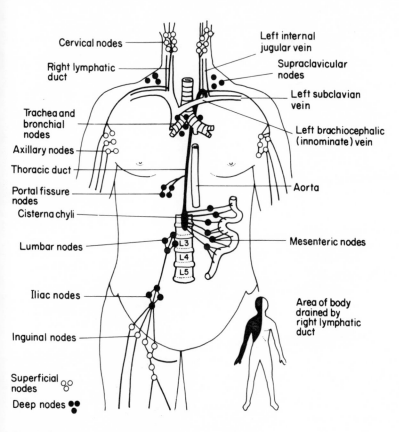

Fig. 11.2 Diagram of the main lymph nodes and ducts.

From the tissue spaces, the tissue fluid passes into **lymph capillaries** which unite to form larger trunks which ultimately rejoin the general circulation. *Lymph* is the name given to the tissue fluid when it has entered the lymphatic vessels and it may be looked upon as a part of the plasma which has filtered through the walls of the blood capillaries. The larger lymphatic vessels resemble small veins and are provided with valves to prevent back flow. A plexus of fine blood vessels accompanies the larger lymph vessels and nourishes them. If the lymph vessels under the skin become inflamed (lymphangitis) the accompanying plexuses of blood vessels become congested and red lines are seen in the skin.

The lymph channels unite to form larger vessels which eventually converge into two large ducts, the **thoracic duct** and the **right lymphatic duct**, which empty their lymph into the left and right brachiocephalic veins respectively. The more important of the two ducts, because it conveys the greater quantity of lymph back into the blood, is the

thoracic duct. This duct is about 40 cm long in the adult and extends from the second lumbar vertebra to the root of the neck. It commences as a dilated sac, the **cisterna chyli**, into which the lymph vessels from the lower limbs and the abdomen, except for part of the convex surface of the liver, empty their lymph. After passing upwards in front of the left side of the thoracic vertebrae, it is joined by the lymphatics from the left arm and the left side of the head and neck before it terminates in the left brachiocephalic vein. The right lymphatic duct, which is about 1 cm in length, conveys lymph from the right side of the head and neck, the right upper limb, the right side of the chest and heart, and part of the convex surface of the liver.

Utilizing the thoracic duct and the right lymphatic duct, all the lymph of the body returns to the blood. There is thus a constant circulation of lymph from the capillaries into the tissue spaces and back again into the blood stream. It has been pointed out in Chapter 2 that the lymphocytes are a recirculating population, migrating from blood to

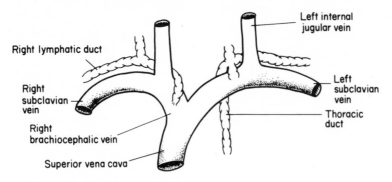

Fig. 11.3   Diagram illustrating the return of lymph to the great veins.

spleen and back and from blood to lymph and back. The *recirculation of lymphocytes* increases the chances that the 'correct' ones will encounter specific antigens and it also ensures the body-wide distribution of memory cells.

The lymph vessels of the small intestine are of special importance because of the digested fat which is absorbed into them. The minute projections, called villi, each contain a small lymph vessel into which the fat, as *chylomicra* (microscopic fatty globules), is absorbed. These channels are called **lacteals** and their contents all ultimately pass to the cisterna chyli. The lymph reaching the thoracic duct from the intestines differs from the clear colourless lymph from other parts of the body in that its fat content makes it cloudy or gives it a milky appearance. This fluid draining from the small intestine is called *chyle*, the term cisterna chyli meaning 'the receiver of the chyle'.

Lymph nodes

Situated in the course of the lymph vessels and generally occurring in groups are the small and oval or kidney-shaped lymph nodes. These are highly organized structures richly populated with lymphocytes and phagocytes and capable of making a controlled response to antigenic stimulation.

The lymph nodes act as *filters* for the lymph, trapping particulate matter and microorganisms. The vessels bringing lymph to the node are called the **afferent vessels**; they enter the gland at points around its periphery and drain into its cortex or outer zone. After percolating through the substance of the node the lymph leaves via the **efferent vessel**; this emerges directly from the medulla, or inner zone, at the small concavity called the **hilum**. Lymph may pass through several groups of glands

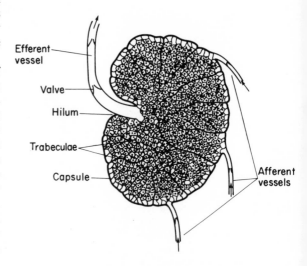

Fig. 11.4   Diagram of a lymph node.

before it is returned to the bloodstream. A lymph node is covered by its **capsule** which is largely composed of collagen fibres. From the capsule, **trabeculae** extend into the node and become continuous with the **reticulum** or meshwork of fine reticulin fibres, which forms the supporting framework for various cells, particularly lymphocytes and macrophages. Numerous lymph channels (sinuses) run through the lymph node, connecting the afferent vessels entering the cortex to the efferent vessel emerging at the hilum. The hilum is also the portal of entry and exit for the blood vessels of the lymph node.

There are many groups of lymph nodes in the body. Among the most important are those in the accompanying table (Table 11.1).

Table 11.1   The major groups of lymph nodes

Nodes	Location	Region drained
Near the surface		
cervical (superficial and deep)	neck	head and neck
axillary	axilla	upper limbs and breasts
inguinal	groin	lower limbs
Deep		
iliac	iliac fossa	pelvic viscera, etc.
lumbar	close to lumbar vertebrae	abdominal wall and viscera
mesenteric	mesentery	alimentary tract
hepatic	portal fissure of liver	liver, gall bladder etc.
pre-aortic	anterior to the abdominal aorta	alimentary tract and accessory organs
thoracic (bronchopulmonary group)	hilum of lung	lungs

Lymph nodes may be enlarged as a result of viral infections such as German measles and glandular fever. The lymph nodes draining an area of suppuration are often swollen and tender, the axillary lymph nodes being thus affected in patients with a septic finger. The axillary nodes also receive lymph from the breast and may be enlarged as a result of metastatic spread of cancer from the breast. Similarly the hilar lymph nodes in the thorax may be enlarged due to secondary malignant involvement in cases of lung cancer. In Hodgkin's disease the lymph nodes are primary sites of the malignant process.

Filariasis, due to a tropical worm, or malignant disease may block all or most of the lymphatic pathways in a limb or in parts of the trunk, resulting in gross swelling of the limb (lymphoedema, elephantiasis) or fluid in the peritoneal cavity (ascites) or pleural cavity (pleural effusion).

Lymphoid tissue elsewhere in the body

Apart from the lymph nodes, there are a number of other collections of lymphoid tissue, namely

(1) those situated in the walls of the alimentary and respiratory tracts and including the tonsils, nasopharyngeal tissues which when enlarged constitute adenoids, Peyer's patches scattered in the small intestine, and the appendix;

(2) the spleen;

(3) the thymus.

The tonsils

The tonsils are two ovoid masses of lymphoid tissue embedded in the side walls of the oral part of the pharynx, between the anterior and posterior pillars of the fauces. Their size varies but they tend to be relatively larger in children than in adults. The lower pole of each tonsil is continuous with lymphoid tissue situated in the base of the tongue. On the surface of the tonsil can be seen a dozen or more small openings which lead into deep narrow recesses called **crypts**.

The efferent lymph vessels of the tonsils (which have no afferent lymph vessels) drain into the upper deep cervical lymph nodes, especially the jugulodigastric group situated below the angle of the jaw. The blood supply of the tonsil is derived from branches of the external carotid artery, the tonsillar branch of the facial artery being the chief artery of supply. The nerves supplying the tonsil are the lesser palatine nerves, which are branches of the pterygopalatine ganglion, and the glossopharyngeal (IXth cranial nerve). As the latter nerve, through its tympanic branch, also supplies the mucous membrane lining the tympanic cavity, tonsillar pain due to inflammation or cancer may be referred to the ear.

The tonsils form part of a ring of lymphoid tissue guarding the entrance of the alimentary and respiratory tracts against bacterial invasion. Inflammation of the tonsils — tonsillitis — is common, particularly in young people. An abscess in the bed of the tonsil is called a quinsy.

The spleen

The spleen is a dark purple-coloured organ lying for the most part in the left hypochondriac region of the abdomen, between the fundus of the stomach and the diaphragm. Its long axis lies in the line of the tenth rib. It varies in size and weight during the lifetime of an individual but in an adult it is usually about 12 cm long, 8 cm broad and 3 or 4 cm thick. The normal spleen is not palpable and any spleen which can be palpated is therefore necessarily abnormally large due to disease of some kind, for example glandular fever, Hodgkin's disease or portal hypertension.

The spleen has diaphragmatic and visceral surfaces. The diaphragmatic surface is in contact with the under surface of the diaphragm, which separates it from the left lung. The visceral surface presents gastric, renal, pancreatic and colic impressions. The gastric impression is in contact with the posterior wall of the stomach and presents, near its lower limit, a long fissure, termed the **hilum**, where the branches of the splenic artery enter the spleen. The major tributaries of the splenic vein emerge from the hilum and this vein eventually joins with the superior mesenteric vein to form the portal vein.

Structure

The spleen has an outer coat of peritoneum which is firmly adherent to the internal fibro-elastic coat or splenic capsule. From the capsule, trabeculae pass into the spleen and branch to form a framework which is continuous with a fine reticulum. Closely associated with the reticulum are reticular cells, or fibroblasts, and macrophages. Two kinds of splenic pulp, red pulp and white pulp, occupy the interstices of the reticulum. The **white pulp** consists of periarteriolar sheaths of lymphatic tissue with enlargements, called splenic lymphatic **follicles** or Malpighian bodies, containing rounded masses of **lymphocytes**. These follicles, which are centres of lymphocyte production, are visible to the naked eye as whitish dots against the dark red background of the red pulp on the freshly cut surface of the spleen. The **red pulp** consists of numerous venous sinusoids, containing blood, separated by a network of perivascular tissue which is referred to as the **splenic cords**. These so-called cords contain numerous **macrophages** and are the sites of intense *phagocytic* activity. They also contain numerous lymphocytes, many of which are derived from the white pulp.

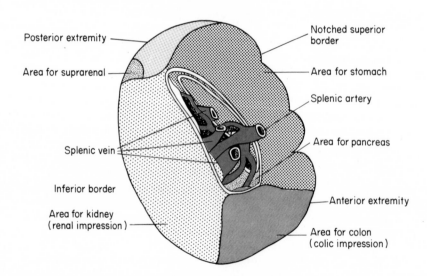

Fig. 11.5   The medial or visceral aspect of the spleen.

Functions

The spleen is a lymphoid organ which is not essential to life and can safely be removed (splenectomy) but which has several functions.

(1) *Phagocytosis* The phagocytes of the spleen, mostly macrophages, remove microorganisms and worn-out red cells, white cells and platelets from the blood. *Iron* is removed from the haemoglobin of the red cell debris and is conveyed to the bone marrow for re-use in red cell production. From the remaining pigment, the spleen forms *bilirubin* which, together with that formed in the liver, is secreted in the bile.

(2) *Immune response* Just as the spleen shares with the other peripheral lymphoid organs the function of phagocytosis, so it joins with them in the immune response. **B-lymphocytes** are sited mainly in the follicles and **T-lymphocytes** are the main cells of the periarteriolar sheaths. These lymphocytes can migrate into other areas of the spleen. B-lymphocytes stimulated by particular antigens, are transformed into plasma cells which produce **antibodies** (immunoglobulins) which are released into the general circulation, so conferring *humoral immunity* against specific bacterial infections. Some of the T-lymphocytes co-operate with the B-cells and are known as helper T-cells. Other T-lymphocytes, which are cytotoxic and able to kill infected circulating cells, are called killer T-cells. *Cellular immunity*, based on T-lymphocytes, provides defence against infection by viruses, fungi and a few bacteria including the tubercle bacillus.

(3) *Haemopoiesis* In the fetus, the spleen is an important haemopoietic organ and precursors of red cells, neutrophils and platelets are found in the red pulp. The white pulp of the mature spleen contributes to the circulating pool of immunocompetent lymphocytes.

(4) *Red cell storage* The spleen can act as a reservoir for red cells, which it discharges into the blood stream in an emergency, such as that created by *anoxia*. This function is probably of relatively minor importance in human beings.

The spleen is a soft organ. A normal spleen may be ruptured by severe external trauma and an enlarged spleen (as in glandular fever) may occasionally be ruptured by even trivial trauma. Rupture of the spleen leads to severe internal haemorrhage and shock. Diagnosis of the condition may be difficult but is essential, if the patient's life is to be saved, because only splenectomy will stop the bleeding.

The thymus

The thymus is situated superiorly in the anterior mediastinum of the thorax. It lies in front of the heart and pericardium and the aortic arch and its branches and on the front and sides of the trachea. It is **a central organ of the lymphoid system**. It controls lymphocyte production in both the peripheral and the central lymphoid organs, including the thymus itself, and is essential for the maturation of lymphocytes into the cells responsible for cellular immunity, **T-lymphocytes**.

The thymus is relatively large at birth and continues to increase in size until puberty, whereafter it usually gradually diminishes as a result of atrophic changes.

Thymectomy is beneficial in some cases of myasthenia gravis, a rare chronic disease characterized by abnormal fatiguability and weakness of certain muscles; this is due to an autoimmune process affecting the receptor proteins of the neuromuscular junctions, antibodies being formed against these receptors.

Questions

1. What do you understand by the terms (a) phagocyte (b) macrophage?
2. Name the lymphoid organs and tissues.
3. What is the thoracic duct and what is chyle?
4. Write short notes on lymph nodes.
5. Describe the tonsils.
6. Describe the structure of the spleen. What are the functions of this organ?
7. What is the major function of the thymus?

12
The Digestive System

Every cell in the body requires food for heat and *energy*, and for *growth* and *repair* of tissues. The purpose of digestion is to alter foodstuffs and convert them into simple constituents that can be absorbed into the blood and utilized by the various tissues according to their requirements.

The alimentary canal is a long tube extending from the mouth to the anus. Although it is one continuous tube it is described in the following parts:

    mouth
    pharynx
    oesophagus
    stomach
    small intestine
    large intestine

Digestion takes place in the alimentary canal and is aided by the accessory organs of digestion, the salivary glands, liver and pancreas (Fig. 12.1), which pour their secretions into the canal to assist the processes of digestion.

The greater part of the alimentary tract and the organs concerned with digestion lie within the abdominal cavity. For convenience of description the abdomen can be divided into nine regions by imaginary lines drawn through certain fixed points (Fig. 12.2). The *regions of the abdomen* are important in a practical sense in identifying the positions of the various abdominal organs and locating the source of pain or disease. For example the pain of peptic ulceration is felt in the epigastrium, whilst the pain of acute appendicitis arises in the right iliac fossa.

Food is physically and chemically complex in composition and two kinds of digestive processes are needed before absorption of food is possible, namely mechanical and chemical processes.

The mechanical process of digestion consists of mastication, swallowing and the peristaltic movements of the alimentary tract, which mix the food with digestive juices and propel it along the tract.

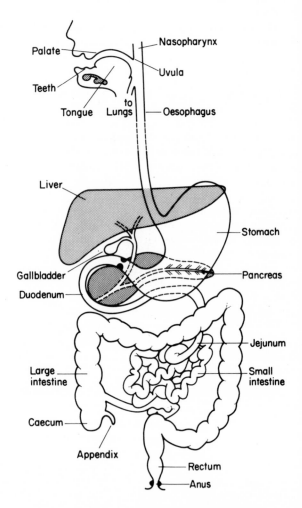

Fig. 12.1   The digestive system and accessory organs.

The chemical process of digestion consists of all the changes brought about by the enzymes of the digestive juices. A digestive **enzyme** is a substance, secreted by the glands of the alimentary tract (including the salivary glands and pancreas), which is able to act on a foodstuff and convert it into a simple form in which it can be absorbed. Each enzyme has a specific action, i.e. acts on a particular type of foodstuff.

The alimentary tract deals with food in four stages:

(*1*) *Ingestion*, or the intake of food.

(*2*) *Digestion*, which commences in the mouth, but is mainly carried out in the stomach and small intestine, involves both mechanical and chemical processes.

(*3*) *Absorption*. The passage of digested food substances through the mucosa of the gastro-intestinal tract into the blood and lymph.

(*4*) *Egestion* ('excretion'). Food substances which cannot be digested (e.g. cellulose) or have no food value are eliminated in the faeces.

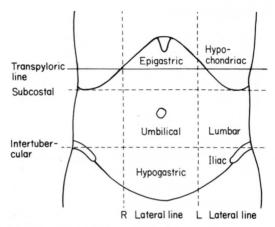

Fig. 12.2 The regions of the abdomen.

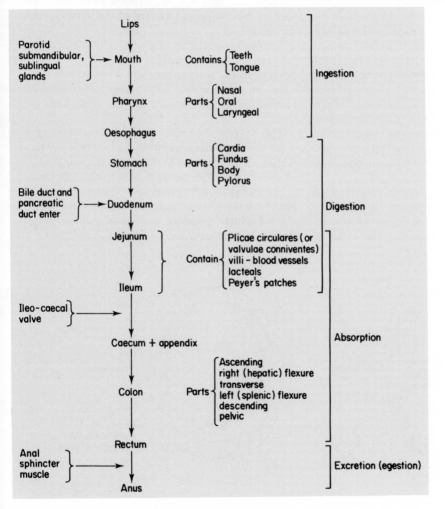

Fig. 12.3 Scheme illustrating the alimentary canal.

The mouth

The mouth, or buccal cavity, is the upper expanded portion of the alimentary tract. It consists of two parts. The outer part, or vestibule, is the recess between the teeth and the inner surface of the cheeks. The inner central part communicates with the oro-pharynx, and contains the tongue.

The anterior opening of the mouth is bounded by the lips, which contain muscle fibres, and are lined on their inner surface by mucous membrane, which is thinner than skin, allowing the colour of the blood in the underlying capillaries to show through and give them their natural red colour. This is replaced by blueness (cyanosis) if the blood is not properly oxygenated. The whole of the interior of the mouth is also lined with mucous membrane which is kept moist by saliva.

The roof of the mouth is formed by the palate. The anterior portion is the **hard palate**, and is formed by the maxilla and the palatine bones; the posterior portion is the **soft palate**. The soft palate is fleshy and muscular, and has a cone-shaped prolongation which hangs down in the back of the mouth, and is called the **uvula**. Curving downwards from either side of the uvula are two folds of mucous membrane, these are the anterior and posterior pillars of the **fauces**, between which lie the palatine **tonsils** (see p. 162).

During the development of the fetus *in utero* the palate may fail to become completely fused, a condition known as cleft palate. The degree of severity may vary from simple bifurcation of the uvula to complete absence of the roof of the mouth, and may be associated with a cleft upper lip (hare lip).

The floor of the mouth is mainly formed by the **tongue**, but the anterior portion under the tongue consists of a sheet of mucous membrane covering muscles and soft tissues. Some drugs (e.g. trinitrin) can be absorbed through the mucous membrane of the mouth if allowed to dissolve under the tongue.

The ducts of the salivary glands open into the mouth.

The tongue

The tongue is a highly mobile organ composed entirely of voluntary muscle. It is essential for mastication, swallowing, speech and taste. The root of the tongue is attached to the hyoid bone in the neck.

The mucous membrane which lines the mouth also covers the free surface of the tongue. On the dorsum of the tongue are numerous minute elevations of the mucous membrane called papillae. Embedded in the papillae are the **taste buds** (see p. 259).

Three types of papillae are found on the tongue.

(a) **Circumvallate papillae**. These are 8–10 large papillae arranged in a V-shaped pattern at the base of the tongue; they contain numerous taste buds.

(b) **Fungiform papillae** are flattened papillae found on the top and sides of the tongue.

(c) **Filiform papillae** are small pointed elevations found all over the surface of the tongue.

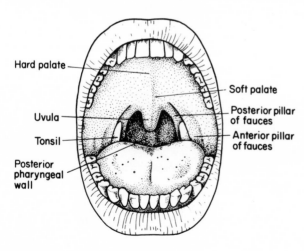

Hard palate
Uvula
Tonsil
Posterior pharyngeal wall
Soft palate
Posterior pillar of fauces
Anterior pillar of fauces

Fig. 12.4   The back of the mouth.

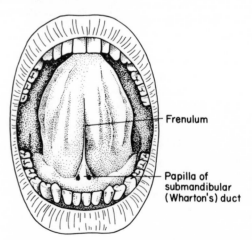

Frenulum
Papilla of submandibular (Wharton's) duct

Fig. 12.5   The floor of the mouth and inferior surface of the tongue.

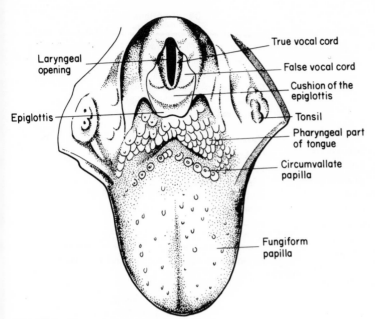

Fig. 12.6 The tongue and opening of the larynx.

The under surface of the anterior part of the tongue is covered by thinner mucous membrane, and this forms a central fold, called the **frenulum**, which attaches the tongue to the floor of the mouth.

Nerve supply

The voluntary muscles of the tongue are supplied by the hypoglossal nerve. The lingual branch of the mandibular nerve supplies the anterior portion of the tongue with ordinary sensation, whilst the facial nerve supplies taste fibres. The posterior portion of the tongue is supplied with both ordinary sensation and taste fibres by the glossopharyngeal nerve.

The teeth

Each individual develops two sets of teeth, the temporary teeth of childhood and the permanent teeth of adult life.

Temporary teeth (also known as deciduous or milk teeth). These are twenty in number, five in each half of each jaw.

Permanent teeth There are thirty two permanent teeth, sixteen in the upper and sixteen in the lower jaw, (or eight in each half of each jaw). According to their shape and position they are classified into 4 groups.

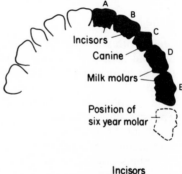

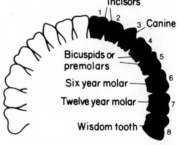

Fig. 12.7 Arrangement of the teeth — (above) temporary and (below) permanent.

Class of teeth	Number on each side of both jaws
Incisors	2
Canines	1
Premolars	2
Molars	3

The **incisors** are the sharp cutting, or biting, teeth situated in front. To the lateral side of the incisors are the **canine** teeth (eye teeth), used for grasping. Behind the canines are the **premolars** and the **molars,** or grinding teeth. The teeth are set in each jaw in the shape of a horseshoe, which is sometimes referred to as the dental arch.

At birth all the temporary teeth and most of the permanent teeth are buried in the alveoli of the jaws. Several of the latter, however, are formed after birth.

The development and maintenance of strong healthy teeth depends on an adequate supply of calcium, phosphates, and vitamins A and D in the diet. If a pregnant woman's diet is deficient in these constituents her own body stores will be depleted to supply the fetus. For this reason emphasis is placed on dental care during pregnancy.

Eruption of the teeth

The eruption of both sets of teeth follows a fairly regular order and as a general rule those of the lower jaw appear before those of the upper jaw.

The infant usually cuts its first teeth, the lower central incisors, about the 6th or 7th month. Others continue to appear and the full set of milk teeth is completed at about the age of 2 years.

The earliest permanent teeth to appear are the first molars. Others appear in turn and the permanent set is complete at about the age of 12 years, except for the third molars, or wisdom teeth, which appear between the 17th and 25th years, or may never erupt but remain buried in the gums. Sometimes these teeth are unable to erupt because there has not been sufficient mandibular growth to accommodate them. They then become impacted and have to be removed.

Structure and appearance of teeth

A tooth consists of a crown, a neck and a root. The **crown** is the portion projecting above the gum.

Dentition table

Temporary teeth	Months to appear
incisors	6–12
first molars	12–14
canines	14–20
second molars	20–24

Permanent teeth	Year
first molars	6 th
central incisors	7 th
lateral incisors	8 th
premolars	9 th and 10 th
canines	11 th
second molars	12th
third molars	17th – 25th

For convenience a dental surgeon may use the following formula (see Fig 12.7)

LEFT (*upper*)	RIGHT (*upper*)
87654321	12345678
87654321	12345678
(*lower*)	(*lower*)

Fig. 12.8 Dentition table.

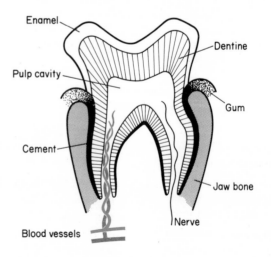

Fig. 12.9 The structure of a tooth.

The slight constriction where it is surrounded by the gum margin is the **neck** and the part buried in the alveolus of the jaw is the **root**.

The main mass of the tooth consists of a substance called **dentine**, a very hard material which resembles bone. The exposed portion of the tooth is covered with a thin layer of **enamel**, which is even stronger than dentine.

The interior of a tooth is hollow and is called the **pulp cavity**. This contains soft connective tissue, small blood vessels and nerves which enter the root of the tooth through a fine canal.

The teeth are firmly fixed in the alveoli of the jaw by a special calcified cement substance and a strong layer of connective tissue called the **periodontal membrane**.

The incisor teeth have a crown consisting of a sharp cutting edge, and a single pointed root. The canines have a crown which terminates in a point, and a single root. The premolars have a single root and a crown consisting of two elevations or cusps (which explains why they are sometimes called bicuspids). The molars have a square-shaped crown with four cusps. The upper molars possess three roots, the lower two roots.

Nerve supply

The nerve supply to the teeth is by branches of the trigeminal nerve; the teeth of the upper jaw are supplied by the maxillary division, and those of the lower jaw by the mandibular division.

The pain from a decayed (carious) tooth is usually felt in the tooth itself, but may sometimes be referred to different parts of the face, forehead, chin or ear.

The **teeth** are essential for efficient mastication. The incisors are the cutting teeth and permit the individual to bite off small pieces of food. The canines are really grasping teeth and, in animals such as dogs, are much larger and longer than in man. They prevent the food slipping while the incisors are cutting it off. The molars and premolars are the grinding teeth and it will be noticed that mastication is a side-to-side movement of the lower jaw so that the food is ground between them.

A film known as 'plaque' normally covers teeth and is difficult to remove, even with a good toothbrush. It contains bacteria which break down carbohydrates into acids which attack the enamel and eventually cause dental decay (caries). Dental caries is reduced when the concentration of

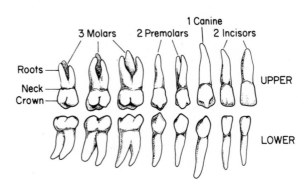

Fig. 12.10 Shapes of human teeth.

fluoride in drinking water is about 1 part per million, which makes enamel more acid-resistant and inhibits the bacterial enzymes in plaque which break down carbohydrate.

The salivary glands

There are three pairs of salivary glands, the parotid, submandibular and sublingual glands. They secrete saliva, the first of the digestive juices.

Parotid glands

These lie below and slightly in front of each ear. They are the largest of the salivary glands, and the deepest part of the gland lies between the upper part of the ramus of the mandible and the mastoid process. Each has a duct (Stenson's duct), which runs diagonally in the substance of the cheek and enters the vestibule of the mouth at a small papilla which can be seen opposite the second upper molar tooth.

Infectious parotitis, or mumps, is an acute viral infection that causes swelling of the parotid glands.

Submandibular glands

Each is situated just under cover of the angle of the mandible. The submandibular, or Wharton's duct passes forwards in the floor of the mouth and enters by a papilla close to the frenulum under the tongue (Figs 12.5 and 12.11).

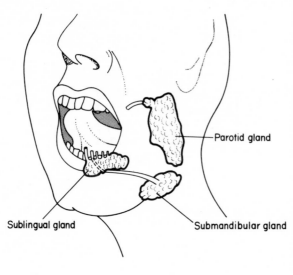

Fig. 12.11 Diagram of salivary glands.

Sublingual glands.

These lie in the anterior part of the floor of the mouth between the tongue and the mandible, close to the midline. They pour their secretions directly into the mouth by a number of small ducts (Fig. 12.11).

Saliva

Saliva is the mixed secretions of the three pairs of salivary glands. It is a fluid consisting of 90% water, and contains the enzyme **ptyalin** (salivary amylase) which begins the digestion of carbohydrate. It also contains mucin, a thick lubricant, and a small amount of calcium salts.

Functions of saliva

(*a*) Saliva constantly moistens the mouth and tongue.

(*b*) It moistens and lubricates food so that it can be rolled into a soft mass, or bolus, suitable for swallowing.

(*c*) Ptyalin (salivary amylase) begins to act on cooked starches and converts them to dextrins and maltose.

(*d*) Saliva dissolves part of the food, which stimulates the taste buds.

(*e*) It has a cleansing action and helps to keep the mouth and teeth free of debris.

Secretion of saliva

The secretion of saliva is under the control of the superior and inferior salivary nuclei in the brain stem. The salivary glands are supplied with both parasympathetic and sympathetic nerve fibres. Parasympathetic nerves stimulate the flow of saliva, whilst the sympathetic nerves inhibit its production.

The sight, smell or thought of food stimulates an initial increase in the secretion of saliva (the sensation of the 'mouth watering').

When food is placed in the mouth there is a further increase in the output of saliva. The teeth are essential for mastication of food so that a greater surface area of the food is exposed to the saliva. The tongue aids mastication by moving the food about in the mouth.

Inhibition of the secretion of saliva is caused by activity of the sympathetic nerve fibres, as in fear or excitement, leading to dryness of the mouth. Dehydration, and some drugs (e.g. atropine), also decrease the amount of saliva.

The pharynx

The pharynx, or throat, is an expanded muscular tube lying behind the nose and mouth. It is about 13 cm (5 in) long and extends from the base of the skull to the level of the cricoid cartilage, where it becomes continuous with the oesophagus. It lies just in front of the cervical vertebrae.

The pharynx is anatomically divided into three parts:
(1) the nasopharynx;
(2) the oropharynx;
(3) the laryngopharynx.

The **nasopharynx** lies immediately behind the nasal cavities and extends to the level of the soft palate (see p. 158).

The **oropharynx** is situated behind the mouth, extending from the soft palate above to the hyoid bone below. Its anterior aspect is open to the mouth and is bounded by the anterior pillars of the fauces. The palatine tonsils and the posterior pillars of the fauces lie within its lateral walls. The posterior wall can be seen through the mouth, and lies in front of the second and third cervical vertebrae.

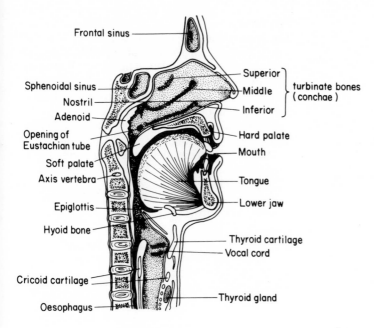

Fig. 12.12   Section through the nose and throat.

The **laryngopharynx** is narrower than the other parts and lies behind the larynx. Above it is continuous with the oropharynx; below it merges into the oesophagus.

The nasopharynx is not a functional part of the alimentary tract, but belongs to the respiratory system. The oropharynx is a shared entrance for the respiratory and alimentary tracts, since air must pass through it to reach the larynx and food to reach the oesophagus.

The walls of the pharynx are composed of muscles arranged in thin overlapping sheets; these form the **constrictor muscles** which contract in the act of swallowing. There is a middle layer of fibrous connective tissue and a lining of mucous membrane similar to that of the mouth.

The nerve supply of the pharynx is from branches of the Vth, lXth, Xth, and Xlth cranial nerves.

The oesophagus

The oesophagus, or gullet, is a collapsible muscular tube extending from the pharynx to the cardiac orifice of the stomach. It is about 25 cm (10 in) long and passes downwards through the neck, the thorax and the abdominal cavity. It lies in the midline in front of the vertebral column and behind the trachea as far as its bifurcation, then behind the left atrium of the heart. The thoracic aorta is situated on the left side of the lower half of the oesophagus. The oesophagus leaves the thorax by piercing the diaphragm and enters the abdominal cavity for a short distance before terminating in the stomach at the cardia.

The oesophagus is composed of four layers of tissue.

(*1*) A lining of mucous membrane with stratified epithelium.

(*2*) A submucous layer containing blood vessels, nerves and mucous glands which secrete a viscid lubricant mucus.

(*3*) A muscle layer consisting of an inner coat of circular muscle fibres and an outer coat of longitudinal muscle fibres.

The muscle of the upper third of the oesophagus is skeletal muscle, whilst that of the middle third is mixed skeletal and visceral muscle. The lower third contains only visceral muscle fibres.

(*4*) An outer covering of connective tissue.

The circular muscle fibres of the oesophagus are thickened to form sphincter muscles at its upper (cricopharyngeal) and lower (cardiac sphincter) ends, which relax when swallowing takes place. The cardiac sphincter prevents regurgitation of stomach contents into the oesophagus.

Swallowing or deglutition

When a mouthful of food has been masticated and well mixed with saliva the movements of the tongue and cheeks convert it into a soft rounded mass called a **bolus**.

The act of swallowing takes place in three stages.

(1) The first stage is under voluntary control. The lips are closed. The tongue is raised against the hard palate and forces the bolus of food past the pillars of the fauces into the oropharynx.

The bolus stimulates **swallowing receptors** situated around the opening of the pharynx, and nervous impulses are transmitted to the brainstem, which activates a series of automatic muscular contractions.

(2) The soft palate rises and closes the naso-pharynx. The vocal cords in the larynx close and the epiglottis moves backwards to cover the glottis; these two actions prevent food entering the trachea.

(3) The cricopharyngeal sphincter of the oeso-phagus relaxes and the larynx is pulled upwards, opening the oesophagus. The pharyngeal muscles constrict to force the bolus of food from the pharynx downwards into the oesophagus.

The bolus is carried down the oesophagus by the contraction of its muscular walls (peristalsis). The cardiac sphincter relaxes and the bolus enters the stomach.

Nervous control of swallowing

The swallowing receptors in the pharynx transmit nerve impulses to the swallowing centre in the medulla via the glossopharyngeal and vagus nerves. The swallowing centre then initiates a sequence of muscular reactions which cannot be stopped.

The nerve impulses to the pharyngeal and laryngeal muscles are transmitted via the glosso-pharyngeal, vagus and accessory nerves.

The process of swallowing occupies about ten to twenty seconds, depending on whether liquids or solids are taken, the passage of the latter being slower. It can be observed by means of X-rays if the patient is given a barium mixture to swallow. In this way any obstruction of the oesophagus can be demonstrated, and if necessary, further inves-tigated by oesophagoscopy. Difficulty in swallow-ing is called *dysphagia*.

Peristalsis

The special method of muscular contraction by which the bolus of food is carried down the oeso-phagus is called peristalsis. This form of muscular action is of great importance and occurs through-out the whole of the alimentary tract for the pur-pose of passing on its contents.

The circular muscle of the tube immediately behind the bolus contracts, whilst that directly in front relaxes. This results in the bolus being forced into the relaxed portion. The contraction of muscle follows closely behind the bolus and further relaxa-tion occurs in front of it, thus the bolus is passed steadily forwards (Fig. 12.13).

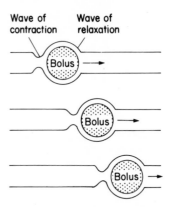

Fig. 12.13 Diagram illustrating peristalsis, showing wave of contraction preceded by wave of relaxation of muscle, resulting in forward movement of bolus of food.

The stomach

The stomach is a dilated portion of the alimentary canal, situated between the oesophagus and the beginning of the small intestine. It lies in the epi-gastric, umbilical and left hypochondriac regions of the abdominal cavity. It is a muscular, dis-tensible organ and its shape and size vary according to the amount and type of its contents, and to the pressure exerted on it by surrounding organs.

The stomach has two openings, the cardiac and pyloric orifices, and is described as having two curvatures, the greater and lesser curvatures, and anterior and posterior surfaces.

The oesophagus enters the stomach at the **cardiac orifice**, which is situated at the upper end of the lesser curvature. To the left of the cardiac orifice is the dome-shaped upper part of the stomach called the **fundus**, which in life generally contains a bubble of air. The main part of the stomach is called the **body**, and this narrows at its lower end to become the **pyloric antrum**. The stomach joins the duodenum (the first part of the small intestine) at the pylorus via the pyloric orifice.

The **greater curvature** begins where the oesophagus enters the stomach and extends over the fundus, then curves downwards to the pylorus.

The **lesser curvature** is shorter and extends from the cardiac orifice to the pylorus (Fig. 12.14).

The anterior surface of the stomach lies partly behind the left lobe of the liver and the under-surface of the left side of the diaphragm, and partly in contact with the anterior abdominal wall. The posterior surface lies in the 'stomach bed' which is formed by the left kidney, the pancreas and the spleen.

The structure of the stomach

The walls of the stomach consist of four layers of tissue.

(1) An outer covering of **serous membrane** (the peritoneum).

(2) A **muscle layer**, which actually consists of three layers of visceral muscle fibres.

(a) Longitudinal muscle fibres are the most superficial, and are continuous with the longitudinal fibres of the oesophagus.

(b) Circular muscle fibres form the middle layer. They are continuous with those of the oesophagus, and are thickened around the pylorus forming the **pyloric sphincter**.

(c) Oblique fibres, the innermost layer, are found mainly in the body of the stomach.

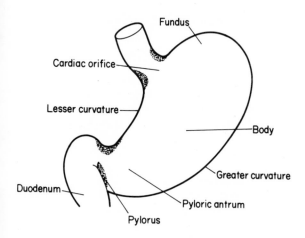

Fig. 12.14 Diagram illustrating parts of the stomach.

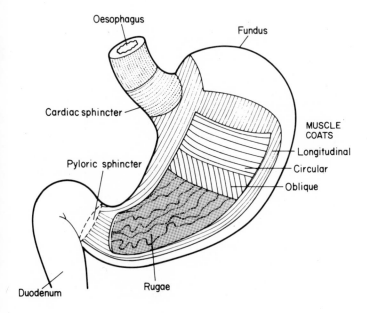

Fig. 12.15 Structure of the stomach.

(3) A submucous layer consisting of loose areolar tissue carrying blood vessels, lymphatics and nerves.

(4) A lining **mucous membrane** which is thick, and has a smooth, soft velvety surface when the stomach is distended. When the stomach is empty it is thrown into numerous irregular folds or **rugae**.

The mucous membrane contains microscopic depressions or alveoli; these are the **gastric pits** and they are particularly numerous in the fundus and body of the stomach. Embedded in the gastric pits are glands which secrete gastric juice.

The glands around the cardiac orifice mainly secrete mucus, whilst those in the fundus and body secrete digestive enzymes and *hydrochloric acid*. The glands in the pyloric antrum secrete mainly mucus and a protein compound called *intrinsic factor*.

Covering the surface of the mucous membrane, including the gastric pits, is a single layer of columnar epithelial cells, the surface mucous cells, which secrete mucus to act as a protective lubricant. In the absence of mucus secretion the enzymes and acid of the gastric juice produce craters in the wall of the stomach; these are called gastric ulcers.

Blood supply

The stomach receives a liberal arterial blood supply from the coeliac axis, the first branch of the abdominal aorta. The blood from the stomach drains into the portal vein and thence to the liver.

Nerve supply

The stomach is under the control of the autonomic nervous system. It receives parasympathetic nerve fibres from the vagus nerves; stimulation of these fibres increases gastric motility and the secretions of the gastric glands. Sympathetic nerve fibres are supplied from the coeliac plexus, and stimulation of these decreases gastric motility and inhibits digestion.

Digestion in the stomach

Movements in the stomach

The stomach, which resembles an elastic bag, acts as a reservoir, and can contain very large quantities of food. The gastric glands of the mucosal lining secrete large amounts of digestive juices which are mixed with the stored food by rippling waves of peristalsis, called *tonus* or *mixing* waves. The action of these waves also moves the food towards the antral part of the stomach, where vigorous peristaltic waves break down or churn the food and gastric secretions to form a milky-white sludge called *chyme*.

The pyloric sphincter usually remains slightly constricted, but when the chyme is sufficiently fluid in consistency strong waves of peristalsis are able to force the chyme through the pylorus into the duodenum.

Gastric secretions

Gastric juice is a clear, watery, strongly acid fluid, containing hydrochloric acid, enzymes, mineral salts and mucus. A total quantity of $1\frac{1}{2}$ to 2 litres is secreted each day.

When food first reaches the stomach, salivary digestion is continued until hydrochloric acid inhibits the action of ptyalin. Hydrochloric acid converts pepsinogen to pepsin. It also kills many microorganisms which may be ingested in food.

There are three **enzymes** in gastric juice.

Pepsin Pepsinogen is formed by the gastric glands, but needs the presence of hydrochloric acid to convert it to the active protein-splitting enzyme pepsin. This reduces proteins to proteoses, peptones and polypeptides; further stages of protein digestion take place in the small intestine.

Rennin This is an enzyme possessed by infants that curdles milk. In the presence of calcium ions, it converts the soluble milk protein caseinogen to the insoluble form, casein, which is then reduced to peptones by pepsin.

Gastric lipase, an enzyme for beginning the digestion of fats, is produced in very small amounts.

Mineral salts Gastric juice contains sodium, potassium, chloride, magnesium, calcium, phosphate and sulphate.

Mucus is secreted by the stomach mucosa to protect it from the digestive action of pepsin.

Very little absorption takes place in the stomach, but the mucosa absorbs some water, glucose, alcohol and certain drugs.

The gastric mucosa also secretes a protein

compound called '*intrinsic factor*' (of Castle), which is needed to facilitate the absorption of *vitamin B₁₂* (sometimes referred to as the 'extrinsic factor'), from the alimentary tract. Vitamin B_{12} is essential for the normal development of red blood cells. Absence of intrinsic factor from the gastric juices and consequent lack of vitamin B_{12} in the body leads to a disease called pernicious anaemia.

A small quantity of gastric juice is produced continuously, but the intake of food is followed by a considerable increase in the output of gastric juice. The secretion of gastric juice occurs in three phases (Fig. 12.16):

Cephalic phase The thought, sight, smell or taste of food causes an increase in the flow of gastric juice by reflex stimulation of the vagus nerve. This prepares the stomach for food that is to be eaten.

Gastric phase Distension of the stomach by food stimulates the mucosa of the pyloric antrum to secrete a hormone called **gastrin**. Gastrin is absorbed into the blood and carried to the gastric glands where it stimulates increased secretion of hydrochloric acid and enzymes.

Certain substances in food, such as meat extracts, partially digested proteins, alcohol and caffeine, even in small amounts, can also cause release of this hormone. Gastrin production continues throughout the time food remains in the stomach.

Intestinal phase Presence of partially digested food in the duodenum causes the release of *intestinal gastrin*, and this also stimulates the stomach to secrete gastric juice.

However, if too much chyme is already in the duodenum, or the chyme is excessively acid, or contains large quantities of proteins or fat, the duodenum initiates strong feedback mechanisms, both nervous and hormonal, to inhibit further gastric emptying. Reflex nervous impulses are transmitted from the duodenum back to the stomach continuously throughout the period of digestion. The duodenum is able to inhibit gastric emptying via the *enterogastric reflex*, which reduces pyloric activity and increases the tone of the pyloric sphincter, thus effectively closing the stomach until duodenal contents have been adequately digested.

The presence of excess chyme, especially when it contains a high percentage of fats, causes the cells of the duodenal mucosa to release several hormones into the bloodstream, among these are **cholecystokinin** and **secretin**. The hormonal mechanism slows gastric peristalsis to allow time for fat digestion in the small intestine.

Vomiting

Vomiting is the method by which the upper part of the alimentary tract rids itself of its contents. It is a reflex action and may be initiated by a variety of stimuli, including over-distention of the stomach, local irritation of the pharynx, oesophagus, stomach or small intestine, disturbance of the vestibular apparatus of the ear, and emotional influences such as fear, anxiety or disgust.

Nervous impulses are transmitted by both the vagus nerve and sympathetic nerve fibres to the **vomiting centre** in the medulla. The appropriate motor reactions are then initiated to cause the vomiting act.

The individual takes a deep breath, the nasopharynx and larynx are closed. The abdominal muscles contract and the diaphragm descends, thus raising the intra-abdominal pressure, and compressing the stomach. The pyloric sphincter contracts, the cardiac sphincter relaxes, and the stomach contents are expelled through the oesophagus by a strong wave of reverse peristalsis.

Vomiting may also be caused by some emetic drugs (e.g. apomorphine and ipecachuanha), or by direct pressure on the vomiting centre by raised intracranial pressure.

The small intestine

The small intestine is a convoluted tube, extending from the pylorus of the stomach to the ileo-caecal valve, where it joins the large intestine. It is approximately 5 m (16 ft) long and consists of three parts:

> duodenum
> jejunum
> ileum

The duodenum

The duodenum is the first part of the small intestine. It is approximately 25–30 cm (10–12 in) long, and lies in a C-shape around the head of the pancreas. It is fixed to the posterior abdominal wall by peritoneum, which covers only its anterior

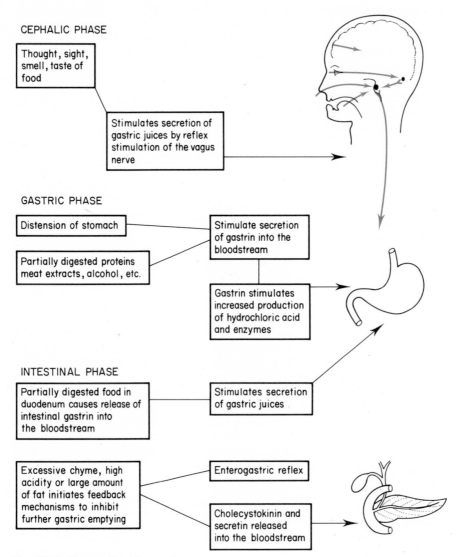

CEPHALIC PHASE

Thought, sight, smell, taste of food

Stimulates secretion of gastric juices by reflex stimulation of the vagus nerve

GASTRIC PHASE

Distension of stomach

Partially digested proteins meat extracts, alcohol, etc.

Stimulate secretion of gastrin into the bloodstream

Gastrin stimulates increased production of hydrochloric acid and enzymes

INTESTINAL PHASE

Partially digested food in duodenum causes release of intestinal gastrin into the bloodstream

Stimulates secretion of gastric juices

Excessive chyme, high acidity or large amount of fat initiates feedback mechanisms to inhibit further gastric emptying

Enterogastric reflex

Cholecystokinin and secretin released into the bloodstream

Fig. 12.16   Phases of gastric secretion.

surface. About halfway along the concave surface of the duodenum the bile duct and the pancreatic duct enter the duodenal lumen together at a small papilla called the **ampulla of Vater**.

Jejunum and ileum

The jejunum forms two-fifths of the small intestine, whilst the ileum forms the remaining three-fifths. They extend from the duodenojejunal junction to the ileo-caecal valve. Although given different names the jejunum passes imperceptibly

into the ileum. They lay coiled in the central and lower abdominal cavity and are attached to the posterior abdominal wall by the mesentery, a fan-shaped fold of peritoneum.

Structure of the small intestine

The small intestine, like the stomach, has four coats of tissue:

(*1*) an outer serous coat (peritoneum);

(*2*) a muscular coat consisting of longitudinal and circular muscle fibres (but no oblique fibres);

(*3*) a submucous coat;

(*4*) an inner coat of mucous membrane. In the small intestine the mucous membrane is arranged in permanent folds, called **plicae circulares** (valvulae conniventes). Their effect is to greatly increase the surface for secretion and absorption. The surface of the mucous membrane is covered with minute finger-like processes called **villi**, which give the surface a soft velvety appearance.

The villi are highly vascular and each villus contains a network of *blood capillaries* and a *central lacteal* (an intestinal lymphatic vessel), supported by loose connective tissue and some smooth muscle fibres. The surface of the villi are covered by a single layer of epithelial cells, which have a 'brush border' forming thousands of microvilli. The microvilli further increase the total absorptive area of the small intestine.

Between the villi lie simple tubular glands, the **crypts of Lieberkuhn**, which secrete alkaline intestinal juice (*succus entericus*). At the base of the crypts lie the cells of Paneth. These secrete the digestive enzymes into the intestinal juice.

The small intestine also secretes mucus from its entire surface as a protective lubricant, but this is an especially important function in the duodenum because of the extreme acidity of the chyme from the stomach. In the first few centimetres of the duodenum are large mucous glands, called **Brunner's glands**, which protect the duodenal mucosa from the powerful effects of hydrochloric acid and pepsin until the acidity of the chyme has been neutralized by pancreatic juice.

Lymphatic tissue is scattered throughout the whole of the small intestine either in solitary nodes or in collected masses called **Peyer's patches**, which are found mainly in the ileum.

Blood supply
The duodenum derives its blood supply from branches of the coeliac axis, whilst the rest of the small intestine receives blood from the superior mesenteric artery.

Nerve supply
The small intestine receives both parasympathetic and sympathetic nerve fibres. The parasympathetic fibres increase peristaltic movements and secretions of the intestinal glands, and relax the sphincter of the ileocaecal valve. The sympathetic fibres reduce peristalsis and close the ileocaecal valve.

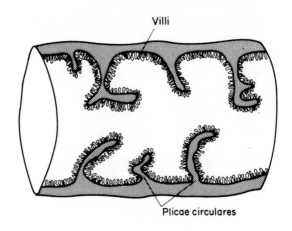

Fig. 12.17 Section of small intestine illustrating mucosal folds and villi.

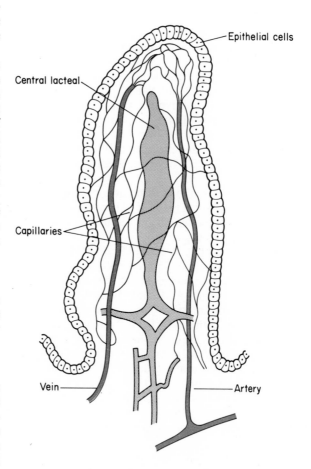

Fig. 12.18 Structure of a villus.

Digestion and absorption in the small intestine

The functions of the small intestine are to complete digestion of food, to absorb the end-products of digestion and to move the chyme along the length of the small intestine.

The presence of the acid chyme in the duodenum distends the intestinal walls and causes the mucosa to secrete mucus and intestinal juice (succus entericus), which contains the enzyme **enterokinase**. At the same time the mucosa also releases into the bloodstream the hormones secretin and cholecystokinin which are carried to the pancreas. Secretin stimulates the pancreas to secrete large quantities of fluid containing sodium bicarbonate. The bicarbonate reacts with the acid in the chyme and neutralizes it; failure to neutralize this acidity allows the protein-digesting enzyme pepsin to erode the walls of the duodenum leading to ulceration. Cholecystokinin causes the pancreas to secrete digestive enzymes; **amylase** for digesting carbohydrates to disaccharides (sugars), **trypsinogen** and **chymotrypsinogen** for digesting proteins and **lipase** for digesting fats. The fluid and enzymes enter the duodenum via the pancreatic duct and the ampulla of Vater.

Trypsinogen and chymotrypsinogen are inactive enzymes. Trypsinogen is converted to an active enzyme **trypsin**, in the presence of enterokinase, and in turn activates chymotrypsinogen to convert it to **chymotrypsin**. These enzymes reduce proteins to polypeptides.

Cholecystokinin also stimulates the gallbladder to empty bile, which is essential for the digestion of fats, into the duodenum via the bile duct and the ampulla of Vater.

Bile contains bile salts, which act as a powerful detergent, emulsifying fats to small globules so that the lipases can act on larger surface areas and convert the fat to fatty acids and glycerol.

The small intestine also secretes digestive enzymes within the epithelial cells covering the villi. Thus digestion takes place inside these cells or in close proximity to them. The enzymes sucrase, maltase and lactase are responsible for splitting disaccharides to monosaccharides (glucose), the end-products of carbohydrate digestion; a group of enzymes called peptidases reduce polypeptides to amino acids, and lipases complete the digestion of fats.

Absorption of the digested food occurs through the epithelial surface of the villi.

Monosaccharides and amino acids are absorbed into the blood capillaries of the villi by the process of active transport (p. 8), which requires the expenditure of energy by the epithelial cells. This allows the absorption to take place even when there are very small concentrations of these substances in the intestine.

Fatty acids and glycerol molecules diffuse easily through the epithelial cells of the villi and are reformed into small neutral fat globules called **chylomicrons**, which are absorbed into the central lacteals. When mixed with the lymph in the lacteals the chylomicrons give it a milky appearance and is called *chyle*. The chyle is carried from the lacteals into the abdominal lymphatic vessels and the cisterna chyli, then via the thoracic duct to be emptied into the bloodstream through the left brachiocephalic vein.

In addition the small intestine absorbs iron, vitamins and calcium (the latter in the presence of vitamin D). Any fluid that is ingested is also absorbed, and almost all the secretions of the alimentary tract, i.e. saliva, gastric juice, intestinal juice, pancreatic secretions and bile, are re-absorbed. This amounts to a total of approximately 8 litres a day, only about 800 ml is passed on to the large intestine. In this way fluid circulates between the body fluids and the alimentary tract continuously.

Movements of the small intestine

When food enters the small intestine as chyme it initiates peristaltic activity called **segmentation**, this is a series of rhythmical alternating contractions and relaxations of the intestinal wall which mix the chyme with the intestinal juice and enzymes, and bring it into contact with the mucosa for absorption. *Propulsive* peristaltic movements push the chyme along the length of the small intestine until it reaches the ileocaecal valve, which opens to allow spurts of the fluid contents of the ileum to enter the large intestine. By this time the absorption of digested foodstuffs is complete.

If the mucosa of the small intestine becomes irritated or inflamed, violent peristaltic waves occur and chyme is passed rapidly into the large intestine before digestion and absorption can take place. Frequent watery stools are passed and may result in dehydration as vast amounts of body fluids are lost.

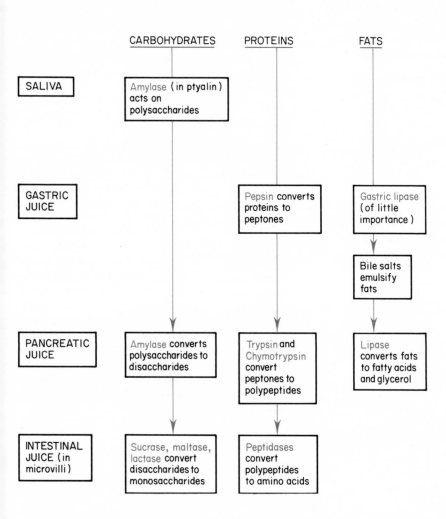

Fig. 12.19 Summary of digestive enzymes.

Large intestine

The large intestine, or colon, is 1.5 m (5 ft) long and extends from the ileocaecal valve to the anus. It is described as having the following parts:

caecum and vermiform appendix
ascending colon
transverse colon
descending colon
pelvic or sigmoid colon
rectum and anal canal

The caecum

This is a blind dilated sac situated in the right iliac fossa, into which the ileum enters at the ileocaecal valve. The mucous membrane is so arranged at this point that together with a sphincter muscle, it acts as a valve permitting the contents of the ileum to pass into the caecum but preventing their return into the ileum.

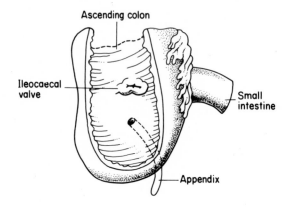

Fig. 12.20 The caecum opened from the front.

The appendix

The appendix (vermiform = worm-like) is a narrow blind tube projecting from the caecum just below the ileocaecal valve. It is usually about 9 cm (3.5 in) in length. Sometimes it lies to the side of the caecum, sometimes tucked up behind it, or its tip may hang down into the pelvic cavity. It is lined with mucous membrane and contains lymphoid tissue in its walls.

Inflammation of the appendix (appendicitis) is a relatively common condition which may affect both children and adults. It is characterized by pain and tenderness in the right iliac fossa. The main danger is that the appendix may perforate and liberate its infected contents into the peritoneal cavity, causing peritonitis.

The ascending colon

This passes upwards from the caecum through the right lumbar region and is held in position on the posterior abdominal wall by peritoneum. When it reaches the under surface of the liver it turns sharply to the left at the right or **hepatic flexure** to become the transverse colon.

The transverse colon

This passes across the abdominal cavity as a loop that may hang well below the umbilicus. It rises as it reaches its left extremity and comes in contact with the spleen, where it turns sharply downwards at the left or **splenic flexure** to continue as the descending colon. The transverse colon lies in a fold of peritoneum which extends downwards from the greater curvature of the stomach called the **greater omentum**. It is therefore freely movable within the abdominal cavity.

The descending colon

This passes downwards in the left lumbar region. It is anchored to the posterior abdominal wall by peritoneum, and like the ascending colon is not movable.

The pelvic or sigmoid colon

This is the continuation of the descending colon, and it makes an S-shaped curve, the *sigmoid flexure*, to enter the pelvic cavity to become the rectum. It is attached to the left side of the pelvis by a fold of peritoneum called the sigmoid mesacolon.

The rectum

The rectum lies in the pelvic cavity and is situated in the concave hollow formed by the anterior surface of the sacrum. It is about 13 cm (5 in) long and narrows to form the anal canal. This is about 3 cm (1.2 in) in length and its lower half is lined with

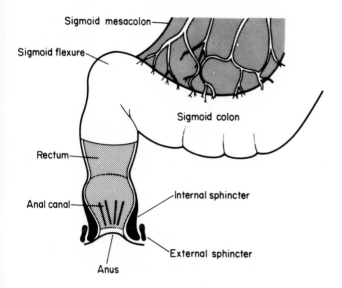

Fig. 12.21 Diagram of sigmoid colon, rectum and anus.

stratified squamous epithelium which is continuous with the mucous membrane lining the rest of the anal canal and the rectum and with the skin where the anus opens to the exterior of the body.

Structure of the large intestine

The large intestine, like the small intestine, has four coats; peritoneal or serous, muscular, submucous and mucous.

The characteristic feature, however, is that the longitudinal muscles do not form a continuous layer over the whole gut but are arranged in three separate longitudinal bands, the **taenia coli**. These bands are somewhat shorter than the length of the large intestine, which accounts for its sacculated appearance.

When the longitudinal fibres reach the rectum they spread out over the whole surface. In the upper half of the anal canal the circular smooth muscle fibres are thickened to form the **internal anal sphincter**. The lower half is composed of striated muscles fibres, which form the **external anal sphincter**.

The mucous membrane lining contains no villi and is not thrown into folds like that of the small intestine. The large intestine, like the oesophagus, does not take any part in digestion but the whole of its mucosa contains mucous glands that secrete large quantities of mucus to protect it from the effects of the digestive enzymes of the small intestine, and to provide lubrication for the passage of faeces.

Blood supply

The superior mesenteric artery supplies the caecum, the appendix, the ascending colon and most of the transverse colon, whilst the inferior mesenteric artery supplies part of the transverse colon, the descending colon, the sigmoid colon and the upper part of the rectum.

The blood supply of the lower rectum and the anus is from branches of the inferior mesenteric, internal iliac and pudendal arteries. The veins of the anus form dilated vessels within the anal ring called the haemorroidal plexus, and this blood is drained into the inferior mesenteric, pudendal and internal iliac veins. Back pressure along the venous system causes enlargement of the veins in the haemorroidal plexus which results in the formation of haemorrhoids (piles).

Nerve supply

This is from the autonomic nervous system. The internal anal sphincter is controlled by the parasympathetic and sympathetic nerve fibres, but the external sphincter receives its nerve supply from the spinal nerves of the sacrum and is under voluntary control.

Functions of the large intestine

The functions of the large intestine are the absorption of water and electrolyes from the chyme which enters it from the small intestine, and the storage of faecal material until it is expelled as faeces.

By the time the contents of the small intestine reach the caecum the digestive process is completed and most of the products of digestion have been absorbed. However the residue is still in a liquid state, amounting to approximately 800 ml per day. Most of the water and electrolytes are reabsorbed by the first half of the large intestine, leaving only a faecal residue of between 100 and 200 ml each day.

The large intestine also contains a multitude of bacteria living in symbiosis. These synthesize vitamins B and K, which are absorbed into the blood by the mucosa. As long as these bacteria remain within the colon they are harmless to the body, but if they reach other organs they become pathogenic (i.e. disease producing). Destruction of the bacteria by the administration of antibiotic drugs can occasionally lead to vitamin K deficiency within a few days.

The walls of the large intestine are able to excrete an excess of calcium, iron and drugs of the heavy metal type, such as bismuth, into the faeces. Iron given by mouth makes the faeces black in colour.

The faeces

The faeces are normally a semi-solid, paste-like mass coloured brown by stercobilin, a pigment derived from the bilirubin and biliverdin of the bile. Water, even after the absorption which takes place in the colon, still forms 65 to 70 per cent of the total bulk of the faeces. The remainder consists mainly of undecomposed cellulose, some fatty acids, protein residue, bacteria and epithelial cells. The surface of the faeces is lubricated by the mucin secreted by the large intestine.

If the faeces are passed too rapidly through the intestine insufficient time will be allowed for the absorption of water, and this results in the watery character of the stool in some cases of diarrhoea.

Movements of the large intestine

Peristaltic movements of the large intestine are generally very sluggish, and consist of mixing movements similar to the segmentation movements of the small intestine, or of mass peristalsis which transfers the faecal material into the descending colon and the pelvic colon. These latter movements occur two or three times a day.

Defaecation is the expulsion of faeces from the anal canal. The rectum is usually empty until mass peristaltic movements propel faecal matter from the pelvic colon into the rectum. The entry of faeces into the rectum distends the walls of the cavity and causes nervous impulses to pass to the lower part of the spinal cord. Reflex signals are transmitted through the sacral parasympathetic nerves causing contraction of the descending colon, sigmoid flexure and rectum, and relaxation of the internal anal sphincter. At the same time impulses are sent to the brain, where they arouse the conscious sensation of the desire to defaecate. If the time is appropriate the external sphincter is relaxed by conscious control and defaecation takes place.

However, since the external sphincter is under voluntary control it can be tightened and the call to defaecate ignored. The rectum accomodates itself to its contents by relaxation of its muscular walls, nerve impulses cease and the desire passes off. The arrival of more faecal material in the rectum causes further distension and initiates the defaecation reflex again.

Repeated failure to respond to the defaecation reflex is a common cause of constipation. The retention of faeces in the rectum allows continued absorption of water and results in a hardened or constipated stool.

Defaecation in the infant is by reflex action and cannot be controlled until the nervous system is fully developed.

During defaecation the following actions occur:

(*1*) the sphincter muscles of the anus relax;

(*2*) the muscular walls of the rectum contract;

(*3*) the muscles of the pelvic floor contract;

(*4*) intra-abdominal pressure is raised by holding the breath and contracting the diaphragm, and by contracting the muscles of the abdominal wall.

The injection of an enema into the rectum has the effect of rapidly distending its walls and so initiating the defaecation reflex. It also helps to soften and break up hard masses of faeces.

The peritoneum

The peritoneum is a serous membrane, and like the other important serous membranes of the body, the pleura and the pericardium, consists of two separate layers. The **parietal layer** lines the walls of the abdominal cavity and the **visceral layer** partly

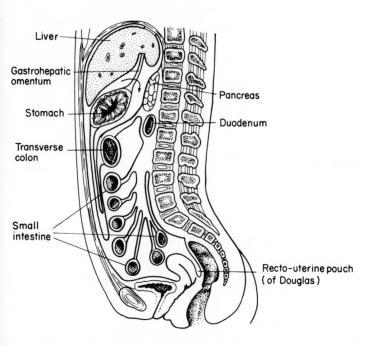

Liver

Gastrohepatic
omentum

Stomach

Transverse
colon

Small
intestine

Pancreas

Duodenum

Recto-uterine pouch
(of Douglas)

Fig. 12.22 Sagittal section of the abdomen showing disposition of the peritoneum. (The arrow passes through the foramen of Winslow into the lesser sac.)

or completely covers the organs. The membrane secretes a small amount of serous fluid to lubricate the surface of the viscera so that they are free to move without friction within the abdomen.

The detailed arrangement of the peritoneum and its various folds is very complicated, but a general conception can be obtained in the following way. Imagine the peritoneum to be a closed bag having front and back surfaces. Then consider the effect if an organ were pushed into the middle of the back surface in such a way that this surface completely surrounded the organ. It will then be understood that one organ may be completely covered by peritoneum while structures like the pancreas, lying on the posterior abdominal wall will be covered by the membrane only on their anterior surfaces.

Thus the organs completely surrounded by peritoneum will be suspended from the posterior abdominal wall by a double fold of the membrane. It is in this way that a **mesentery**, or fold of peritoneum, by which the intestine is attached to the posterior abdominal wall is formed. It is between the two layers of this fold that the blood vessels reach the organs, for the abdominal aorta and its branches lie outside the peritoneal cavity.

The stomach, intestines (except for the duo-denum and rectum), the liver and spleen are almost completely surrounded by peritoneum, which therefore forms an outer coat for these structures. The duodenum, rectum and pancreas are covered only on their anterior surfaces.

Peritoneal ligaments
The liver, uterus and other organs are partly maintained in position by means of double folds of peritoneum which form suspensory ligaments.

The omenta
These are folds of peritoneum connected to the stomach. The **great omentum** hangs from the lower border of the stomach like an apron in front of the small intestines; in its posterior portion lies the transverse colon. The **lesser omentum** stretches from the lower border of the liver to the lesser curvature of the stomach.

The mesentery
This is the fold of peritoneum which encloses the small intestine and anchors it to the posterior abdominal wall. The attachment to the abdominal wall is relatively short, whereas the intestinal part is many feet long, so that the mesentery can be described as a fan-shaped structure.

The pelvic peritoneum

The peritoneum in the pelvis is continuous with that of the rest of the abdominal cavity. It covers the front aspect of the rectum. In the male it passes forwards over the posterior and upper surfaces of the bladder to become continuous with that on the anterior abdominal wall. In the female it passes from the rectum over the posterior and anterior surfaces of the uterus before reaching the bladder. The sac between the rectum and the uterus in the female is called the **recto-uterine pouch** (*of Douglas*). On either side, in the female, the membrane covers the uterine (*Fallopian*) tubes. These tubes have an opening into the peritoneal cavity where their mucous membrane is directly continuous with the peritoneum. It is in this way that the ova are able to pass from the ovaries, which lie within the peritoneal cavity, into the uterus. In the male the peritoneum is a completely closed sac.

Functions of the peritoneum

(*1*) It is a serous membrane which enables the abdominal contents to glide over each other without friction.

(*2*) It forms a partial or complete covering for the abdominal organs.

(*3*) It forms ligaments and mesenteries which help to keep the organs in position.

(*4*) The omenta and mesentery contain a considerable amount of fat, and act as important fat stores for the body.

(*5*) The omentum can move about inside the cavity and in the event of inflammation tends to wrap itself round the affected part of the alimentary tract and prevent the infection from spreading to the rest of the peritoneum. It is very common in cases of acute appendicitis to find the appendix totally surrounded by the omentum. On account of its protective mobility it has been called the 'abdominal policeman'. It is shorter and less well-developed in infancy and childhood, so that appendicitis in these cases may be very serious.

(*6*) It has the power to absorb fluids in large quantities.

Peritoneal dialysis

The peritoneum is a membrane through which some electrolytes and simple substances can be exchanged with others in the blood. For example, if a suitable glucose-electrolyte solution is introduced into the peritoneal cavity its strength will equalize with the blood, i.e. urea will pass from the blood into the dialysis fluid which lacks this substance. The fluid is then removed from the peritoneal cavity, the blood urea having been reduced. The procedure is therefore used in the treatment of some cases of renal failure (uraemia).

The accessory organs of digestion

The liver, biliary system and pancreas form the accessory organs of digestion.

The liver

The liver is the largest organ in the body; it weighs between 1.0–2.5 kg (2.2–5.5 lb) and is heavier in the male than the female. It is a wedge-shaped organ lying immediately below the diaphragm in the right hypochondrium and epigastrium.

The liver is described as having right and left lobes, and superior, inferior, anterior and posterior surfaces.

The lobes

The right lobe is much larger than the left; the division between them is marked by the **falciform ligament** on the anterior surface, and by the ligamenta teres and venosum on the inferior and posterior surfaces. The under surface of the right lobe is further subdivided into the quadrate and caudate lobes.

The surfaces of the liver

The **superior surface** is in contact with the under surface of the diaphragm; the potential spaces between the liver and the diaphragm are called the subphrenic spaces.

The **inferior surface** is related to other abdominal viscera, including the kidney and right (hepatic) flexure of the colon on the right and the fundus of the stomach on the left.

The **anterior surface** is separated from the right lower ribs and costal cartilages by the margin of the diaphragm and, in the midline is related to the anterior abdominal wall.

The **posterior surface** crosses the vertebral column in the midline and is also related to the aorta, inferior vena cava and lower end of the oesophagus.

In the centre of the inferior surface, lying between the quadrate and caudate lobes, is the hilum or **porta hepatis** (the door of the liver). All the vessels and nerves entering and leaving the liver, with the exception of the hepatic vein, pass through the porta hepatis.

Attached to the under surface of the right lobe is the gall-bladder.

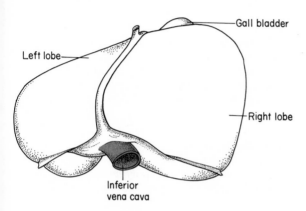

Fig. 12.23   The liver seen from above.

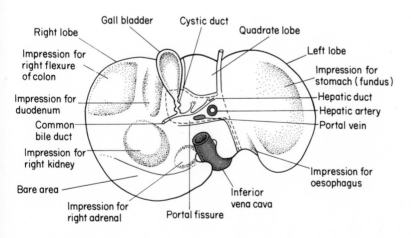

Fig. 12.24   The under-surface of the liver.

Blood supply

The liver receives blood from two sources and is an extremely vascular organ (approximately one-fifth of the liver volume is blood).

The **hepatic artery**, which is a branch of the coeliac axis from the abdominal aorta, conveys oxygenated blood to the liver cells.

The **portal vein** conveys venous blood, poor in oxygen but rich in nutrients, from the stomach and intestines.

Venous drainage from the liver is by the **hepatic veins** which empty into the inferior vena cava.

Due to its great vascularity, lacerations of the liver are very dangerous and result in profuse haemorrage.

Structure of the liver

The liver is composed of a large number of hexagonal **lobules**, each about 1 mm in diameter. A small branch of the hepatic vein extends through the centre of each lobule. The liver cells are arranged in plates, or sheets, one cell thick, around the central vein. The plates of cells form an irregular anastomosing system throughout the liver; between the plates of cells lie spaces which contain the **sinusoids**. The sinusoids are blood vessels with incomplete walls, and are irregular in shape and wider than blood capillaries. They are lined by thin endothelium and Kupffer cells, which are phagocytes and remove cellular debris and bacteria from the blood.

Arranged around the periphery of each lobule are branches of the hepatic artery, the portal vein and the hepatic bile ducts. The blood vessels form small branches which pass between the lobules and enter the sinusoids. Thus the sinusoids receive oxygenated blood from the hepatic artery and blood rich in nutrients from the portal vein. The sinusoids drain into the central vein which joins the veins from adjacent lobules to form interlobular veins, these in turn unite to form the hepatic veins, which drain the blood from the liver into the inferior vena cava.

The hepatic cells are polyhedral (have many sides) and perform many metabolic activities. Each

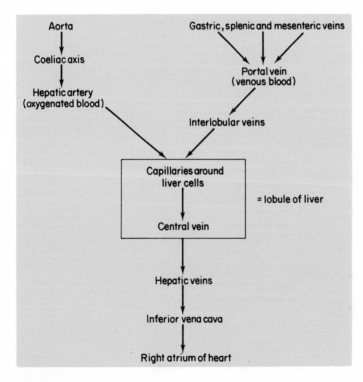

Fig. 12.25   The blood supply of the liver.

cell has two surfaces that face sinusoids, and several surfaces in contact with other cells. As the blood passes through the sinusoids, the liver cells remove absorbed nutrients from the circulation. Bile is secreted by the liver cells and drains into minute **bile canaliculi** which lie between the walls of the cells. The bile canaliculi join to form **intralobular ductules,** which unite to become the right and left **hepatic ducts**.

The liver has a thin fibrous capsule (Glisson's capsule), almost completely covered by peritoneum.

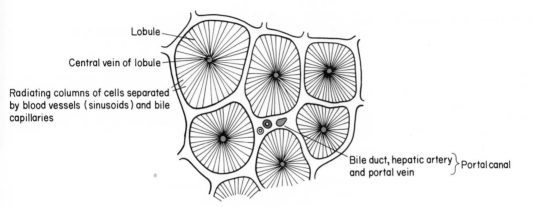

Lobule
Central vein of lobule
Radiating columns of cells separated by blood vessels (sinusoids) and bile capillaries
Bile duct, hepatic artery and portal vein } Portal canal

Fig. 12.26 Liver, showing liver lobules and radial arrangement of the cells from the centre of each lobule.

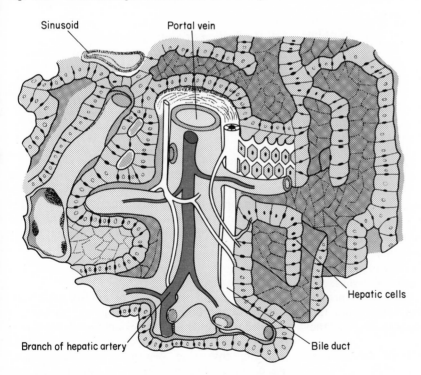

Sinusoid
Portal vein
Hepatic cells
Branch of hepatic artery
Bile duct

Fig. 12.27 Liver lobule, illustrating plates of liver cells and sinusoids.

Functions of the liver

The liver performs a variety of different functions, and in addition to the secretion of bile, plays a vital role in body metabolism.

(1)   Secretion of bile

The secretion of bile is an *exocrine* function. All the liver cells continually form a small amount of bile, which is secreted into the bile canaliculi and passes via the hepatic ducts to the gall-bladder for storage until required to assist the digestive processes.

(2)   Storage of glycogen

Carbohydrate taken in the food is broken down by the digestive juices into the monosaccharide, glucose, and in this form is absorbed into the tributaries of the portal vein and conveyed to the liver. Here it is converted into a polysaccharide, glycogen, and stored. Glycogen is reconverted into glucose whenever the blood glucose level begins to fall too low.

   The liver is also capable of converting amino acids and glycerol to glucose (the process of gluconeogenesis), should the need arise (p. 186).

(3)   Metabolism of fats

Fats are split to fatty acids and glycerol by the lipase of the pancreatic juice in the presence of bile salts, and are absorbed in this form. In the villi of the small intestine these substances are recombined into neutral fats and carried by the thoracic duct to the blood which conveys them to the fat depots of the body. The liver metabolizes fats, whether these reach it from the intestine, or are mobilized from the fat stores, to a form in which it can be used by the tissues of the body. The liver also converts excess carbohydrate and amino acids into fat for storage.

(4)   Deamination of amino acids

The end-products of protein digestion are amino acids which are absorbed into the portal circulation through the villi of the small intestine. Many of the amino acids pass through the liver and are used by the tissues for growth, and repair of tissues. Any remaining amino acids are oxidized to provide energy, or converted to carbohydrates and fats. However before they can be used in this way they must be deaminated.

   Deamination is the removal of the nitrogen containing portion of the amino acids and its conversion to ammonia, which in turn combines with carbon dioxide to form urea. The urea is excreted in the urine by the kidneys.

(5)   Production of the plasma proteins

The liver forms 90–95% of the plasma proteins, albumin, globulin and fibrinogen. The remaining 5–10% are gamma globulins formed by the plasma cells of the lymphoid tissue.

(6)   Storage of vitamins

Large quantities of vitamins A, D and B_{12} are stored in the liver. Sufficient amounts of vitamin A can be stored to prevent deficiency for up to two years, whilst vitamins D and B_{12} can be stored to prevent deficiency for up to four months. The liver is also capable of synthesizing vitamin A from carotene, found in tomatoes, carrots, and other vegetables.

(7)   Storage of iron

The liver stores iron in the form of a protein compound called **ferritin**. The iron is derived from the haemoglobin of worn-out red blood cells which have been destroyed in the spleen, and from iron absorbed by the small intestine. This reserve of iron is re-used for the formation of haemoglobin.

(8)   Production of clotting factors

The liver plays an important role in blood coagulation by forming many of the substances involved in the clotting process. These include fibrinogen, prothrombin and factor VII. Vitamin K is needed for the formation of prothrombin and factors VII, IX, and X.

(9)   Production of heat

The metabolic functions of the liver involve the expenditure of large amounts of energy, which is accompanied by the production of heat. This excess heat is distributed to the body by the blood stream and helps to maintain normal body temperature.

(10)   Detoxification

The liver is able to destroy or modify toxic substances in the body. Many drugs are chemically reduced to simpler non-toxic compounds for excretion, or are totally destroyed. Several hormones, including thyroxine, oestrogen and aldosterone, are either chemically altered or excreted by the liver.

The biliary tract

The biliary tract is the excretory apparatus of the liver and consists of:
 the common hepatic duct
 the gall-bladder
 the cystic duct
 the common bile duct

The common **hepatic duct** is formed by the junction of the right and left hepatic ducts which drain bile from the right and left lobes of the liver. It runs downwards from the porta hepatis for about 3 cm, where it is joined at an acute angle by the **cystic duct** from the gall-bladder and continues downwards closely related to the hepatic artery and the portal vein as the **common bile duct**. This passes behind the first part of the duodenum and then is buried in the head of the pancreas. It enters the duodenum at a small papilla called the ampulla of Vater, where it is joined by the pancreatic duct.

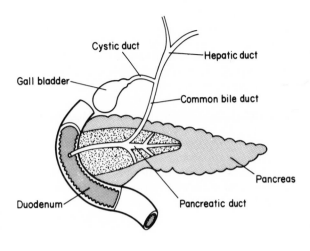

Fig. 12.28 Diagram of biliary tract and pancreas.

The gall-bladder

This is a pear-shaped sac, with a capacity of approximately 60 ml, attached to the under surface of the right lobe of the liver. The rounded end of the sac, the fundus, projects from beneath the inferior border of the liver.

Structure of the gall-bladder

The gall-bladder consists of three coats:

(*1*) An outer peritoneal coat continuous with the peritoneum covering the liver, which binds the gall-bladder in position on the under surface of the liver.

(*2*) A muscular coat, which contracts to enable the gall-bladder to empty its contents into the common bile duct.

(*3*) An inner coat of mucous membrane of columnar epithelial cells which secrete mucus. The mucous membrane is highly vascular and is thrown into rugae.

The neck of the gall-bladder is continuous with the cystic duct. The mucous membrane lining the neck and cystic duct projects into the lumen in oblique folds which form a spiral valve.

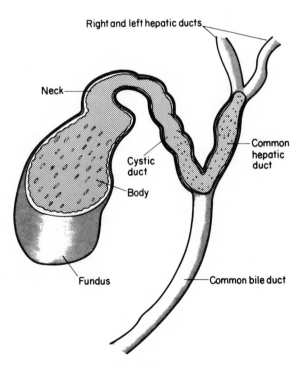

Fig. 12.29 Structure of gall-bladder.

Functions of the gall-bladder

The gall-bladder acts as a reservoir for the storage of bile, which after leaving the liver by the hepatic duct, passes up the cystic duct to the gall-bladder. The liver secretes between 600–1000 ml of bile per day, but the mucosal lining of the gall-bladder reabsorbs the fluid and electrolytes, thus concentrating the other bile constituents.

When food enters the duodenum, especially fat-containing food, the hormone **cholecystokinin** is secreted by the duodenal walls and passes in the blood to the gall-bladder, where it causes the muscular walls to contract and expel the bile. At the same time, duodenal peristalsis inhibits the sphincter of Oddi and causes it to relax and allow the bile to enter the duodenum.

Bile

Bile is the external secretion of the liver, and is produced in a dilute form, which is then concentrated by the gall-bladder to a viscid, greenish fluid. It is composed of:

>   water
>   bile salts (sodium glycocholate and sodium taurocholate)
>   bile pigment (bilirubin)
>   cholesterol
>   mucus

Bile salts have important functions in assisting the digestive action of the pancreatic enzymes, and in aiding the absorption of fats and fat-soluble vitamins from the small intestine.

These salts, by lowering surface tension, cause fats to break up or emulsify into small droplets, allowing the fat-digesting enzymes to work more efficiently and convert them into fatty acids and glycerol, in which form they are absorbed.

Bile salts do not appear in the faeces as they are reabsorbed from the small intestine and returned to the liver.

Bile pigments are derived from the breakdown of the haemoglobin of worn-out red blood cells, and give the bile its characteristic colour. The bile pigments are converted in the bowel to **urobilinogen** by bacterial action. Some urobilinogen is reabsorbed into the blood and is excreted by the kidneys into the urine. Exposure of urine to the air causes urobilinogen to be oxidized to urobilin. In the faeces, urobilinogen is altered and oxidized to form stercobilin which gives the faeces a dark brown colour.

If there is an obstruction to the excretion of bile, the bile pigments accumulate in the blood, giving the skin and mucous membranes a yellow colour (jaundice). At the same time they appear in the urine, which is turned a dark brown. Absence of bile pigments from the faeces results in pale clay-coloured stools, whilst absence of bile salts leads to an excess of foul-smelling fat in the stools.

The pancreas

The pancreas is a gland which has both exocrine and endocrine functions. Its exocrine function is the secretion of digestive enzymes. The endocrine secretions of the pancreas are two hormones, **insulin** and **glucagon**.

It is a greyish-pink gland lying transversely across the posterior abdominal wall at the level of the first and second lumbar vertebrae and is situated behind the stomach.

The pancreas is described as having a head, a body and a tail. The **head** is situated to the right and fits into the C-shaped curve of the duodenum. Buried in the substance of the gland is the termination of the common bile duct as it joins the pancreatic duct to form the ampulla of Vater, which is guarded by the sphincter of Oddi.

The **body** lies in front of the lumbar vertebrae, behind the stomach; the **tail** of the gland is in contact with the gastric surface of the spleen.

The anterior surface of the pancreas is covered by peritoneum.

Structure of the pancreas

The pancreas consists of a number of lobules, supported by fine loose connective tissue. Each lobule contains masses of secretory acini (cells arranged in a grape-like formation), lined with columnar epithelium. From the lobules small ducts emerge which unite to form larger ducts until they reach the main **pancreatic duct**, which extends from left to right in the centre of the organ and, in the head of the pancreas, joins the termination of the common bile duct to enter the duodenum at the ampulla of Vater.

Embedded between the acinar cells of the pancreas lie clusters of cells differing in character and appearance from those of the secretory epithelium. These are the **islets of Langerhans**,

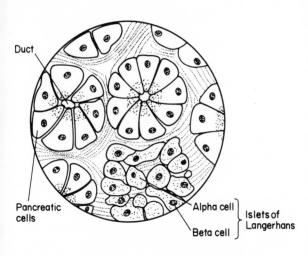

Duct

Pancreatic cells

Alpha cell
Beta cell

Islets of Langerhans

Fig. 12.30 Diagram of pancreatic cells.

which secrete the hormones insulin and glucagon directly into the blood.

Functions of the pancreas

Exocrine The acinar cells of the pancreas secrete digestive enzymes in an inactive form, which only become activated after their release into the pancreatic juice and the small intestine. When chyme enters the duodenum it causes the release of secretin and cholecystokinin into the blood which pass to the pancreas and stimulate the secretion of large quantities of fluids containing sodium bicarbonate and the release of the **digestive enzymes**, amylase, lipase, trypsinogen and chymotrypsinogen (p. 170). The pancreatic juice enters the duodenum by the pancreatic duct at the ampulla of Vater with the bile released from the gall-bladder.

The *endocrine* function of the pancreas is the secretion of insulin and glucagon. Insulin is important in the metabolism of carbohydrates, fats and protein (p. 186). Glucagon is concerned with the breakdown of liver glycogen (glycogenolysis), and with increased gluconeogenesis. It is also lipolytic (breaks down fat) and has numerous other actions.

Questions

1. What are the functions of the alimentary tract?
2. What is a digestive enzyme? List the enzymes of the alimentary tract together with their actions.
3. Give an account of the teeth in the child and the adult.
4. Describe the tongue and its functions.
5. State what you know of the functions of the mouth. Describe what you see when it is open.
6. Give an account of the salivary glands. What is the value of their secretions?
7. Describe (a) the structure of the oesophagus and (b) the act of swallowing.
8. Describe the anatomy of the stomach. What are its functions?
9. Describe the structure of the small intestine and comment on its functions.
10. Describe the large intestine and its functions.
11. Give an account of the peritoneum and mention any clinically important points related to it.
12. What is the largest organ in the body and what are its functions?
13. What do you know of the functions of the pancreas?
14. Describe the biliary tract. What is bile?

13
Metabolism and Nutrition

Energy is needed for the multitude of activities performed by the body. It is also required for growth, and repair of tissues. This energy is obtained from ingested food, which is first digested and absorbed, and finally metabolized.

Metabolism is the total of the chemical reactions which occur in the whole body. Metabolism consists of two major processes, catabolism and anabolism.

Catabolism is the *breaking down* of large molecules to smaller units to release energy and heat.

Anabolism is the *building* or synthesis of new compounds, and this process is energy consuming.

In the healthy adult there is a balance between catabolism and anabolism, which is called the energy-balance, (i.e. energy produced equals energy used). Both processes occur continuously and simultaneously, and consist of a series of precisely regulated chemical reactions.

The principal nutrients of the body are carbohydrates, fats and proteins. During the process of digestion, **carbohydrates** are converted to glucose, **fats** are converted to fatty acids and glycerol, and **proteins** are converted to amino acids. These molecules are able to enter the cells of the body tissues. Inside the cells the nutrients react chemically with oxygen under the influence of enzymes to release energy.

The energy liberated from the nutrients by oxidation is utilized to form a high energy compound, *adenosine triphosphate (ATP)*. This is synthesized and stored in the mitochondria of the cell, and thus provides an *energy reserve* which is available for cellular metabolism.

Attached to the nucleus of the ATP molecule are three phosphate groups, two of which are connected by high-energy bonds. ATP releases its energy by splitting one or both of these bonds. The splitting of the first bond yields energy and reduces ATP to adenosine diphosphate (ADP). Splitting of the second bond releases further energy and the ADP is degraded to adenosine monophosphate (AMP).

The ADP and AMP are rapidly reconverted to ATP by the mitochondria, using energy obtained

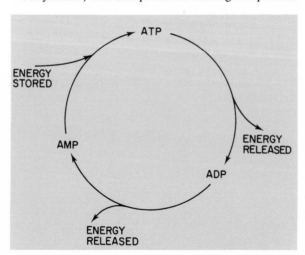

Fig. 13.1   ATP is degraded to ADP and AMP which are rapidly reconverted to ATP in a continuous cycle.

from the oxidation of nutrients. This ATP/ADP cycle can be repeated again and again.

The oxidation of the nutrients releases energy and heat. The heat is distributed to the whole body by the bloodstream and helps to maintain normal body temperature. Excess heat is dissipated from the body by the skin, by respiration and in the urine and faeces.

The *metabolic rate* is the rate at which energy is released in the body. A convenient method of estimating the energy requirements of the body is to measure it in terms of heat. The unit of heat used for this purpose is the **Calorie** (abbreviated to C). A Calorie (or kilocalorie) may be defined as the amount of heat required to raise 1 litre of water through 1 degree Centigrade.

The energy value of food is also measured in Calories:

1 gram of carbohydrate	= 4 Calories (16 kJ, or kilojoules)
1 gram of protein	= 4 Calories (17 kJ)
1 gram of fat	= 9 Calories (37 kJ)

Basal metabolic rate

The basal metabolic rate (BMR) is the rate of the body's energy expenditure under 'basal conditions'. This means the individual is at rest, mentally and physically, has not eaten for at least 12 hours (i.e. is in the post-absorptive state), and is in a warm comfortable environment. Under these conditions the metabolic needs of the body are at their lowest, energy being used only to sustain vital functions (e.g. breathing, the beating of the heart, maintenance of normal body temperature).

The basal metabolic rate can be calculated by estimating the amount of oxygen consumed in a given time. Since individuals vary greatly in size, the BMR is expressed in Calories per square metre of body surface per hour. The body surface area is calculated from measurements of an individual's weight and height.

Men oxidize food faster than women and therefore have a higher basal metabolic rate. For example, a male in his twenties has a BMR of about 40 C per square metre of body surface per hour, whilst a woman of the same age has a BMR of about 37 C per square metre of body surface per hour.

Factors influencing metabolism

Age The metabolic rate of children is relatively greater than that of adults due to high rates of cellular activity and growth. The BMR decreases with increasing age.

Exercise requires energy. Strenuous physical exercise can increase the metabolic rate as much as a hundred times that of the BMR of an individual for a few seconds at a time.

Body temperature An increase in body temperature increases the BMR. A decrease in body temperature results in a decrease in the metabolic rate and in oxygen consumption.

Environmental temperature The average metabolic rate of individuals living in tropical countries is considerably lower than that of people living in cold climates.

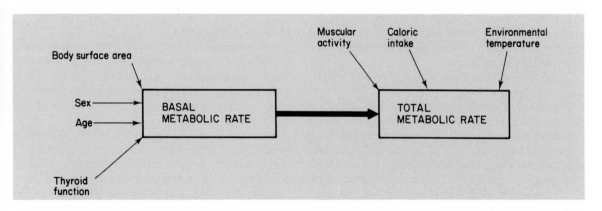

Fig. 13.2 Relationship between basal metabolic rate and total metabolic rate.

Thyroid hormone plays an important part in metabolism. When an excess of the hormone is secreted (thyrotoxicosis) the basal metabolism is increased; when there is a deficiency of thyroid hormone (myxoedema and cretinism) basal metabolism is slow.

Stimulation of the sympathetic nervous system as in fright or acute anxiety, causes a temporary increase in the metabolic rate in order that the body may cope with an emergency.

Drugs Certain drugs, such as the amphetamines, or caffeine, can increase the BMR.

Carbohydrate metabolism

Carbohydrate is the term used to describe starches and sugars. They are compounds of carbon, hydrogen and oxygen ($C_6H_{12}O_6$). Carbohydrates are absorbed through the villi of the small intestine as glucose, fructose and galactose. They are the primary energy source of the body. The fructose and galactose are converted to glucose by the liver and returned to the bloodstream.

Glucose is used to provide energy for cellular activity. Before glucose can be oxidized in the cells it must be transported across the cell membrane by a process called facilitated diffusion, which requires the use of a carrier substance. The presence of the hormone insulin accelerates this process. Failure of the pancreas to secrete sufficient insulin, as in diabetes mellitus, means that only very small amounts of glucose are able to enter the cells.

The maintenance of the blood glucose concentration at a relatively constant level is essential for the survival and function of brain cells. A fall in the blood glucose level rapidly results in disturbance of the central nervous system and, if uncorrected, leads to loss of consciousness and death within a few hours.

The blood glucose level is regulated by the liver and hormones.

Following a meal, glucose is rapidly absorbed and the blood glucose level rises above normal. As the blood passes through the liver, excess glucose is transported with the aid of insulin into the liver cells, where it is converted to glycogen and stored. This process reduces the blood glucose concentration to normal. A fall in the blood glucose level stimulates the pancreas to secrete the hormone glucagon, which causes the liver to reconvert glycogen to glucose and release it into the blood, thus raising the blood glucose to normal.

Several other hormones are also involved in accelerating the conversion of glycogen to glucose, these include adrenalin, glucocorticoids, ACTH, and thyroxine.

Gluconeogenesis
Glucose is stored mainly in the liver and muscles of the body, but these stores are limited. During periods of starvation the liver converts proteins or glycerol to glucose in order to maintain blood glucose levels at normal concentration.

Catabolism of glucose
The catabolism of glucose by cells to release energy occurs in two stages, an anaerobic stage called glycolysis, which does not require oxygen, and an aerobic stage for which oxygen is essential.

Glycolysis
Each molecule of glucose is split by a series of cellular enzymes into two molecules of pyruvic acid, with the release of a small amount of energy. This process is especially important when oxygen is in short supply. For example, during strenuous exercise the respiratory and circulatory systems may not be able to deliver sufficient oxygen to the muscle cells for the oxidation of glucose. However this process can only be used for very limited periods of time because it results in the accumulation of lactic acid, and the muscles incur an '*oxygen debt*'. Immediately following such exercise the oxygen debt must be repaid. Respiratory stimulation increases the respiratory rate to provide enough oxygen to reconvert lactic acid to pyruvic acid which is either oxidized to carbon dioxide and water for excretion, or converted back to glucose.

Oxidation of glucose
When adequate oxygen is available the pyruvic acid is converted to acetyl coenzyme A, which then combines with oxaloacetic acid to enter the *citric acid cycle* (Krebs cycle). The citric acid cycle completely oxidizes the glucose to carbon dioxide, water and ATP.

The conversion of one molecule of glucose to pyruvic acid (glycolysis) yields four molecules of ATP, whilst the conversion of pyruvic acid to acetyl coenzyme A and the citric acid cycle yield a further thirty-four molecules of ATP. Thus the greatest amount of energy is released from glucose by the oxidative process.

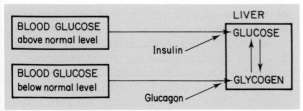

Fig. 13.3   Blood glucose levels are regulated by the liver and hormones.

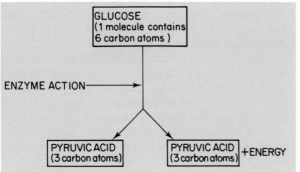

Fig. 13.4   Glycolysis (an anaerobic process) is the first stage of carbohydrate metabolism.

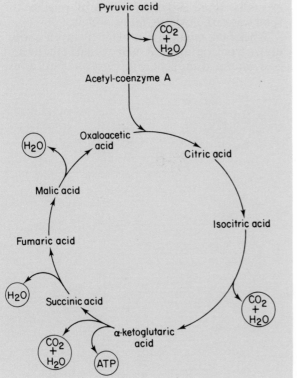

Fig. 13.5   Simplified diagram of citric acid cycle (Krebs cycle).

Fat metabolism

Fats are absorbed as fatty acids and glycerol into the lacteals of the villi in the small intestine, where they are recombined to neutral fats and transported to the bloodstream by the lymphatic vessels. Fats are compounds containing carbon and hydrogen with very little oxygen. They constitute a more concentrated energy source than carbohydrates since catabolism of 1 gram of fat yields 9 Calories of heat and energy, whilst 1 gram of carbohydrate yields only 4 Calories.

Catabolism of fats

Neutral fats are split by the liver to glycerol and fatty acids. Glycerol is similar to the breakdown products of glucose, and can be converted to a compound that enters the glycolytic pathway. Fatty acids are oxidized to acetyl coenzyme A and, provided that carbohydrates are being metabolized, can enter the citric acid cycle. Both carbohydrates and fats enter the citric acid cycle by combining with oxaloacetic acid, which is manufactured from carbohydrate. Thus carbohydrates can supply their own oxaloacetic acid, but fats cannot, and must use that produced by carbohydrates to be fully oxidized.

If no carbohydrate is available, as in starvation, or when carbohydrate cannot be metabolized as in diabetes mellitus, then the acetyl coenzyme A molecules from fatty acids accumulate in the body. The liver cells condense the acetyl coenzyme A molecules to form acetoacetic acid, some of which is converted to beta-hydroxybutyric acid and acetone. These three substances are collectively known as **ketone bodies** or ketoacids. Accumulation of ketone bodies in the blood is called *ketosis*, and when severe can cause coma and death. The presence of ketone bodies can be detected on the breath and in the urine as acetone.

Any fat which is not immediately required for heat and energy is stored in the fat depots of the body. The fat depots are the body's largest energy reserve source, and are found under the skin, in the folds of the omentum, between the muscles and around various organs such as the kidneys. This adipose tissue also provides heat insulation for the body.

Excess glucose and amino acids are converted by the liver to fat, which is also stored in the fat depots. Thus when an individual takes in more Calories than are required for metabolic needs, the excess is stored as fat regardless of whether the food was carbohydrate, fat or protein.

When fat is required to provide energy, it is withdrawn from the fat depots and carried to the liver to be desaturated and split to glycerol and fatty acids.

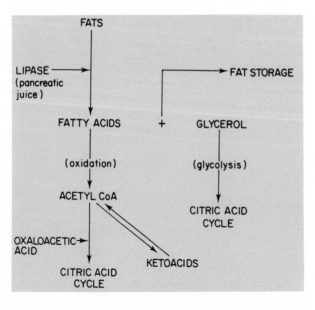

Fig. 13.6 Fat catabolism.

Phospholipids and cholesterol

These two substances are present in large quantities in the body, and are synthesized in all the cells of the body. Both are fat-soluble and only slightly water-soluble.

Phospholipids are complex compounds which appear to be important in the transport of fats in the blood, and in the formation of cell membranes and the structure of the myelin sheaths of the nerve fibres.

Cholesterol is composed mainly of a sterol nucleus synthesized from acetic acid. Cholesterol is found in cell membranes, and is also used by the adrenal glands in the formation of adrenocortical hormones, by the ovaries to form oestrogen and progesterone, and by the testes to form testosterone.

A large amount of cholesterol is present in the horny layer of the epidermis, where, with other lipids, it contributes to the 'water tightness' of the skin.

Atherosclerosis is a disease of the arteries in which fatty deposits containing large quantities of cholesterol and other lipids develop in the arterial walls. Fibrous tissue grows into the deposits and they become calcified, leading to narrowing of the lumen and 'hardening of the arteries'.

Protein metabolism

Proteins are complex organic compounds containing carbon, hydrogen and oxygen with the addition of nitrogen, sulphur and phosphorus. Large protein molecules consist of many amino acids linked together to form chains known as **peptides**. The links between the chains are called *peptide bonds*.

Proteins are essential to the body, for they provide most of the structural elements of the cells and the enzymes which are necessary for all biochemical reactions.

Twenty-one amino acids have been identified and named. Some of these are known as **essential amino acids** because they cannot be synthesized in the human body and therefore must be supplied in food. The other amino acids are known as non-essential because, although they are vital to life, they do not need to be supplied in food, but can be synthesized by the body itself.

Every cell in the body is capable of manufacturing the proteins it needs, and this process is controlled by the genes in the cell nucleus. Protein anabolism thus produces many substances necessary for healthy body function. Among these are the plasma proteins synthesized by the liver, which are essential for the maintenance of fluid balance and osmotic pressure. The plasma proteins

Table 13.1

Essential amino acids (those which must be supplied in food)	Non-essential amino acids (those which do not need to be supplied in food since they can be synthesized in the body)
Arginine	Alanine
Histidine	Aspartic acid
Leucine	Cystine
Isoleucine	Glutamic acid
Lysine	Glycine
Methionine	Hydroxyproline
Phenylalanine	Proline
Threonine	Serine
Tryptophan	Tryosine
Valine	Cysteine
	Hydroxylysine

prothrombin and fibrinogen play vital roles in blood coagulation. Protein synthesis is also important for the production of the gamma globulin of antibodies essential in the defence of the body against infection, and for growth and the replacement of cells destroyed by daily wear and tear.

When the blood concentration of amino acids is high, some are absorbed into the liver cells and stored. As the amino acid concentration of the blood falls to below normal the stored amino acids are released from the liver back into the blood. To some extent most of the other body cells can also store amino acids. The amino acids stored in this way are referred to as the amino acid pool.

Surplus amino acids are utilized for heat and energy. The first stage of protein catabolism is *deamination*, and takes place in the liver. The nitrogen portion is split off from the amino acid molecule to form ammonia, which in turn combines with carbon dioxide to form urea in the liver. The urea is excreted by the kidneys into the urine.

The deaminated amino acids can be directly oxidized to form carbon dioxide and water with the release of heat and energy, or they may be converted to fat or carbohydrate.

Normally a state of *nitrogen balance* exists in the healthy adult body, i.e. the rate of protein intake equals the rate of protein utilization. A negative nitrogen balance occurs when protein utilization is greater than the protein intake, and results in a state of 'tissue wasting'. Malnutrition and debilitating diseases such as thyrotoxicosis may cause a negative nitrogen balance.

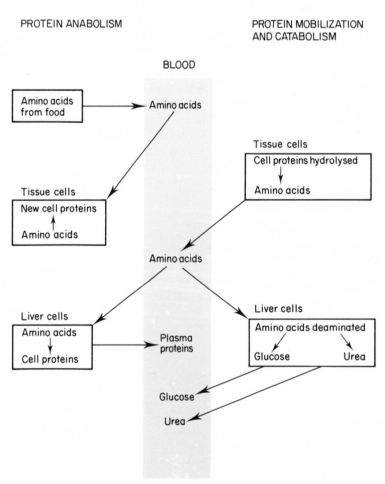

Fig. 13.7 Diagram illustrating the main features of protein metabolism.

A positive nitrogen balance occurs when protein intake is greater than protein utilization. This happens during growth, pregnancy, and recovery from a debilitating illness.

Protein metabolism is controlled by hormones. Growth hormone and testosterone have a stimulating effect on protein synthesis and are sometimes known as anabolic hormones. Glucocorticoids have a profound effect on protein catabolism. Thyroid hormone is also necessary for protein anabolism, but excessive thyroxine secretion stimulates protein mobilization and catabolism.

Mineral metabolism

A number of mineral salts play an important part in the vital processes of metabolism. Some are present in relatively large amounts, whilst others are needed only in minute quantities, and are known as *trace elements*. These inorganic compounds must be derived from food.

Calcium is found mainly in dairy products and green vegetables. In the UK, calcium carbonate is added to flour and is therefore also present in bread and cakes.

Growing children need more calcium than adults, and deficiency results in poor growth, rickets and badly formed teeth. Pregnant women also require extra calcium to form the developing bones and teeth of the fetus.

Calcium is most plentiful in the body in the form of calcium phosphate in the bones. A small constant concentration of calcium is maintained in the blood and plays an important role in blood coagulation and normal muscle function. The bones act as a reservoir for calcium so that when the blood level falls below normal, calcium is withdrawn from the bones.

The presence of *vitamin D* is necessary for the absorption of calcium from the intestine. Large amounts of undigested fats in the intestine inhibit calcium absorption because the calcium forms salts with the fat which cannot be absorbed.

The metabolism of calcium is controlled by parathormone from the parathyroid glands and the thyroid hormone calcitonin. Damage to the parathyroid glands may lead to low blood levels of calcium which increase the excitability of nerve fibres and result in a condition called tetany.

Phosphorus is present in many foods, especially dairy products, liver and kidney. Phosphate is extremely important in the body, for not only is it combined with calcium in the formation of bones and teeth, it is the principal anion of intracellular fluid and is essential in a large number of chemical reactions in the body (e.g. in adenosine triphosphate).

Magnesium is present in many foods, especially those of vegetable origin. It is an essential constituent of all cells but most of the body's

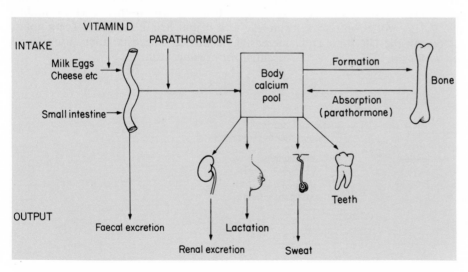

Fig. 13.8 Diagram illustrating the main features of calcium metabolism.

magnesium is in the bones. It is required for the functioning of some of the enzymes involved in the utilization of energy. Deficiency of magnesium is rare but may result from losses in the stools, in patients with diarrhoea or, together with potassium, in the urine of patients taking diuretics.

Sodium is present in most foods, and is also added to food as salt (sodium chloride) during cooking and at the table. Sodium is the principal extracellular cation in the body and maintenance of normal concentrations are essential to life, for it is involved in the electrolyte and fluid balance of the body, in the transmission of nerve impulses and muscular contraction.

Sodium is lost from the body in urine and sweat. Excretion of sodium in the urine is controlled by hormones from the pituitary gland and the adrenal glands, but there is less control over the loss of sodium in sweat. Thus extra salt intake may be needed in conditions which cause increased sweating (e.g. in strenuous physical exercise, high environmental temperatures).

Potassium is the principal intracellular cation, and is present in most foods. It is required for the chemical activities of cells and, like sodium, is needed for the maintenance of electrolyte and fluid balance, transmission of nerve impulses and muscular contraction. Potassium is excreted in urine but, unlike sodium, is not lost in sweat.

Iron is found in foods of animal origin (especially liver, kidney and beef), in egg yolks, potatoes, green vegtables and carrots. In the UK flour and white bread are fortified with iron.

Iron is essential for the manufacture of haemoglobin in the red blood cells, and is needed by all body cells in small amounts. The body is very economical in its use of iron, for when red blood cells are destroyed the iron is extracted and recycled for use in new erythrocytes. Iron is lost from the body in cells which are shed from the skin, mucous membranes and gastro-intestinal tract, and the loss amounts to about 1 mg per day. It is also lost when bleeding occurs, and women in the child-bearing years have a greater loss of iron than men due to menstrual blood loss, pregnancy and childbirth.

Absorption of iron occurs mainly in the duodenum and jejunum, and is aided by hydro-chloric acid and the presence of vitamin C (ascorbic acid).

Iodine is a *trace element*, needed only in minute quantities but essential for the formation of thyroid hormones, which in turn are essential for normal metabolism. Iodine is found mainly in sea-fish and shell-fish, and in vegetables grown in soil containing iodine. In areas where soil and water are deficient in iodine it is added to table salt to prevent goitre.

Fluorine is a trace element found in bones and teeth, where it helps to prevent dental caries (dental decay). The only important nutritional sources are drinking water, tea and fish, especially those fish of which the bones are eaten. The natural content of fluorine in water is often below the optimal level of 1 part per million (1 mg per litre) and some water authorities bring it up to this level by adding fluoride as a dental health measure. Teeth apart, increasing evidence suggests that fluorine is essential for life and health.

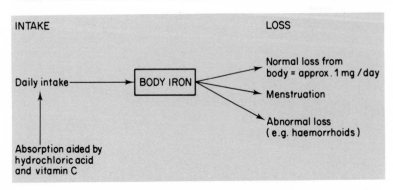

Fig. 13.9 Intake and loss of iron.

Some other trace elements are essential for human metabolism and include copper, zinc, manganese, cobalt and chromium. These elements and needed only in minute amounts.

Vitamins

Vitamins, or accessory food factors, are organic compounds which are needed in small quantities for normal metabolism. They are not oxidized to supply energy and do not form part of the body structure. Although some vitamins can be manufactured in the body, the amounts are insufficient and these vitamins must be supplemented by the vitamins taken in the diet.

Vitamins A, D, E and K are fat-soluble, whilst vitamins of the B group and vitamin C are water-soluble.

Vitamin A (retinol)

This is found in animal foods such as liver, dairy produce, eggs and fish-oils. Yellow and green vegtables contain substances called carotenes, which can be converted by the liver to vitamin A.

Vitamin A is used to synthesize the light-sensitive pigments of the retina, and is necessary for normal growth of most cells of the body, especially epithelial cells.

Lack of vitamin A manifests in (a) night-blindness; (b) dry, scaly skin which becomes susceptible to infection; (c) hardening or keratinization of the cornea which may result in blindness; (d) atrophy of the germinal epithelium of the gonads.

Vitamin D

This increases calcium absorption from the gastro-intestinal tract and is concerned with the deposition of calcium and phosphorus in bones. It is particularly important in infants and children for development and growth of bones and in the pregnant woman for development of the bones and teeth of the fetus.

Several vitamin D compounds exist. Vitamin D_2 (ergocalciferol) and vitamin D_3 (cholecalciferol) are closely related. A substance called 7-dehydroxycholesterol, present in the skin, is converted in the skin to vitamin D_3 on exposure to the ultraviolet rays of sunlight. Foods containing vitamin D include fatty fish, margarine, butter and eggs. The vitamin D compounds must be altered to 1,25-dihydroxycholecalciferol by a complicated process that involves first the liver and then the kidneys, under the influence of parathormone. 1,25-dihydroxycholecalciferol is the active form of vitamin D. In the absence of the kidneys the active form of the vitamin is not produced.

Lack of vitamin D results in rickets in children and osteomalacia in adults.

Vitamin E (tocopherols)

This includes several related compounds, found mainly in wheat germ, vegetable oils, peanuts, milk, butter and eggs. In animals, lack of vitamin E causes muscle weakness and paralysis (especially of the hind quarters), anaemia and infertility in both sexes. The effects of this vitamin in the human body are not yet clear, but deficiency prevents normal growth.

Vitamin K

This is necessary for synthesis of clotting factors by the liver (including prothrombin, and factors VII, IX and X). Vitamin K consists of a number of naturally occurring substances, which are present in many foods, and are synthesized by bacteria in the large intestine.

Absorption of vitamin K requires the assistance of bile salts, and any obstruction which prevents bile from being produced or entering the duodenum leads to deficiency of the vitamin within a few days. Administration of oral antibiotics may destroy the bacteria in the gut and may lead to vitamin K deficiency.

Vitamin B complex

This comprises a number of chemically distinct compounds, which are often but not always present in the same foods (see Table 13.2) Some of these vitamins act as coenzymes in the body (i.e. they work closely with an enzyme and are vital to its function).

Thiamine (vitamin B_1) is essential for many biochemical reactions, but is especially important for the conversion of pyruvic acid to acetyl coenzyme A. Without thiamine the oxidative processes in the metabolism of carbohydrate and fats become deficient. This results in widespread abnormalities in the body which particularly affect the nervous system, the gastrointestinal tract and cardiac function.

Riboflavin (vitamin B$_2$) is required for the oxidation of foodstuffs. Lack of this vitamin is associated with soreness of the lips and keratitis (inflammation of the cornea).

Nicotinamide (vitamin B$_3$), sometimes also called niacin, is essential for carbohydrate oxidation and the citric acid cycle. The body is able to manufacture nicotinamide from the amino acid tryptophan. Deficiency causes a disease called *pellagra*, which is characterized by gastrointestinal disturbances, lesions in the central nervous system with mental changes, and scaly, pigmented skin in areas exposed to sunlight or mechanical irritation.

Pyridoxine (vitamin B$_6$) functions as a coenzymes for many chemical reactions in amino acid and protein metabolism.

Cyanocobalamin (vitamin B$_{12}$)

Unlike other vitamins, cyanocobalamin (vitamin B$_{12}$) contains a metal, *cobalt*. Vitamin B$_{12}$ performs several metabolic functions, acting as a coenzyme. It is essential for the maturation of red blood cells, and deficiency of the vitamin leads to pernicious anaemia and demyelination of the nerve fibres of the spinal cord (a condition known as subacute degeneration of the spinal cord).

The cause of vitamin B$_{12}$ deficiency is rarely lack of the vitamin in the diet, but often lack of a substance called *intrinsic factor* produced by the mucosal cells of the stomach, which combines with vitamin B$_{12}$ in the food. Intrinsic factor combined with the vitamin is bound to the wall of the small intestine and the vitamin is then absorbed.

Folic acid

Folic acid is essential for the synthesis of nucleoproteins and for cell division. It is widely distributed in food, and is also produced by bacteria in the large intestine.

Pantothenic acid

This is used by the body to form acetyl coenzyme A, thus deficiency of this vitamin leads to depressed metabolism of carbohydrates and fats.

Vitamin C (ascorbic acid)

Vitamin C is essential for many oxidative reactions of metabolism, and maintains normal intercellular substances in the body, including the formation of

Table 13.2

Vitamin	Source
(1) Fat soluble	
A (retinol)	Liver, dairy produce, eggs and fish-oils
D	Fish, margarine, butter and eggs, sunshine
E (tocopherols)	Wheatgerm, vegetable oils, peanuts and dairy products
K	Present in many foods; synthesized by bacteria in large intestine.
(2) Water soluble	
B$_1$ (thiamine)	Cereals, wholemeal flour, yeast, peas and beans
B$_2$ (riboflavin)	Meat, milk and wholemeal flour
B$_3$ (nicotinamide)	Meat, liver and wholemeal flour
B$_6$ (pyridoxine)	Present in many foods
B$_{12}$ (cyanocobalamin)	Liver, meat and animal products
Folic acid	Present in many foods; synthesized by bacteria in large intestine
Pantothenic acid	Liver, meat, eggs and milk
C (ascorbic acid)	Fresh fruit and vegetables

connective tissue, and the intercellular matrix of bones and teeth.

Deficiency of vitamin C leads to a disease called *scurvy*. Wounds fail to heal because connective and fibrous tissue are not formed. Bone growth ceases, and blood capillaries become fragile, resulting in bleeding into the gums, skin and joints.

Vitamin C is also required for the absorption of iron, and the normal function of folic acid; thus anaemia is another feature of deficiency of this vitamin.

Water

Water forms approximately 60% of the body weight and forms the basis of the body fluids (blood plasma, lymph and tissue fluids). Water is an essential constituent of all the body tissues and cells. Many chemicals are dissolved in water, while others are in suspension, and thus can be transported between the blood plasma and the tissue fluids, and then between the tissue fluids and the cells.

Water is a very important part of the diet, because, although it has no nutritional value, a lack of water can lead to dehydration and death within a few days. Many foods consist largely of water, but in addition 1.5 litres of fluids a day are necessary to ensure an adequate fluid intake.

Water passes readily through the walls of the stomach and intestines by diffusion and osmosis (p. 8). During metabolism, the oxidation of nutrients results in the production of carbon dioxide and water, and in the healthy adult this amounts to about 1–1.5 litres of water a day.

The control of body temperature is assisted by the evaporation of water as sweat. Water is also excreted by the kidneys as urine, by the large intestine in faeces and by the lungs as water vapour.

Fibre

Adequate peristalsis in the bowel only occurs when there is a sufficient food residue for the muscle in its walls to act upon. This is because the normal stimulus to peristaltic action is the stretching of the walls by the bowel contents. The undigested faecal residue which performs this function consists of fibre and in the normal diet it is mainly provided by the cellulose found in vegetables, salads, fruit and wholemeal bread. Insufficient fibre leads to sluggish peristalsis and a tendency to constipation.

Nutrition

Nutrition is the supply of food to the tissues and its absorption and metabolism for growth, energy, maintenance and repair of the living body.

For maintenance of a healthy body, food must include carbohydrates and fats, which can be oxidized to provide heat and energy, in sufficient quantities to suit an individual's metabolic needs. Proteins are required to supply essential amino acids to provide materials for growth and repair of body tissues. Minerals and vitamins are needed to provide the special compounds required for the chemical reactions of metabolism.

The nutrients described above are present in varying proportions in different foods, and to be satisfactory any diet must contain adequate proportions of all the nutrients required by the body. Thus the requirements of a balanced diet will vary throughout an individual's life-cycle. A child grows most rapidly in the early years of life and growth can only occur if the organs and tissues receive the nutrients needed for synthesis of their protein and cellular structures. A newborn infant needs about five times as much protein as the adult per unit of body weight. As the child grows older the rate of growth slows down, and after puberty the need for nutrients gradually changes, although a greater protein intake is needed until growth ceases. An inadequate supply of energy and protein are the commonest causes of failure to grow.

Each individual requires food to supply sufficient Calories to maintain the basal metabolic rate and to supply energy for the activities of everyday life. The average adult needs about 25 Calories per kilogram of body weight per 24 hours for basal metabolism, i.e. a man weighing 63 kg (10 stone) needs 1575 Calories per day (63 kg × 25 C = 1575 C). Additional Calories are needed for his daily activities, the amount of which will depend on his occupation and physical activity; thus if he has a sedentary job he will need a total of approximately 2700 Calories per day, but if his occupation is moderately active this would increase to 3000 C, and if very active to 3600 C per day.

Women, because they have a lower basal metabolic rate than men, need between 2200 and 2500 Calories depending on their physical activity.

The energy requirements of the individual are thus dependent on body size and composition, age, sex, physical activity, climate and environment.

The nutritional value of a protein depends on the

amino acids that it contains. Proteins which contain all the essential amino acids are called *complete* or **first class proteins**, whilst those that lack one or more of the essential amino acids are called *incomplete* or **second class proteins**. First class proteins are mainly those derived from foods of animal origin such as meat, fish, milk and eggs. Second class proteins are mainly derived from vegetables.

Fats are the most concentrated source of energy in the diet, providing more than twice as much as proteins or carbohydrates. Fats are of animal or vegetable origin. Animal fats are found in dairy products, meat and oily fish. Vegetable fats are found in olive oil, vegetable oils and margarine. Carbohydrates represent the cheapest and most commonly occurring class of food in the diet. They are widely distributed in vegetable foods and fruit and include sugars, starches and cellulose.

Dietary intake of food varies widely, due partly to the availability of some foodstuffs. For example, in the UK protein forms approximately 15 per cent of food intake, with approximately 30 per cent fats and 55 per cent carbohydrates. However in the underdeveloped areas of the world the intake of protein and fats may be as little as half these values.

Religious, cultural and traditional practices also affect the diet of many people throughout the world.

Control of food intake

Within the lateral hypothalamus are a group of neurone that function as an **appetite centre**, and stimulation of these cells results in sensations of hunger. Hunger means a craving for food. When a person has not eaten for some hours the stomach is stimulated to contract rhythmically, causing 'gnawing' sensations, which are called *hunger pangs*. These are relieved when food is eaten.

Appetite is the term used when there is a desire for specific types of food.

A cluster of neurones in the medial hypothalamus is thought to function as a **satiety centre**, and stimulation of these cells inhibits food intake.

Both centres may be affected by blood levels of nutrients and by hormones.

Obesity

Obesity results from an imbalance between energy intake and energy output. When more energy, in the form of food, is taken into the body than is used in energy expenditure the surplus energy is converted to fat and stored as adipose tissue (for each 9 Calories of excess energy 1 gram of fat is stored). Thus most cases of obesity are caused by overeating. This may be due to habit rather than hunger, but rarely may also arise due to an abnormality of the appetite centre in the hypothalamus.

Starvation

During a period when very little or no food is being taken into the body, stores of carbohydrate (i.e. the glycogen stored in the liver and muscles) are used up within the first 24 hours. After this the individual starts to break down stored fats to provide the energy necessary for cellular function. When the store of fats is almost depleted, the body begins to use protein to supply energy. However proteins are needed for cellular structure and chemical activities, and continued depletion of protein for use as energy results in death when the proteins of the body have been reduced to one-half their normal level.

Questions

1. How is the energy value of food expressed? Give examples.
2. What is meant by metabolic rate and which factors may influence it?
3. What is a carbohydrate? What is its importance in human nutrition?
4. What are essential amino acids? What is meant by the term nitrogen balance?
5. What is a trace element? Give four examples and state the clinical importance of two of them.
6. What are vitamins? Mention some of the sources and the value to health of any two of them.
7. What are the essential constituents of a normal diet? Discuss the importance of the substances you mention.
8. Name good sources of the following and explain their importance in normal nutrition: (a) first-class protein, (b) vitamin C, (c) iron, (d) calcium.

14
The Endocrine System

Many of the activities of the body are controlled by the nervous system and one of the characteristics of this system is the rapidity of its response to various kinds of stimuli. There is, however, a second major system which exercises control over the body's activities, especially those of a slower character, such as growth, which are manifest over long periods. The organs of this second system are called the **endocrine** or **ductless glands**. They produce special chemical substances called *hormones* which they secrete directly into the bloodstream. They are, therefore, sometimes referred to as the organs of internal secretion.

A hormone may be defined as a chemical messenger, secreted by a ductless gland, which is transported by the circulation to its target cells in distant organs, which it is thereby able to influence. An alternative, simpler definition of a hormone, is that it is a circulating substance which acts on organs distant from its site of origin.

In some instances an organ may have both an internal secretion which enters the blood directly and an external secretion which leaves it by a duct. The pancreas (p. 00) is an example. The internal secretions of the pancreas, insulin and glucagon, pass into the blood while the pancreatic juice reaches the duodenum via the pancreatic duct. The pancreas is therefore said to have both *endocrine* and *exocrine* functions.

Numerous hormones have been isolated and their chemical structure elucidated. An increasing number can be prepared artificially in the laboratory either by direct chemical synthesis or by recombinant DNA technology, which is already used commercially to produce human insulin on a large scale.

Some hormones, such as adrenalin, have an immediate action. Others, especially the pituitary growth hormone, exercise their influence over many years.

The function of the ductless glands has been studied in a number of ways. Before any hormones were isolated, some knowledge of their action was gained by observing the effects of disease. It was found that two sets of symptoms existed, namely those produced by excessive activity of the gland and over-secretion (*hypersecretion*) of hormones and those resulting from under-activity and under-secretion (*hyposecretion*). Clearly, therefore, the action of a gland in health is to maintain a balance between these effects.

Another method of study is to obtain the hormone itself either by recovering it from the gland or manufacturing it in the laboratory and to consider the effects produced by its administration both in health and in disease.

A third method is to study the results of the destruction or removal of the gland in animals.

The major endocrine glands may be arranged in two groups:

(1)   (a) the anterior pituitary gland
        (b) the adrenal cortex
        (c) the thyroid gland
        (d) the sex glands or gonads

(2)   (a) the posterior pituitary gland
        (b) the adrenal medulla
        (c) the parathyroid glands
        (d) the pancreas

The anterior pituitary gland controls the other members of the first group via the intermediary of *trophic* hormones (such as ACTH or corticotrophin), which it secretes. The anterior pituitary gland is itself under the control of the hypothalamus (see p. 233) which secretes *releasing* and *inhibiting* hormones (e.g. corticotrophin-releasing factor).

The glands of the second group are controlled by other stimuli, both chemical and neural.

The gastrointestinal tract also produces certain internal secretions, all of which are peptides and some of which, such as gastrin and secretin, fulfill the criteria necessary for their inclusion as hormones. Some of them are present in both the gastrointestinal tract and within the nervous system, where they appear to act as neurotransmitters. These are known as the 'brain-gut peptides' and include enkephalins (see p. 242), somatostatin and vasoactive intestinal peptide (VIP).

Disorders of the thyroid gland are common and relatively easy to understand; this gland is therefore considered first.

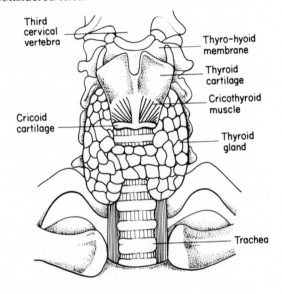

Fig. 14.1 The larynx, the thyroid gland and the cervical portion of the trachea.

The thyroid gland

The thyroid gland is situated in the lower part of the neck and when enlarged it forms the once familiar goitre. It consists of **two lobes**, one on either side of the trachea, joined together by an **isthmus** which passes in front of the trachea just below the cricoid cartilage. The lobes are conical and have upper and lower poles, the upper pole extending to the side or wing of the thyroid cartilage. It receives its plentiful blood supply from the superior and inferior thyroid arteries which are respectively branches of the external carotid and subclavian arteries. In the groove between the trachea and oesophagus on each side of the neck, the recurrent laryngeal nerve lies in close relationship to the thyroid gland and may be damaged by a carcinoma of the gland or by thyroid surgery, resulting in a change in the voice.

Microscopically, the thyroid contains two types of hormone-producing cell, the **follicular cells** which produce *thyroid hormones* and the **C (clear) cells** which produce *calcitonin*. The shape of the follicular cells depends upon whether or not they are being stimulated by thyrotrophin (thyroid stimulating hormone, TSH), derived from the pituitary and circulating in the blood.

The thyroid gland stores large amounts of thyroid hormones in an inactive form, called *colloid*, within compartments called **follicles**, which are lined by follicular cells. The colloid consists of *thyroglobulin* and is produced by the follicular cells. These cells are also responsible for

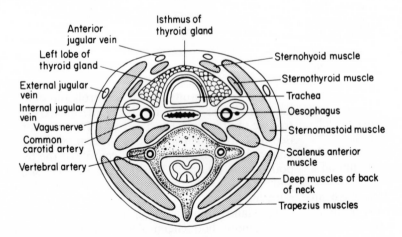

Fig. 14.2 Transverse section of the neck at the level of the seventh cervical vertebra, showing the position of important structures.

converting it into the thyroid hormones, *tri-iodothyronine (T₃)* and *thyroxine (T₄)* and releasing them into the bloodstream. Thyroxine is the main hormone secreted by the thyroid but is itself relatively inactive. The active hormone is tri-iodothyronine and, by its action on cells, it regulates the basal metabolic rate and influences growth and maturation, particularly of nerve tissue.

Both thyroid hormones contain a high proportion of *iodine* which the thyroid gland 'traps' from the blood. It is particularly this stage (the trapping of iodide) of thyroid hormone production that is influenced by TSH. In turn the release of TSH from the anterior pituitary is stimulated by *thyrotrophin-releasing hormone (TRH)* secreted by the hypothalamus. There are feedback mechanisms to inhibit the secretion of TSH and TRH (see Fig. 14.3).

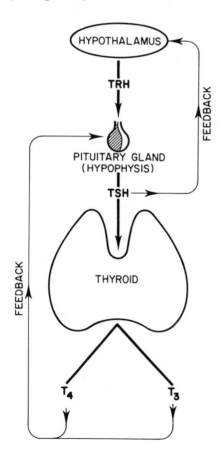

Fig. 14.3 The hypothalamus − hypophyseal control of thyroid function.

The iodine required for thyroid hormone synthesis is obtained from foods, especially sea foods, and drinking water. In all continents, there are regions in which the soil is deficient in iodine. If these are inland regions, in which sea fish are rarely eaten and the population lives mainly on local foods, the iodine deficiency will manifest itself among the inhabitants by enlargement of the thyroid gland to form a *goitre*. At one time, this was prevalent in the Peak District of England and was known as 'Derbyshire Neck.' Such 'endemic' goitres may still be found in the Andes and the Himalayas.

A goitre may also be found in some patients who have no dietary deficiency of iodine but who have thyroid disease. It is most often found with hypersecretion of the gland, which causes thyrotoxicosis, in which there is loss of weight despite a good appetite, indicating that the rate of general metabolism is increased. Other clinical features are nervousness and restlessness, increased sweating, hot moist hands, a rapid pulse, a tremor, and protrusion of the eyeballs (exophthalmos).

Underactivity (hyposecretion) of the thyroid gland produces cretinism in infants and myxoedema in adults. In contrast with the appearance of extreme nervousness seen in hyperthyroidism, the patient often has a mentally dull appearance and there is slowness of speech and movement. The skin is dry and thick and the pulse is slow. The body temperature tends to be subnormal and these patients are at risk from hypothermia. The metabolism of the body is slowed down and this is reflected in the reduced level of bodily activity. Myxoedema and cretinism are both treated by administering thyroxine by mouth, thereby supplying the missing factor; this is called substitution therapy.

The C cells, or parafollicular cells of the thyroid gland produce the hormone *calcitonin* which lowers blood calcium levels. Its secretion is controlled solely by the level of calcium in the blood, providing another example of a feedback mechanism.

The parathyroid glands

These are small ovoid glands, smaller than peas, which lie on the posterior surface of the thyroid gland. Usually there are two pairs of parathyroid glands, a superior pair and an inferior pair. The chief cells of the parathyroid glands secrete a hormone called *parathormone*, which raises serum calcium levels. The activity of the parathyroid gland is controlled by alterations in the level of calcium in the blood. A raised blood calcium level inhibits the secretion of parathormone and a low blood calcium level stimulates secretion of the hormone. This is the main feedback mechanism for the regulation of blood calcium levels, thyroid calcitonin secretion playing a lesser role. The maintenance of blood calcium levels within narrow limits is essential for the normal function of cells, most obviously those of the heart, skeletal muscles and nerves.

Decreased activity of the parathyroid glands (hypoparathyroidism) results in a low blood calcium level and the condition known as tetany. This is characterized by muscular spasms and increased irritability of the nervous system. Increased activity of the glands (hyperparathyroidism) causes an increase in the levels of calcium in the blood and urine and results in renal calculi and bone disease (osteitis fibrosa).

The suprarenal glands

The suprarenal or adrenal glands are two small flattened yellowish bodies situated on the upper pole of each kidney. Each is about 5 cm (2 in) high, 3 cm (1.2 in) wide and 1 cm (0.4 in) thick and is plentifully supplied with blood by three arteries, derived from the aorta, the renal artery and the inferior phrenic artery. The profuse sympathetic nerve supply from the coeliac plexus goes almost entirely to the medulla, which is the dark interior of the adrenal gland. The outer part of the gland, the cortex, is yellowish, because it is rich in lipids, and has entirely separate functions from the medulla.

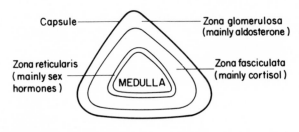

Fig. 14.4 The suprarenal gland – general structure and function.

Functions of the suprarenal cortex

The hormones of the suprarenal cortex are derived from cholesterol and belong to a class of fatty or wax-like substances called *steroids*. They fall into three main groups corresponding to their main effects:

(1) The mineralocorticoids These regulate the body sodium and potassium ion levels and therefore help to maintain the fluid and electrolyte balance of the body. The principal mineralocorticoid is *aldosterone* and of lesser importance is corticosterone. They act on the collecting tubules of the kidney in such a way that sodium is retained in the body and any excess of potassium is excreted.

(2) The glucocorticoids These derive their name from their influence on carbohydrate metabolism. The principal glucocorticoid is *cortisol (hydrocortisone)* and others, of lesser importance, are corticosterone and 11-deoxycortisol. Among their main effects are:

   (*a*) increased output of glucose from the liver into the blood;
   (*b*) increased breakdown (catabolism) of protein;
   (*c*) liberation of lipid from tissue stores and redistribution of adipose tissue;
   (*d*) suppression of growth hormone release and activity;
   (*e*) diminution of eosinophils and lymphocytes in the blood;
   (*f*) immunosuppression;
   (*g*) suppression of inflammation;
   (*h*) enhancement of water diuresis.

Cortisone (which is converted to hydrocortisone in the body) and many similar synthetic substances (analogues), such as prednisone and prednisolone, are used in clinical medicine for many different purposes but mostly to suppress inflammation not due to infective agents. In high dosage they may have many side-effects including a mineralocorticoid effect which causes a disturbance of salt and water balance.

(3) Sex hormones Small quantities of sex hormones, *androgens* and *oestrogens*, are produced by the suprarenal gland. These influence sexual development and growth. In females, the suprarenal glands are the principal source of androgens, which are required by both sexes for normal pubertal and skeletal development.

Unlike the secretions of the adrenal medulla, those of the cortex are not regulated by nervous impulses. The adrenal production of glucocorticoids and sex hormones is controlled by pituitary adrenocorticotrophic hormone (ACTH) or *corticotrophin*. The pituitary secretion of this hormone is, in turn, stimulated by corticotrophin releasing factor (*corticotrophin-RF*). The whole control structure is known as the hypothalamic-pituitary-adrenal (HPA) axis. Large doses of cortisone will suppress the pituitary secretion of corticotrophin by a negative feed-back mechanism (see Fig. 14.5). This effect may be very persistent in

patients after steroid therapy has been discontinued. Such patients, with partial suppression of pituitary activity, have a diminished ability to respond to stress such as that of an infection, an accident, or a surgical operation. Booster doses of corticosteroids are necessary for these patients at such times to prevent a hypoadrenal crisis, which could be fatal.

Hypoadrenalism may also result from destructive disease of the adrenal glands by an autoimmune process or tuberculosis, for example. The condition is known as *Addison's disease* and is characterized by a low blood pressure (hypotension), brown pigmentation of the skin and mucous membranes, and excessive loss of sodium from the body, with hyponatraemia (a low serum sodium level) and often dehydration.

Over secretion of cortisol occurs in *Cushing's syndrome*, which is characterized by rounding ('mooning') of the face, obesity with a 'lemon on tooth-picks' distribution and a 'buffalo-hump', hypertension, diabetes and osteoporosis.

Aldosterone secretion is largely independent of ACTH but is influenced by changes in the volume of extracellular fluid, a decrease in the latter causing greater aldosterone secretion and vice versa. Aldosterone also plays a part in blood pressure regulation through the renin-angiotensin mechanism controlled by the juxtaglomerular apparatus of the kidney (see Fig. 14.6).

Functions of the suprarenal medulla

The medulla secretes *adrenalin* and *noradrenalin*. The secretion of these catecholamines is not controlled by the anterior pituitary gland but by the sympathetic nervous system. They cause a constriction of the arterioles of the body, resulting in a rise of blood pressure, and also an increase in the rate and force of the heartbeat. They also relax the involuntary muscle of the bronchi and stimulate the liver to convert glycogen into glucose, which is liberated into the bloodstream.

Adrenal secretion of catecholamines constitutes a reserve mechanism that comes into action at times of stress. These hormones are poured into the bloodstream during fear or anger and they are responsible for many of the changes which accompany these emotions. Blanching of the skin due to constriction of the arterioles favours diversion of the blood to the muscles, where it is most needed. The increased blood pressure and force of the

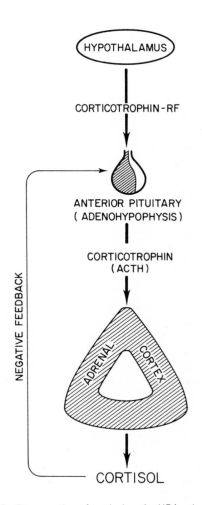

Fig. 14.5 The secretion of cortisol – the HPA axis.

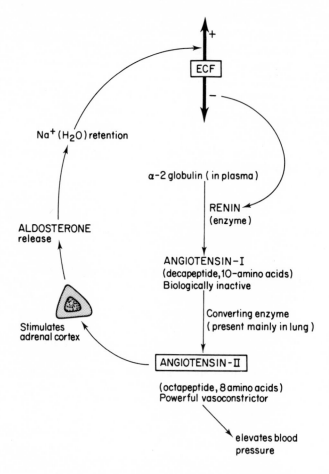

Renin is secreted by specialized smooth muscle cells in the afferent arteriole close to the **macula densa**, which is composed of specialized cells of the distal tubule and, with the afferent arteriole, makes up the juxtaglomerular apparatus.

Stimuli for release of renin:

(a) fall in renal perfusion pressure;
(b) stimulation of autonomic nervous system;
(c) sodium content of distal tubule (diuretics alter this).

Fig. 14.6   The renin-angiotensin-aldosterone mechanism.

heartbeat result in better circulation both in the muscles and the brain, while the liberation of glucose supplies the muscles with the necessary fuel for increased activity.

A tumour of the adrenal medulla, known as a *phaeochromocytoma*, is a rare but important cause of hypertension.

The pituitary gland (hypophysis)

This is a gland about 1 cm (0.4 in) in diameter, situated at the base of the brain in the saddle-shaped depression in the sphenoid bone known as the sella turcica. Its attachment to the brain is by a short stalk placed just behind the optic chiasma where the optic nerves from each eye meet.

The pituitary gland consists of two parts, the anterior and posterior lobes, which have different modes of development and entirely different functions. Both are under the control of the hypothalamus but by different mechanisms. A *neural* mechanism between the hypothalamus and pituitary controls the secretion of hormones by the posterior lobe. The secretion of anterior pituitary hormones is controlled by stimulatory and inhibitory factors or *hormones* which are secreted by the hypothalamus. These factors are secreted into the blood of the portal venous system which runs in the pituitary stalk and they are thereby carried to the anterior pituitary gland. The control of secretion of hypothalamic, anterior pituitary and dependent peripheral glands is by *feedback loops*

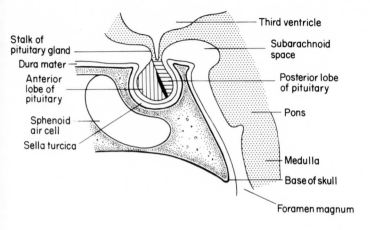

Labels on diagram:
- Third ventricle
- Subarachnoid space
- Stalk of pituitary gland
- Dura mater
- Anterior lobe of pituitary
- Posterior lobe of pituitary
- Sphenoid air cell
- Pons
- Sella turcica
- Medulla
- Base of skull
- Foramen magnum

Fig. 14.7 Diagram of the pituitary gland.

which 'turn off' the trophic hormone when the peripheral gland has secreted enough hormone to achieve normal plasma levels.

Functions of the anterior pituitary (adenohypophysis)

Six hormones are produced by the anterior lobe of the pituitary gland. They are the thyroid stimulating hormone (TSH), corticotrophin or adrenocorticotrophic hormone (ACTH), growth hormone (GH), follicle-stimulating hormone (FSH), luteinizing hormone (LH) and prolactin (PRL).

Thyroid-stimulating hormone

This stimulates the growth and activity of the thyroid gland. The release of TSH into the circulation is stimulated by thyrotrophin-releasing hormone (TRH) from the hypothalamus. Its production is inhibited by a high circulating triiodothyronine (T_3) level.

Corticotrophin

ACTH promotes the production of steroid hormones, especially cortisol, in the suprarenal cortex. The circulating level of cortisol provides feedback to the pituitary and to the hypothalamus, which controls the anterior pituitary by its secretion of corticotrophin-releasing factor (CRF). There is diurnal variation in the secretion of ACTH, resulting in plasma cortisol levels being at their peak at about 6 a.m. and at their lowest at about midnight.

This fact forms the basis of a screening test for Cushing's disease, in which there is hypersecretion of cortisol and the diurnal variation is lost.

Melanocyte pigmentation is also under the control of ACTH and patients with Addison's disease, in whom there is adrenal insufficiency (hypofunction), are abnormally pigmented as a result of loss of the negative feedback normally provided by cortisol.

Growth hormone (GH) (somatotrophin)

This influences the synthesis by the liver of a group of proteins known as somatomedins and thereby exerts its effect on growth. The release of GH from the anterior pituitary is inhibited by somatostatin, which occurs in the hypothalamus and in the pancreatic islets.

Growth hormone promotes protein synthesis and antagonizes the actions of insulin on carbohydrate metabolism. Over-secretion of growth hormone in childhood leads to excessive growth in the length of bones and the condition known as *gigantism*. Hypersecretion of growth hormone in adult life, when the length-wise growth of long bones has ceased, results in enlargement of bone and soft tissues, producing a condition known as *acromegaly*, in which the bones of the face, hands and feet enlarge while the lips become thick and the facial features become coarse. Under-secretion of GH in children results in failure of growth, with normal proportionment and normal mental development. As soon as the diagnosis is made, injections of growth hormone are begun with the aim of achieving as much normal growth as possible before the epiphyses fuse under the influence of the sex hormones.

The gonadotrophins (FSH and LH)

FSH brings about the ripening of ovarian follicles, with attendant oestrogen production, in the female, and stimulates spermatogenesis in the male.

LH causes ovulation and formation of the corpus luteum, which secretes progesterone, in women, and stimulates the testis to produce testosterone in men.

The release of gonadotrophic (gonad-stimulating) hormones from the anterior pituitary is stimulated by the gonadotrophin-releasing hormone (GnRH). Production of the latter is inhibited by oestrogens and to a lesser extent by progestogens (progesterone-like substances), a fact upon which the effectiveness of the contraceptive pill depends.

Prolactin

This hormone promotes lactation. Its secretion is under the control of the hypothalamus, TRH acting as a prolactin-releasing factor and prolactin-inhibiting factor (PIF) inhibiting the secretion of prolactin.

The serum prolactin level rises normally during pregnancy and is very high during lactation. Infertility is the principal effect of pathologically elevated serum prolactin levels which may result from any one of a number of causes, including a pituitary tumour and certain drugs, including phenothiazines.

The actions of the various hormones have been individually described but in health they all work together and in disease there may be multiple deficiencies, as in Simmond's disease (panhypopituitarism).

Functions of the posterior pituitary (neurohypophysis)

The hormones of the posterior lobe of the pituitary gland are vasopressin and oxytocin. They are synthesized in the hypothalamus and use neural pathways to reach the posterior pituitary, where they are stored prior to release into the circulation.

Vasopressin (antidiuretic hormone, ADH)

This causes the reabsorption of water into the blood from the collecting ducts of the kidneys, thereby concentrating the urine and reducing its volume. The function of vasopressin is to maintain a normal plasma osmolality (270–290 mOsm/kg) and plasma volume. Its release depends upon signals from osmoreceptors in the hypothalamus and volume (stretch) receptors in the walls of the atria of the heart and in the great veins. A rise in plasma osmolality and a low plasma volume both result in an increase in vasopressin release, which causes retention of water by the kidney until the osmotic pressure and volume of the plasma are again within the normal range. Vasopressin release is also affected by pain, emotional stress, alcohol and certain drugs.

Oxytocin

This is released from the posterior pituitary by a neural mechanism and causes contraction of the smooth muscle of the uterus and breast. Suckling stimulates neural pathways and the consequent release of oxytocin from the pituitary causes ejection of milk from the lactating breast.

Oxytocin is used only in Obstetrics. Vasopressin is used in the treatment of diabetes insipidus of pituitary type, in which large volumes of dilute urine are passed.

The pineal gland

The pineal gland is a small reddish-grey structure, about the size of a pea, situated in the midline of the brain immediately behind the third ventricle and under the posterior end of the corpus callosum. In adults it may be calcified and identifiable on skull X-ray films, in which its displacement to one side would indicate the presence of a space-occupying lesion within the cranium.

The pineal gland was considered by Descartes to be the seat of the soul and more recently it was considered to be an evolutionary relic. There is now evidence that it is an active endocrine gland and, although its function is poorly understood, it is thought to have a regulatory role in modulating the activity of the pituitary and other glands. Its secretions generally have an inhibitory effect and there is evidence that a pineal hormone inhibits the growth and maturation of the gonads until puberty. The pineal has a circadian rhythm of endocrine activity and may have an extensive role in co-ordinating circadian and diurnal rhythms throughout the

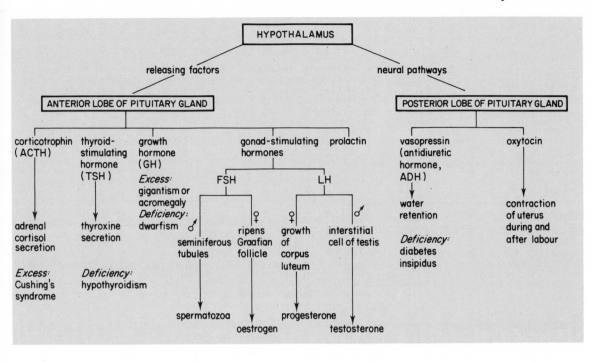

Fig. 14.8 Functions of the pituitary gland.

body, acting via the hypothalamus and pituitary gland.

The sex glands or gonads

See Chapter 18 on the Reproductive System.

The pancreas

This organ and its secretions are described in Chapter 11 on the Digestive System. The endocrine tissue of the pancreas is in the form of clumps of secretory cells known as the **islets of Langerhans**. The islet cells are of three types, alpha, beta and delta. *Insulin* is secreted by the beta cells and, like the other hormones, passes directly into the bloodstream. Insulin acts on a cell membrane receptor and enhances the entry of glucose into most cells, thereby diminishing the plasma glucose concentration. Within the cell, insulin promotes the conversion of glucose to glycogen (glycogenosis), promotes protein synthesis and inhibits fat breakdown (lipolysis). Insulin deficiency results in breakdown of glycogen, protein and fat, leading to a raised plasma glucose level (hyperglycaemia) and ketoacidosis. The ketone bodies are acetoacetic acid, beta-hydroxybutyric acid and acetone, which are derived from the free fatty acids released during fat breakdown or lipolysis.

Normally the blood glucose level remains within a certain range. Any excess of glucose is stored in the liver and muscles as glycogen. If the blood glucose rises above its normal or 'threshold' level the excess of glucose is excreted by the kidneys and glucose appears in the urine (glycosuria). Some people, however, have a low renal threshold, so that they have glycosuria while their blood glucose level is within the accepted limits of normal.

Deficiency of insulin due to disease of the islets of Langerhans results in diabetes mellitus, a condition in which the blood glucose is high and glucose is passed in the urine. In severe cases the disturbed metabolism of fat results in ketoacidosis and the presence of ketone bodies in the urine.

The alpha cells of the pancreas secrete *glucagon*, the metabolic effects of which are the opposite of those of insulin. It causes the breakdown of liver glycogen, thereby releasing glucose into the bloodstream, and promotes lipolysis.

The third hormone of the pancreas is *somatostatin*, which is secreted by the delta cells of the islets of Langerhans. It is able to inhibit the secretion of many hormones and other substances and is one of the 'brain-gut peptides', which act as neurotransmitters. Somatostatin is also secreted by the hypothalamus and, because it inhibits the pituitary release of growth hormone, is also known as growth hormone release inhibiting hormone (GHRIH).

Questions

1. What do you understand by the term hormone? Give three examples.
2. Which other glands are under the hormonal control of the pituitary gland and hypothalamus? Give an example in the form of a schematic illustration.
3. Which hormones are produced by the thyroid gland and what are their actions?
4. How is the secretion of parathormone regulated?
5. Write short notes on the suprarenal glands.
6. Name the six hormones of the anterior lobe of the pituitary gland. Write briefly about two of them.
7. How does the posterior lobe of the pituitary gland regulate the water content of the blood?
8. What are the endocrine secretions of the pancreas? Which cells are concerned with their secretion?

15

The Skin and Regulation of Body Temperature

The skin

The skin is the outer covering of the body and is continuous with the mucous membrane lining the body orifices. It is an extensive and diverse organ, with functions which are vital for survival, including protection, excretion, sensation and temperature regulation. The skin consists of two layers: the epidermis and the dermis.

The epidermis

The epidermis is the most superficial layer of the skin and is derived from the embryonic ectoderm. It is composed of many layers of stratified squamous epithelium.

The **basal** or **germinative** layer is the deepest, and consists of a single layer of columnar cells firmly attached by a basement membrane to the underlying dermis. These cells are constantly dividing to produce new prickle cells which push the older cells upwards towards the surface. **Prickle cells** derive their name from minute intercellular bridges which hold the cells together. As they approach the surface of the epidermis these cells become flattened to compose the **stratum granulosum** (granular layer). The nuclei of cells in the granular layer contain granules of keratohyaline, which is a precursor of **keratin**, a tough fibrous protein. Under the influence of enzymes the nuclei disintegrate and the keratohyaline is converted to keratin.

In the thickest areas of skin over the palms of the hands and the soles of the feet the granular layer transforms into the **stratum lucidum** (or clear layer), but elsewhere it transforms directly into the **stratum corneum** or horny layer. The cells of the horny layer are thin, flat non-nucleated cells composed mainly of keratin. They are bound together to form a strong, pliable membrane which is relatively impermeable to water and prevents loss of tissue fluids from the body. The cells of the horny layer are constantly being shed and replaced by cells from the deeper layers.

The life of an epidermal cell, from its formation as a prickle cell by the germinative layer until it is shed from the surface of the skin is between 28 and 30 days, approximately 14 days being spent in the stratum corneum. In some skin diseases, cell division in the germinative layer is greatly increased and in psoriasis the time taken for new cells to be produced and shed is reduced to 7–10 days. Disease or trauma which destroys the intercellular bridges of the prickle cells results in the separation of the cell layers and the formation of blisters.

Between the cells of the basal layer are **melanocytes**. These are specialized cells, derived from embryonic nervous tissue, which secrete the pigment melanin. *Melanin* is produced from tyrosine under the influence of the enzyme tyrosinase. This pigment is secreted into the cells of the epidermis and protects the deeper layers of the skin from the effects of the ultra-violet rays in sunlight. Production of melanin is partly controlled by genetic inheritance and partly by the melanocyte-stimulating hormone (MSH) secreted by the anterior pituitary gland. Adrenocorticotrophic hormone (ACTH) has a similar structure to MSH and in excess can cause an increase in melanin secretion. Prolonged exposure of the skin to sunlight also stimulates increased pigment production.

Negroes have no more melanocytes than fair-skinned people, but the negro melanocyte produces far greater quantities of melanin.

The dermis

The dermis, or corium, is composed of dense connective tissue containing collagen fibres which strengthen the skin, and some elastic fibres that give the skin its resilience and pliability.

The surface of the dermis is thrown into ridges, or papillae, which interlock with the epidermis.

The configuration of the papillae are responsible for the skin patterns of the palmar aspects of the hands and fingers, and the soles of the feet and toes. The patterns formed by the papillae are so characteristic for each individual that they can be used as a means of identification.

The undersurface of the dermis merges into the subcutaneous tissues which contain a varying number of fat cells and white and yellow connective tissue. The adipose tissue acts as an insulating layer between the skin and the deeper underlying structures of the body.

The dermis contains blood vessels, lymphatics, nerves and some specialized structures derived from the epidermis known as skin appendages. These are the sweat glands, hair, sebaceous glands and nails.

Blood vessels

The circulation of blood through the dermis nourishes the skin and plays an important role in the regulation of body temperature.

The dermis contains an extensive network of blood capillaries arising from arteries in the subcutaneous layers. Each papilla of the dermis is supplied with a loop of capillaries which provides nourishment to the germinative layer of the epidermis and the structures of the dermis. These capillaries drain into veins which join to form venous plexuses a few millimetres below the skin surface. The venous plexuses of the skin contain large quantities of blood that can heat the surface of the skin.

The rate of blood flow through the skin is regulated according to body temperature. The cutaneous arterioles are supplied with sympathetic nerve fibres which are under the control of the heat-regulating centre in the hypothalamus. When body temperature is low the arterioles constrict, blood flow through the skin is reduced and heat loss from the skin is minimal. When body temperature is high the arterioles dilate, blood flow through the skin is increased and heat carried in the blood from the internal organs is lost from the skin surface.

In some areas of the body, such as the hands,

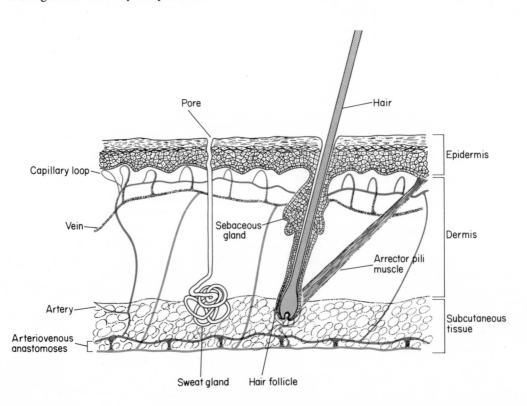

Fig. 15.1 Section through the skin (diagrammatic).

feet, lips and ears, large vascular connections are found between the arteries and the venous plexuses, called arteriovenous anastomoses; these contain powerful muscle fibres which, when relaxed, allow blood to pass from the arteries to the veins without entering the capillary loops and thus help in the conservation of heat by the body.

Lymphatics

The dermis has a rich network of lymphatic vessels. They begin as lymph spaces at the tip of each papilla and pass between the connective tissue fibres to form lymph capillaries in the subcutaneous tissue.

Nerves

The skin supplies much of the body's sensory information and is provided with both sensory nerves and sympathetic nerve fibres. The sensory nerve endings are widely distributed in a fine network of myelinated and unmyelinated fibres throughout the skin and respond to touch, pressure, pain and temperature. Recent evidence suggests that itching should also be regarded as a specific skin sensation, and may be invoked by a variety of physical stimuli. Exogenous chemicals, such as weak acids or alkalis, are thought to cause itching by releasing histamine, while substances containing proteolytic enzymes (e.g. biological washing powders) probably act directly on the nerve endings of the skin.

Nerve impulses from the cutaneous nerves are relayed via the sensory tracts of the spinal cord to the sensory centres in the cerebrum.

Sympathetic nerve fibres are supplied to the blood vessels, the smooth muscle of the hair follicles and the secretory cells of the sweat glands.

Sweat glands

Sweat glands develop as downgrowths from the epidermis. In humans there are two varieties: the eccrine and the apocrine. **Eccrine glands** are found over the entire skin surface, but are not present in mucous membrane. They are most numerous on the palms of the hands and soles of the feet.

The sweat glands have a **glomerulus**, or secreting coil, buried deep in the dermis, and a **duct** which conveys the sweat on to the surface of the skin through a minute opening known as a sweat pore.

The glands are formed of cubical epithelium, and the glomerulus is surrounded by a layer of smooth muscle enclosed in a thin fibrous capsule. Sweat glands are supplied with blood capillaries that form a network around the secreting coils. They are also profusely supplied with sympathetic nerve fibres.

Sweat is a clear watery fluid containing 0.5 per cent of solids. These include sodium chloride, small amounts of other mineral salts and urea.

Sweating is one of the mechanisms by which humans regulate their body temperature. The secretion of sweat is controlled by the heat-regulating centre in the hypothalamus via the sympathetic nervous system.

At normal body temperature water is lost continuously through the skin as *insensible perspiration*; this water loss is not sweat, but is related to evaporation of water passing through the tissue spaces of the epidermis. A rise in body temperature such as occurs in a warm environment or during strenuous exercise leads rapidly to stimulation of the sweat glands and the appearance of sweat over the entire body surface.

Sweating may also be induced by emotional stress such as anxiety or fear, and occurs mainly in the palms, soles and axillae. Excessive production of sweat is called hyperhidrosis.

Apocrine glands

These are large sweat glands whose ducts open into the hair follicles above the sebaceous glands. They are found in the genital and anal regions, the axillae, nipples and areolae, and only begin to function at puberty.

The apocrine glands produce a viscid, milky secretion which is odourless until contaminated with bacteria. Apocrine secretion is stimulated by stress, pain, fear or sexual activity.

The wax-secreting ceruminous glands found in the external acoustic meatus are modified apocrine glands.

Hair

Each hair consists of a free **shaft** extending above the skin surface and a **root** embedded in the skin. The hair root is enclosed in a **hair follicle** which is formed by an invagination of the epidermal cells into the dermis. The lower end of the follicle expands to form the **hair-bulb**. At the base of the

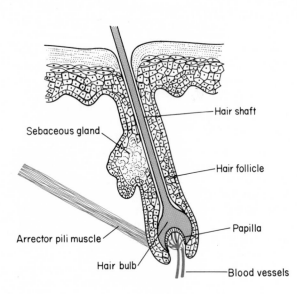

Fig. 15.2   Diagram of hair follicle.

hair-bulb is a small projection of cells derived from the dermis called the **papilla**, which contains blood vessels, nerve endings and melanocytes.

The surface of the papilla is covered by a layer of epidermal cells, the germinal matrix. Hair is formed by proliferation of these cells, which as they are pushed further and further away from the papilla die and become keratinized.

A hair consists of three layers:

(*1*) a cuticle of thin flattened horny cells;

(*2*) a cortex of pigmented spindle-shaped cells;

(*3*) a medulla of cells which lose their nuclei as they are pushed upwards away from the papilla.

Inserted into the walls of the hair follicle is a minute bundle of smooth muscle fibres, the **arrector pilorum**, which is innervated by sympathetic nerve fibres. Contraction of the arrector pilorum causes the hair to 'stand on end' (gooseflesh).

Hair growth occurs in alternating cycles of growth and rest. The growth phase of the human scalp is between 2 and 4 years, while the rest phase lasts about 3 months. Neighbouring hair follicles are at different stages in the cycle.

At the end of a growth cycle the hair becomes detached from the papilla and the hair-bulb passes into a rest phase. When growth is resumed the new hair pushes out the old dead hair.

Loss of hair or balding may occur as a result of mechanical trauma (e.g. excessive brushing or the application of very tight curlers), autoimmunity or the systemic administration of certain anticancer drugs. The common male type baldness, also known as androgenetic alopecia, is the result of a dominant hereditary trait in which the hair follicles are excessively sensitive to androgens.

Sebaceous glands

The sebaceous glands are found all over the body except for the palms of the hands and soles of the feet. They are most numerous on the scalp, forehead, cheeks and chin. The glands are small saccular structures composed of epidermal cells which open into the hair follicles, to form a pilosebaceous unit.

Sebaceous glands secrete *sebum*, which is formed by the disintegration of large nucleated cells within the gland. Sebum contains fatty acids, cholesterol and other substances, and is discharged through the sebaceous duct into the hair follicle.

Sebum is a natural lubricant, which keeps the hair supple, and the skin soft and pliant, protecting it from the effects of moisture and heat.

Modified sebaceous glands, such as those of the areolae and labia minor, and the Meibomian glands of the eyelids, open directly on to the skin surface.

The activity of sebaceous glands is low until puberty, when it increases rapidly due to the effects of the sex hormones.

Hypersecretion of sebum is involved in the aetiology of acne vulgaris, a skin condition occurring in adolescents and young adults.

Nails

The nails are solid plates of modified horny cells, and form a protective covering on the dorsal surfaces of the fingers and toes, corresponding to the claws and hoofs of animals.

The **nail plate** or body is firmly attached to the underlying **nail bed**, which consists of modified epidermal cells. Each nail has a **root** embedded in a fold of skin called the nail groove, and is flanked by the nail wall which extends to form the **cuticle**. The

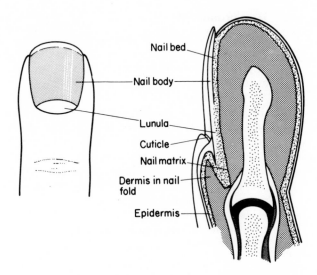

Fig. 15.3 Diagram of nail and nail bed.

body of the nail plate is semi-translucent and derives its pink colour from the blood vessels of the nail bed, while in the region of the lunula (half-moon) at the root the nail is more opaque. The nail plate is free at its distal border.

Fingernails grow more rapidly than toenails, taking between 5 and 6 months to grow from the nail matrix to the fingertip. Nail growth accelerates in summertime, and is increased by nail-biting. Growth may be inhibited by disease or malnutrition.

Functions of the skin

Protection

The skin forms a protective covering for the internal structures of the body. It acts as a barrier preventing the entry of microorganisms and other harmful agents. It helps to maintain a stable internal environment by preventing excessive loss of water, electrolytes, proteins and other substances.

Sensation

The sensory nerve endings of the skin constantly convey information about the external environment to the brain, and serve as an important protective mechanism for the body, besides giving an awareness of immediate surroundings and acting as

an organ of expression. The skin is able to express fear by its pallor, anxiety by sweating and embarassment by blushing.

Storage

The skin and subcutaneous tissues act as a store for water and fat. Approximately 15% of the total water content of the body is contained in the skin, while the subcutaneous tissue serves as one of the main fat depots of the body.

Secretion

The secretion of sebum by the sebaceous glands helps to maintain the integrity and suppleness of the skin.

Sweat produced by the eccrine and apocrine gland plays an important part in the regulation of body temperature and aids in the excretion of metabolic waste products.

Formation of vitamin D

Vitamin D is formed by the action of sunlight on 7-dehydrocholesterol, a fatty substance widely distributed in the skin (p. 193).

Absorption

Substances can be absorbed through the intact skin into the bloodstream and advantage is taken of this fact in the *percutaneous* administration of certain drugs. An example is trinitrin, a nitrate used in the treatment of angina. This drug is most commonly taken under the tongue (sublingually) because it is quickly absorbed through the mucous membrane of the mouth to bring rapid relief from the chest pain. Its effect is brief, however, and if continuous action is required the percutaneous route may be used. The drug may be incorporated into an ointment (e.g. Percutol) or into a 'patch' (e.g. Transiderm-Nitro) which presents a steady dose to the skin through a rate-limiting membrane.

The skin and systemic disease

There are many diseases of the skin alone but often skin lesions represent only one manifestation of a more generalized (systemic) disease. Measles, for example, is not regarded as a skin disease although the rash may be its most obvious feature. It is an example of an *exanthem*, a condition with eruptions on the skin. Many infectious febrile illnesses, especially the 'childhood fevers', are characterized by a rash and a variety of other systemic diseases also have a skin component. Skin lesions associated with deep-seated cancer are known as dermal markers of malignancy.

Regulation of body temperature

Man is a warm-blooded animal and is capable of maintaining a relatively constant central body temperature which is independent of the surrounding environment.

The central, or core, temperature is the temperature of the internal organs and, in health, is maintained at an almost constant level, while the temperature of the skin and subcutaneous tissues is variable, depending on environmental temperature.

The average normal body temperature is 36.8°C (98.2°F) when measured in the closed mouth and approximately 0.5°C (1.0°F) higher when measured in the rectum. The normal body temperature varies during each 24 hour period, being at its highest in the early evening and at its lowest in the early morning.

The constant level of the body temperature is maintained by a balance between the heat produced and the heat lost by the body. Failure to maintain this balance results in a rise or fall in the body temperature.

Heat production

All body tissues produce heat as a by-product of the chemical processes of metabolism, but the greatest heat production is from the most active tissues such as the liver, endocrine glands and muscles.

Muscular activity accounts for about 30% of the body's heat production even at rest. During physical exercise the amount of heat produced is greatly increased. When insufficient heat is produced or when the body is exposed to extreme cold shivering occurs. Shivering is the involuntary contraction of skeletal muscles and is a very powerful means of increasing heat production.

Heat may also be gained by the body when the environmental temperature exceeds body temperature.

Heat loss

Heat is lost from the body principally through the skin by radiation, conduction, convection and evaporation. A small amount of heat is lost in expired air, urine and faeces.

Radiation
Heat is transferred from the body surface to nearby objects that are cooler than the skin, and is transferred to the skin by objects which are warmer than the skin.

Heat is conveyed from the internal organs to the skin by the bloodstream, and radiation of heat from the skin surface can be greatly increased by dilatation of the cutaneous blood vessels.

When the environmental temperature is higher than that of the body, heat cannot be dissipated by radiation.

Conduction
Heat is transferred to any object or substance in direct contact with the body. A rapid loss of heat occurs to an object which is a good conductor of heat, such as metal or light cotton clothing. Heat loss may be minimized by wearing clothing made of wool or fur, which are poor conductors of heat.

Convection
Heat is transferred away from the body surface by movement of air. The air close to the skin is warmed and, as this warm air becomes less dense it rises, allowing cold air to take its place next to the skin.

Evaporation

Heat must be expended to change water into water vapour on the skin surface. There is a small continuous loss of heat from the body in this way due to insensible perspiration (p. 209). This process cannot be controlled for the purpose of temperature regulation but, when body temperature rises, larger amounts of heat can be lost by sweating. The evaporation of sweat is a very efficient way of cooling the body, and is particularly important when environmental temperature is higher than 37°C (98.6°F), since under these conditions the body will gain heat by radiation, conduction and convection. Sweat which drips off the body is ineffective as part of the cooling process. However if the humidity of the surrounding air is high, sweat cannot evaporate and even environmental temperatures of 27°C (80.6°F) become physically uncomfortable.

Control of temperature regulation

Body temperature is controlled by the **temperature regulating centres** of the hypothalamus. These contain two groups of neurons; one group is heat-sensitive, while the other is cold-sensitive. The temperature regulating centres behave like a thermostat, operating in response to the information received through nervous feedback mechanisms from temperature receptors in the body.

A fall in body temperature causes the temperature regulating centres to conserve heat by increasing sympathetic nerve impulses, resulting in constriction of cutaneous blood vessels, and to stimulate heat production by shivering.

A rise in body temperature causes the temperature regulating centres to inhibit sympathetic nerve impulses to cutaneous blood vessels resulting

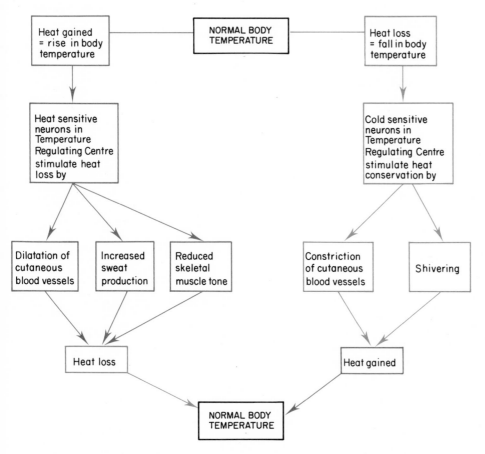

Fig. 15.4 Control of body temperature.

in dilatation of the blood vessels and increased blood flow through the skin. Sweat glands are stimulated to secrete sweat which is evaporated from the skin surface with consequent loss of heat. Skeletal muscle tone is reduced to slow down the rate of heat production.

Pyrexia

Pyrexia means a body temperature higher than normal and may be caused by toxins produced by infecting organisms or by protein breakdown products from rapid tissue destruction. These substances act on the temperature regulating centres, causing the thermostatic control to be reset at a higher level. As a result, the body is stimulated to produce more heat until the body temperature equals that of the new level, and is then maintained at this level as long as the abnormal substance is present in the body.

Substances which affect the temperature regulating system in this way are called *pyrogens*.

Hyperpyrexia
A body temperature of 40.5°C (105°F) or more is called *hyperpyrexia*, and is extremely dangerous as

cellular metabolism is increased so much and so rapidly that the temperature regulating mechanisms are unable to dissipate heat rapidly enough to overcome the rate of heat production. Hyperpyrexia causes degeneration of many body cells (because they literally 'burn themselves out'), and a rectal temperature exceeding 42°C (108°F) causes irreversible brain damage.

Hypothermia

A fall in body temperature below 35°C (95°F) is called *hypothermia*. Hypothermia causes a slowing in the chemical reactions of metabolism and reduces the blood flow and oxygen requirements of the tissues. If the condition remains untreated, and the body temperature continues to fall, the temperature regulating mechanisms are lost. The individual becomes increasingly drowsy, and then comatose due to cerebral ischaemia; heart and respiratory rates become slower as metabolism is progressively reduced. Death may occur when body temperature falls below 26°C (78.8°F), usually from ventricular fibrillation.

Newborn babies and the elderly are particularly susceptible to hypothermia.

Questions

1. Describe the structure and functions of the skin.
2. Give an account of the mechanisms concerned with the regulation of body temperature.

16
The Urinary System

The function of the urinary system is the formation and excretion of urine, thereby eliminating waste products, substances (e.g. drugs) which would be toxic if allowed to accumulate in the body, and excess water. The urinary system consists of the following structures:

the kidneys − the excretory organs
the ureters − the ducts draining the kidneys
the urinary bladder − the urinary reservoir
the urethra − the channel to the exterior

The kidneys

The kidneys are a pair of organs, each of which is about 11 cm ($4\frac{1}{2}$ in) long, 6 cm ($2\frac{1}{2}$ in) wide and 3 cm ($1\frac{1}{4}$ in) thick. They lie obliquely rather than vertically, with their upper poles nearer than their lower poles to the midline, behind the peritoneum of the posterior abdominal wall. The right kidney is about 1.25 cm ($\frac{1}{2}$ in) lower than the left and its lower pole may sometimes be felt on examination of normal subjects during full inspiration. The kidneys move up and down with respiration and are about 2.5 cm (1 in) lower in the standing position than in the recumbent position. The average weight of the adult kidney is about 150 g (5 oz) in the male and 135 g ($4\frac{1}{2}$ oz) in the female. The kidneys are dark red in colour. Although they may be described as bean-shaped, their shape is so characteristic that the term kidney-shaped is frequently used to describe other objects.

Each kidney has anterior and posterior surfaces, superior and inferior poles, and a lateral convex and a medial concave border. In the centre of the medial border is a notch known as the **hilum**, which contains the renal blood vessels and nerves and the renal **pelvis**, which is the funnel-shaped upper end of the ureter. The centre of the hilum is opposite the lower border of the spine of the first lumbar vertebra, about 5 cm from the median plane. The suprarenal glands are sited one on the upper pole of each kidney.

Relations

The kidneys lie embedded in fat (the perirenal or perinephric fat) retroperitoneally on the posterior abdominal wall. Posteriorly the kidneys are related to the diaphragm (which separates it from the pleura), the psoas major and quadratus lumborum muscles and three nerves. Anteriorly the right kidney is related to the liver, the duodenum, the small intestine and the right or hepatic flexure of the colon. In front of the left kidney lie the stomach, the pancreas, the spleen, the jejunum, the left or splenic flexure of the colon and the commencement of the descending colon. The medial border of the right kidney is related to the inferior vena cava and the right ureter. The medial border of the left kidney is related to the aorta and the left ureter. It is not necessarily intended that the student should learn these relations by rote but rather that, with the aid of diagrams or an anatomical model, a mental picture may be obtained.

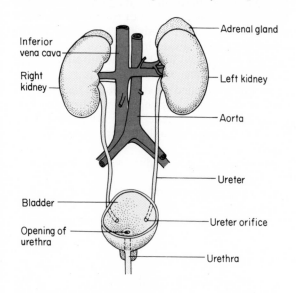

Fig. 16.1 Urinary tract − front view. The top of the bladder is cut off to show the openings of the ureters and urethra.

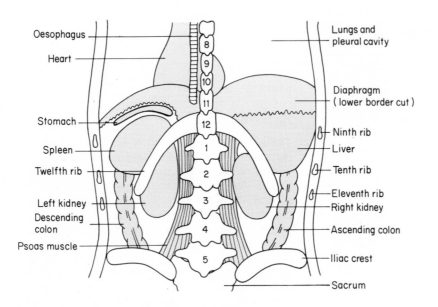

Fig. 16.2   Diagram showing position of kidneys, liver and spleen from behind.

Structure

Surrounding each kidney is a thin smooth fibrous **capsule**. When the kidney is cut in half lengthways by a knife passing through its medial and lateral borders, it is seen to consist of two layers, an outer **cortex** and an inner **medulla**. The pelvis of the ureter is seen to divide into two or three major calyces, each of which subdivides into a number of minor calyces. The renal medulla is arranged in conical masses called **renal pyramids**, the apices of which end in nipple-like papillae which project into the minor calyces. The terminal uriniferous ducts open on the papillae and discharge urine into the ureter.

Microscopically the kidney substance is seen to be composed of:

(*1*)   renal corpuscles (Malpighian bodies);
(*2*)   renal tubules;
(*3*)   blood vessels and supporting tissue.

The renal corpuscle

Each tubule begins in the cortex in a blind expanded end, known as the **glomerular capsule** or **Bowman's** capsule, which is indented by a lobu-

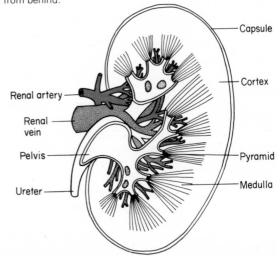

Fig. 16.3   Section through the kidney.

lated tuft of convoluted capillary blood vessels called a **glomerulus**. The two parts together constitute the renal corpuscle or Malpighian body (Fig. 16.4). Its function is filtration of the plasma. The *glomerular filtrate* passes down the tubules, in which its composition is modified by processes of reabsorption and secretion so that urine is finally formed. Water is conserved in the body by its reabsorption from the renal tubules; amongst solutes which are selectively reabsorbed are electrolytes, glucose and amino acids.

The renal tubule

This consists of
* (*a*) the glomerular (Bowman's) capsule;
* (*b*) the proximal convoluted tubule, situated in the cortex;
* (*c*) the loop of Henle in the medulla, with a descending limb, a 'U' bend and an ascending limb;
* (*d*) the distal convoluted tubule;
* (*e*) the junctional tubule;
* (*f*) the collecting duct.

The collecting ducts commence in the cortex and unite with one another to finally open into the terminal uriniferous ducts or papillary ducts, known as the ducts of Bellini, which open on the papillae of the renal pyramids.

Selective reabsorption of solutes occurs mainly in the proximal convoluted tubule. Sodium and chloride are selectively absorbed in the ascending limb of the loop of Henle and in the distal convoluted tubule under the influence of aldosterone. Because of this, the filtrate reaching the end of the distal convoluted tubule is hypotonic in comparison with the blood. However, the filtrate is finally concentrated in the collecting duct, the walls of which have a variable permeability to water under the influence of the antidiuretic hormone (ADH). Up to 95% of the water in the glomerular filtrate is reabsorbed in the renal tubules so that the end product, the urine, is hypertonic to the blood. The concentrations of osmotically active particles are expressed in milli-osmoles (mOsm) per litre or per kilogram and human urine can contain 1400 mOsm/ℓ, achieving almost five times the osmolality of plasma, which is normally 270–300 mOsm/kg of plasma water.

The normal osmolality of the urine varies with diet and fluid intake between 300 and 1000 mOsm/kg of water. Measurement of the urine and plasma osmolalities is often performed in cases of acute renal failure, where a urine/plasma ratio of more than 1.4 indicates a prerenal cause and a ratio of 1.1 or less indicates acute tubular necrosis.

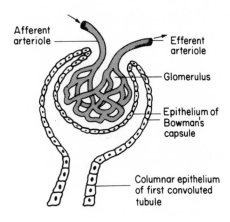

Fig. 16.4 A renal corpuscle, highly magnified.

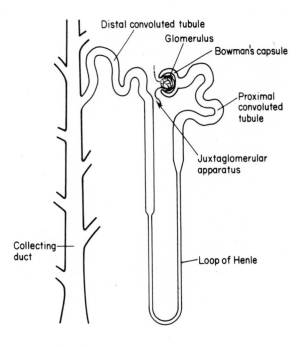

Fig. 16.5 A renal tubule.

Selective reabsorption is not the only function of the tubules. There is *secretion* of various substances into the lumens of the tubules. The secretion of hydrogen ion (H$^+$) by the cells of the proximal and distal convoluted tubules is important in the maintenance of acid-base balance.

Each unit of a renal corpuscle and its renal tubule is called a **nephron**. Each kidney contains between one and two million nephrons, the number diminishing with increasing age. All of the tubules of the kidney are lined by a single layer of cuboidal epithelium and are surrounded by capillary blood vessels.

Near the glomerulus of each nephron, is the **juxtaglomerular apparatus** which is formed principally by cells of the distal convoluted tubule and neighbouring arterioles. The juxtaglomerular cells secrete an enzyme called *renin* into the bloodstream. Working in a feedback mechanism, renin helps to maintain the constancy of the extracellular fluid volume through regulation of the secretion of aldosterone by the cortex of the suprarenal gland. A variety of stimuli, such as sodium depletion and dehydration, increase renin secretion. Some patients with hypertension have high renin levels in their blood.

Apart from being an enzyme, renin may be regarded as a hormone and its secretion is one of the *endocrine activities of the kidney*. Its other endocrine function is the production of a factor which leads to the formation of *erythropoietin* in the blood. This hormone stimulates erythropoiesis, the production and release of red blood cells from the bone marrow.

Blood supply

The kidneys are plentifully supplied with blood from the renal arteries, which are branches of the abdominal aorta. The vascular architecture of the kidneys is complicated but the blood eventually leaves the organs via the renal veins, which enter the inferior vena cava.

Constriction of a renal artery causes increased renin secretion with consequent angiotensin II formation which may account, in part, for the hypertension found in some patients with renal artery stenosis.

Functions of the kidneys

The primary function of the kidneys is to help in keeping the composition of the blood constant by the excretion of abnormal constituents and of normal constituents which are present in excess. In serving this important function, the kidneys perform the following important work:

(*1*) the excretion of water;
(*2*) the excretion of the end-products of protein metabolism;
(*3*) the excretion of ions and regulation of the acid-base balance and pH of the blood;
(*4*) the excretion of drugs, toxins and chemical substances which may be harmful.

(*1*) *Excretion of water* A certain quantity of water is necessary to maintain the normal composition of the blood and tissue fluids. In addition to that taken by mouth, water is an end-product of the metabolism of proteins, carbohydrates and fats. Water is excreted not only by the kidneys but also by the skin, lungs and bowels.

(*2*) *Excretion of end-products of protein metabolism* Urea, which forms about 2% of urine, is the most important of these end-products and forms about a half of the total of the solid constituents of the urine. It is formed in the liver by the removal of the nitrogen-containing fraction of the amino acids, a process called deamination.

Uric acid and urates are derived from the protein found in the nuclei of cells and originate partly from food (exogenous – from without) and partly from the tissues of the individual (endogenous – from within) which are affected by the general processes of wear and tear.

(*3*) *Excretion of ions* Many ions, derived mainly from food, are excreted in the urine. They include sodium, potassium, magnesium, calcium, chloride, phosphate, sulphate and oxalate ions. The correct balance of these ions in the circulating blood is vital. An excess of potassium in the blood, for example, can stop the heart. Hydrogen ion excreted into the renal tubule partly combines with buffers such as ammonia, phosphate and creatinine but also passes into the urine as free hydrogen ion (H$^+$). Bicarbonate ion is not normally found in the urine, having been reabsorbed in the renal tubules. These movements of ions are important in helping to maintain the acid-base balance of the body and keeping the pH of arterial blood at 7.40. Values lower than this represent acidosis and values higher than 7.40 indicate alkalosis.

(*4*) *Excretion of drugs, toxins and other chemical substances* Many drugs and toxins are excreted by the kidneys. This enables a number of drugs to be given over long periods without the risk of cumulative poisoning. In other words, the kidneys help to maintain a drug at a fairly constant level in the blood during the period over which it is being administered and finally eliminates it completely from the circulation when it is discontinued.

Normally no glucose is found in the urine but in diabetes mellitus the level of glucose in the blood, and consequently in the glomerular filtrate, may be so high that the renal tubules cannot reabsorb all of the glucose presented to them. Glycosuria then occurs, glucose being passed in the urine.

It would be very wasteful if glucose was normally lost in the urine and it is only when the plasma concentration of glucose rises above a 'threshold value' that the normal kidney will excrete glucose. The kidneys have a threshold value for many other substances and this factor is important in keeping the composition of the blood constant.

The formation of urine

Three processes are employed by the nephrons of the kidney in the production of urine:

(*1*) filtration;
(*2*) reabsorption (resorption);
(*3*) secretion.

Filtration
This is a simple physical process which occurs in the renal corpuscles. Water, salts and other substances are filtered from the blood in the glomerulus into Bowman's capsule and they quickly pass into the first convoluted tubule.

Absorption
If all of the fluid, salts and other solutes filtered from the glomeruli into the tubules were allowed to pass to the exterior in the urine there would be a serious loss of valuable materials from the body and the composition of the blood would be affected. In order to prevent this, some of the water and solutes are reabsorbed into the circulation by the renal tubular cells, especially those in the loop of Henle.

Secretion
This takes place mainly in the convoluted tubules and is an active, vital process performed by the cells of the cuboidal epithelium lining the tubules. These cells select either abnormal substances or normal substances from the blood when their concentration exceeds their threshold value, and pass them into the lumen of the tubules.

In other words, filtration is a purely mechanical process which takes no account of the requirements of the body, whereas the cells of the renal tubules look after the needs of the body (*a*) by returning much of the filtered material from the glomerular filtrate back into the blood, thereby retaining substances which are useful to the body and (*b*) by excreting those substances which the body does not require.

The composition of the urine

Normal urine is usually a clear yellow or amber-coloured fluid having a specific gravity varying between 1015 and 1025. Specific gravity continues to be measured because, although it is not as useful as osmolality, it is easier to measure. It is usually slightly acid (pH 6) and, when fresh, has an aromatic odour. Although the urine is normally clear at body temperature, when it is allowed to stand and cool it is quite common for a deposit of phosphates or urates to form, especially if the urine is concentrated. Urates dissolve when the urine is warmed. Phosphates can be dissolved by the addition of dilute acetic acid. Another change which quickly occurs in urine on standing is that it acquires an ammoniacal odour because of bacterial decomposition of urea.

Urine consists of water and dissolved solids. Chief among the organic compounds are urea, creatinine and urates, which are the end-products of protein metabolism. Ammonia, secreted by the renal tubules, combines with hydrogen ion to form the ammonium ion. The main ions of inorganic salts in urine are sodium, potassium, magnesium, calcium, chloride, phosphate, sulphate and oxalate. The salts are derived to a large extent from vegetable foods and from common salt (sodium chloride) taken in the diet. Up to 100 mg of protein, mainly albumin, may normally be excreted in the urine in 24 hours. Many hormones are passed out in the urine and, if in excess, may indicate overactivity of an endocrine gland.

The volume of urine passed is usually about $1-1\frac{1}{2}$ litres per 24 hours. There is, however, very considerable variation in health, depending upon

(*1*) fluid intake; and

(*2*) fluid loss from the skin by evaporation of sweat.

In hot weather the urine is decreased in quantity and of deeper colour and higher specific gravity as a result of being more concentrated. In cold weather, when the secretion of sweat is scanty, the converse is true.

The ureter

The ureter is the duct which conveys the urine from the kidney to the bladder. The pelvis of the ureter is its upper expanded portion, which lies mainly in the interior of the kidney. It commences as a number of funnel-shaped channels, which surround the papillae of the pyramids and are called calyces (singular = calyx).

The ureter is a tube about 26 cm (10 in) long, consisting of an outer fibrous coat, a middle muscular layer and an inner layer of transitional epithelium. The upper or abdominal part passes downwards on the posterior abdominal wall on the surface of the psoas major muscle. Crossing the common iliac artery, in front of the sacroiliac joint, it passes forward to enter the bladder. The lower part is called the pelvic portion and, in the female, it is related to the side of the cervix of the uterus. The intravesical portion of the ureter passes obliquely through the wall of the bladder for 2 cm ($\frac{3}{4}$ in). There is no actual valve in the lower end of the ureter but the obliquity of its path through the muscle of the bladder wall normally prevents the reflux of urine from the bladder into the ureter (*vesico-ureteric reflux*). This mechanism may be disturbed in cases of infection of the urinary tract. The passage of urine along the ureters is assisted by the peristaltic contractions of their walls. The urine enters the bladder from the ureteric openings in a regular series of squirts and remains there until it is voided.

The ureter is the source of pain in so-called *renal colic*. This is due to spasms of muscular contraction, usually due to a calculus (stone) in the ureter. The pain is spasmodic and severe and is referred to the areas of skin innervated from those segments of the spinal cord (T11-L2) which also innervate the ureter. Thus the pain starts in the loin and radiates forwards down into the groin and scrotum, or labium majus, and even into the upper part of the thigh anteriorly. Trauma due to the passage of the stone down the ureter may cause the appearance of some blood in the urine.

Pyogenic inflammation of the ureteric pelvis is called 'pyelitis' but the kidney itself is invariably involved to some degree and the preferable term is *pyelonephritis*. In chronic pyelonephritis, one or both kidneys are irregularly distorted, with surface depressions due to underlying scarring.

The urinary bladder

The bladder is the reservoir for the urine received from the kidneys via the ureters. It is a muscular sac lined by mucous membrane covered with transitional epithelium like that of the ureters. The superior surface of the bladder has an additional coat, an outer serous coat of peritoneum.

The muscular coat of the bladder wall is known as the **detrusor** muscle and it is this which is primarily responsible for emptying the bladder during micturition; it may be aided in this action by a rise in intra-abdominal pressure brought about by contraction of muscles of the abdominal wall. The muscle of the bladder is of the unstriated or involuntary type and there are three layers although these cannot be clearly separated because the fibres intermingle. The muscle bundles are arranged in a criss-cross fashion and when hypertrophy occurs, secondary to bladder neck obstruction, a typical trabeculated 'open weave' appearance can be seen through a cystoscope or discerned on an intravenous urogram.

The **neck** is the lowest part of the bladder and in the male it rests on the base of the prostate, with which it fuses. The fundus or **base** of the bladder is directed backwards and downwards and in the female is related to the anterior wall of the vagina. In the male, it is separated from the rectum by the rectovesical pouch of the peritoneum above and by the seminal vesicles and deferent ducts below. The apex of the bladder is its highest point and is directed forwards towards the upper part of the pubic symphysis. The empty bladder also has a superior surface and two inferolateral surfaces.

The bladder must be considered in its two extreme states, empty and full. The empty bladder lies behind the pubic symphysis. As it fills, however, the neck remains fixed whilst the superior surface rises above the symphysis pubis and the inferolateral surfaces become the anterior surface of the now ovoid bladder, resting against the anterior abdominal wall. As the superior surface of the bladder raises with it the serous coat of peritoneum, the anterior walls of the bladder and the abdomen are in direct contact with one another. Because of this, it is possible to pass a trocar and cannula into the full bladder immediately above the pubic symphysis without risk of injuring the peritoneum or infecting the peritoneal cavity. With an average degree of bladder distension, the superior surface of the bladder commonly comes to lie about 5 cm above the pubic symphysis. When it is over-distended, the bladder may reach as high as the umbilicus, or even higher, and can easily be felt as a smooth rounded tumour arising out of the pelvis. Here the word tumour is being used according to its Latin derivation to denote a swelling and not to imply a neoplastic process.

The neck of the bladder opens into the urethra, which can be seen on examination of the interior of the bladder. The two other orifices which can be seen are those of the ureters and the three orifices form the points of a triangle, called the **trigone** of the bladder, with its apex below at the urethral opening (Fig. 16.1).

The urethra

The urethra is the canal conveying the urine from the bladder to the exterior. It differs in the two sexes and its main function in the male is reproductive.

The female urethra is a short tube 4 cm ($1\frac{1}{2}$ in) in length, which leaves the base of the bladder at the trigone and reaches the exterior between the labia minora immediately in front of the opening of the vagina. As it leaves the bladder it is surrounded by the external or **urethral sphincter**, the internal sphincter being in the bladder neck itself.

The male urethra is a channel about 20 cm (8 in) long leading from the bladder to its external orifice, the meatus of the urethra at the extremity of the penis. It has three parts:

(*1*) the prostatic portion;
(*2*) the perineal or membranous portion;
(*3*) the penile or spongy portion.

The first part, as it leaves the bladder, is guarded by the internal or **vesical sphincter** composed of condensed circular muscle fibres of the bladder. Below this, it is surrounded by the prostate. The second part pierces the external or **urethral sphincter**, which is the voluntary sphincter, essen-

tial for the continent state. This membranous part of the male urethra is situated in the perineum and is especially liable to injury, including rupture, by falling astride a structure such as a bar, gate or chair.

Micturition

Micturition is the act of emptying the bladder or passing urine.

Three anatomical facts of importance are:

(*1*) the bladder is a muscular sac;
(*2*) two sphincters surround the urethra;
(*3*) the bladder is supplied with parasympathetic nerves which are motor for the detrusor muscle and inhibitory for the internal (vesical) sphincter. Sensory information from the bladder is conveyed in both the sympathetic and parasympathetic nerves. In the spinal cord, impulses subserving awareness of fullness of the bladder ascend in the posterior columns whilst the pain fibres ascend in the anterolateral white columns.

Micturition is primarily a *reflex* act which, after infancy, can be consciously controlled by impulses from the *higher centres* of the brain which either facilitate or inhibit it.

As urine accumulates in the bladder, the muscle fibres of its walls become gradually stretched. At first, however, the increase in pressure within the bladder (intravesical pressure) is slight because it behaves according to the law of Laplace (the pressure being equal to twice the wall tension divided by the radius) and because the detrusor muscle undergoes modification of its tone. When the pressure reaches a certain level, the sensory or afferent nerves of the bladder are stimulated and an impulse passes to the spinal cord and thence to the higher centres where it is interpreted as the desire to micturate (the sensory side of the act). At this point, the individual can control the act and, if necessary, postpone emptying the bladder. The desire to micturate is first felt at a bladder volume of about 150 ml. If the urge is neglected, it is followed by a sensation of fullness and eventually of pain.

When micturition is performed, the perineal muscles are first relaxed and then the detrusor contracts and the internal and external sphincters relax in turn. It is over the external sphincter, which is composed of striated muscle, that voluntary control is exercised.

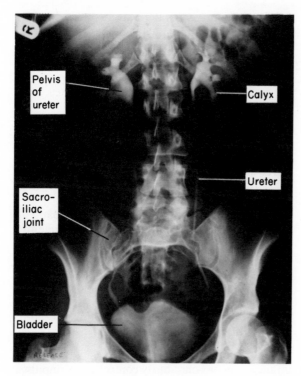

Pelvis of ureter

Calyx

Ureter

Sacro-iliac joint

Bladder

Fig. 16.6   An intravenous urogram (IVU).

In infants, before the higher centres are fully functioning, the act of micturition is a reflex one, the afferent stimulus passing to the spinal cord being followed immediately by the motor response of micturition. A similar state of affairs may occur in some cases of disease of the brain and spinal cord and in unconscious patients.

In altered states of consciousness (semi-consciousness) the patient may react to fullness of the bladder by becoming restless. The nurse who observes this and supplies a urinal may be able to prevent incontinence and keep such a patient dry.

A defect in the nervous mechanism may result in the bladder failing to empty. In such instances retention of urine occurs and the bladder becomes over-distended, a condition which can only be relieved by the passage of a catheter. Retention of urine may also be caused by mechanical obstruction to the outlet from the bladder by enlargement of the prostate or by a narrowing (stricture) of the urethra which impedes the passage of urine to the exterior.

Imaging techniques

The size and shape of the kidneys can be assessed from a plain X-ray film or, failing that, by ultrasonography, using ultrasound waves, or computerized tomography (CT scanning). The urinary tract can be visualized throughout by radiographic techniques, using contrast media consisting of iodine-containing compounds. If these are injected intravenously, an *intravenous urogram (IVU)* is produced. If radio-opaque fluid is injected through a ureteric catheter introduced by cystoscopy, a *retrograde pyelogram* is produced. Similarly the bladder and urethra may be outlined by retrograde injection of contrast media.

Questions

1. What are the characteristics of normal urine?
2. Describe the functions of the kidneys.
3. Draw a diagram of a nephron.
4. Name the three mechanisms by which the renal tubules perform their functions. Give examples.
5. Draw a diagram of the urinary tract.
6. Describe the bladder and the mechanism of micturition. How may micturition be affected by disease?

17
The Nervous System

The nervous system is the most complex and highly developed of all the systems of the body. It makes possible a range of adaptive responses to changes (stimuli) in the environment in the interests of survival of the individual and of the race. Such responses are central to the behaviour of all living organisms and are known as *homeostatic responses*. Simple organisms are simply structured and organized but a large animal with a complex structure requires a nervous system to overcome the problems of recognition of stimuli, storage of information (memory), communication between the various parts of its body, and the execution of effective responses.

The nervous system may be divided into three main portions.
(1) The **central nervous system**, consisting of the brain and spinal cord.
(2) The **peripheral nervous system**, consisting of the nerves between the central nervous system and the muscles and various organs.
(3) The **autonomic nervous system**, which is subdivided into sympathetic and parasympathetic systems.

The brain is the most complex and most important part of the nervous system. In it are the sites of consciousness, thought, memory, creativity, speech, vision, hearing, smell, control of endocrine glandular secretions and autonomic (involuntary) functions, and the will to carry out purposeful actions. It is only because of the brain (his or hers and our own) that we know anything of the personality of an individual.

The brain receives impulses or sensations which are interpreted and stored in the mind. The accumulation of these stored impulses forms the basis of memory. Parts of the brain particularly concerned with memory are the temporal lobes and the hippocampus and its connections. Bilateral destruction of the ventral hippocampus results in loss of recent memory whilst remote memory remains intact (but cannot be added to).

It is thought that memory might be stored as a biochemical change in the neurons. Protein synthesis seems to be involved in some way. It is suggested that the learning process causes a stable alteration in the ribonucleic acid (RNA) in the neurons.

The brain receives impulses from various parts of the body; they are called sensory or *afferent* impulses and are said to travel towards the brain along afferent pathways. Impulses travelling away from the brain and destined to result in movement or action of some sort are termed motor or *efferent* impulses.

The tissue of which the nervous system is constructed has already been described (p. 19).

The meninges

The central nervous system lies within the skull and vertebral column. In addition to the hard bony protection afforded by the axial skeleton, the brain and spinal cord have the following coverings, called meninges:
(1) the **dura mater** or outer layer;
(2) the **arachnoid mater** or middle layer;
(3) the **pia mater** or inner layer.
Between the arachnoid mater and pia mater is a fluid called the *cerebrospinal fluid* which supports the brain and spinal cord and ensures that the pressure upon them is evenly distributed.

The **dura mater** is a strong, thick fibrous membrane consisting largely of white collagen fibres. It lines the interior of the cranium and the vertebral canal, its outer layer actually forming the periosteum at these sites. Two large rigid folds of the inner layer of the dura mater project into the cranial cavity and help to support the brain and to maintain it in position.

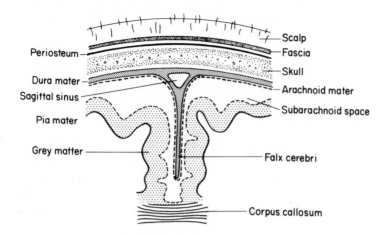

Periosteum —
Dura mater —
Sagittal sinus —
Pia mater —
Grey matter —

Scalp
Fascia
Skull
Arachnoid mater
Subarachnoid space
Falx cerebri
Corpus callosum

Fig. 17.1   Diagram of the scalp and meninges.

(1) The **falx cerebri** is a sickle-shaped fold lying vertically in the mid-line and separating the right and left cerebral hemispheres.

(2) The **tentorium cerebelli** is a crescentic arched sheet which lies horizontally and forms a tent-like roof for the posterior cranial fossa, thereby separating the cerebrum above from the cerebellum below.

The venous sinuses, for example the sagittal, transverse and cavernous sinuses (p. 145), are contained between layers of the dura mater. The **arachnoid mater** (arachnoid = spider-like) is a delicate transparent membrane situated between the dura mater and the pia mater. It is closely applied to the dura mater but is separated from the pia mater by the **subarachnoid space**, a narrow but extensive space which contains the cerebrospinal fluid. Between the under surface of the cerebellum and the medulla oblongata this space is enlarged to form the **cisterna magna**, of importance because a needle is sometimes passed between the occiput and atlas vertebra into this great cistern in order to withdraw cerebrospinal fluid (cisternal puncture). Great care must be taken not to insert the needle too far as it will damage the medulla oblongata, possibly with fatal results.

The **pia mater** is the innermost of the meninges and is closely applied to the brain and spinal cord. It is a very delicate membrane and carries numerous small blood vessels which supply the surface of the brain and spinal cord. It closely follows the surface of the brain and dips into all the fissures between the convolutions, whereas the arachnoid mater forms only a bag-like covering for the central nervous system. The pia mater is invaginated to form the tela choroidea and **choroid plexuses** of the third and fourth ventricles and the choroid plexuses of the lateral ventricles. It is the choroid plexuses which secrete the cerebrospinal fluid.

At the lower end of the spinal cord, below the conus medullaris, the pia mater continues downwards as a long slender filament, the **filum terminale**, to become attached to the dorsum of the coccyx. The subarachnoid space is particularly capacious around the filum terminale, making this the site of election for needling (lumbar puncture) and withdrawal of cerebrospinal fluid.

Cerebrospinal fluid

This is a clear, colourless, slightly alkaline fluid with a specific gravity of 1007. It occupies

(*1*) the ventricles of the brain, into which it is secreted by the choroid plexuses, and

(*2*) the subarachnoid space, into which it is secreted by the plexuses in the lateral recesses of the fourth ventricle.

The choroid plexuses are very vascular structures, consisting largely of a network of fine blood vessels. The cerebrospinal fluid (CSF) has an inorganic salt content similar to that of blood plasma, slightly less glucose and a small amount of protein. The glucose concentration of the fluid depends on the concentration in the blood and is abnormally high in uncontrolled diabetes mellitus. It is reduced in cases where the glucose is being utilized by bacteria and leucocytes, as in pyogenic meningitis. The protein content of the CSF is elevated in many pathological conditions because of increased permeability of the walls of the small blood vessels in the choroid plexuses, i.e. a reduced blood-brain barrier. Thus the CSF protein is increased, for example in cases of meningitis, of various types, and in cases of polyneuritis. Some diseases of the nervous system cause characteristic changes in the proportions of albumin and of the various globulins in the CSF, and these can be brought to light by electrophoresis after suitable concentration of the fluid.

The median aperture of the fourth ventricle (foramen of Magendie), situated in the roof of that ventricle, provides communication between the ventricular system of the brain and the subarachnoid space. Cerebrospinal fluid flows out of the fourth ventricle, through the median aperture, into the subarachnoid space at the base of the brain. It then flows upwards over the brain towards the superior saggital sinus and is reabsorbed into the bloodstream through the arachnoid villi associated with this venous sinus. If the reabsorptive capacity of the villi is diminished, the fluid accumulates, resulting in a condition known as *communicating* or *external hydrocephalus*. Blockage of the foramen of Magendie or obstruction within the ventricular system, by a tumor for example, also causes cerebrospinal fluid to accumulate and in this case the condition is called *internal hydrocephalus*.

The cerebrospinal fluid acts as a 'water cushion' to support the brain and spinal cord and protect it from jars and shocks due to body movements and from blows to the head.

Lumbar puncture

Owing to the intimate association of the cerebrospinal fluid with the nervous tissues and the meninges, disease of these structures is often associated with changes in the fluid, which may be removed for examination by the procedure known as lumbar puncture. This consists of passing a special needle (a fine trocar and cannula) through the skin and between the spines of the second and third or third and fourth lumbar vertebrae into the subarachnoid space. The spinal cord does not extend below the level of the first lumbar vertebra and is

therefore not vulnerable to injury from this procedure.

The patient lies on his or her side (usually on the left side if the doctor is right-handed) and the spine must be fully flexed so as to widen the interspinous spaces as much as possible. The needle is inserted in the mid-line, under local anaesthesia, and passed forwards and upwards through the skin and supraspinous and interspinous ligaments until it reaches the vertebral canal. The dura mater and arachnoid mater are then penetrated in succession. When the trocar is removed from the cannula, cerebrospinal fluid drips out. Its pressure is measured using a simple manometer which is connected to the cannula by a short length of flexible tubing and a male to female connector. Queckenstedt's test is performed by applying pressure to the neck so as to compress the jugular veins. This obstructs the venous drainage of blood from the cranium and raises the intracranial pressure. The increased pressure is transmitted to the spinal subarachnoid space and the cerebrospinal fluid level in the manometer rises briskly, unless there is a spinal block due, for example, to a spinal tumour.

The colour and clarity of the CSF should be noted. It may be turbid in cases of pyogenic meningitis and is uniformly bloodstained in cases of subarachnoid haemorrhage. Non-uniform bloodstaining, unequal in intensity in serial samples, may result from minor trauma during lumbar puncture.

Samples of cerebrospinal fluid are taken for laboratory examination – usually a cell count, glucose and protein estimations, and bacteriological examination. Substances may be injected into the subarachnoid space (intrathecal injection) if necessary. These include antibiotics, of which *the intrathecal dosage must be carefully checked*, in some cases of meningitis; anti-cancer drugs, such as methotrexate in cases of acute lymphoblastic leukaemia; and radio-opaque dye to outline a suspected spinal tumour or other obstruction. For spinal anaesthesia, lignocaine is introduced by means of lumbar puncture into the *epidural* (extradural) space of the vertebral canal. Epidural analgesia as used, for example, to relieve pain in labour, has superseded subarachnoid injection because it is safer.

The patient is nursed lying flat after a lumbar puncture because the head of pressure created by sitting or standing facilitates the further loss of cerebrospinal fluid through the puncture hole in the arachnoid mater. Sufficient fluid may be lost for traction on nerve roots and blood vessels to occur, resulting in a severe post-lumbar puncture headache. The finer the lumbar puncture needle the less likely is this complication to occur. Should it occur, the patient is advised to lie flat and encouraged to drink extra fluids.

The brain

The brain is that portion of the central nervous system which lies within the cavity of the skull. It consists of the following parts.

(1) The **forebrain**, consisting of the two cerebral hemispheres (the cerebrum) and the diencephalon or 'interbrain', which largely corresponds to the third ventricle and the structures bounding it.

(2) The **midbrain** or mesencephalon.
(3) The **hindbrain**, comprising
    (a) the pons
    (b) the medulla oblongata
    (c) the cerebellum.

The midbrain, pons and medulla oblongata together constitute the brainstem. When there is irreversible cessation of function of the brainstem, *brain death* is diagnosed. The heart may continue to beat but no spontaneous breathing occurs and respiration depends upon mechanical ventilation. Before the ventilator may be turned off, the diagnosis of brain death has to be substantiated, with due safeguards, by the repeated demonstration of *absence of all brainstem reflexes* by two doctors, one of whom has to be appropriately experienced and to have been registered for five years or more.

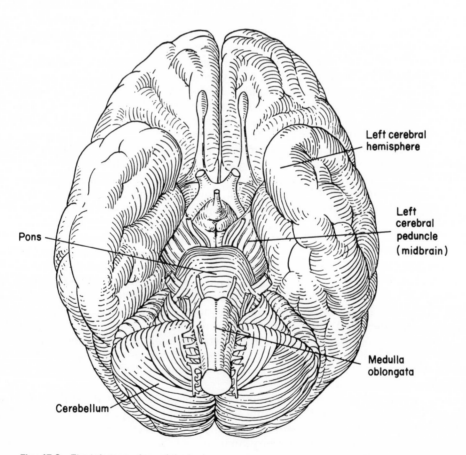

Fig. 17.2 The inferior surface of the brain.

The cerebrum

This is the largest part of the brain and consists of right and left **cerebral hemispheres**, which are separated in the mid-line by a fold of dura mater, the falx cerebri (p. 224). Below the central portion of the falx, the cerebral hemispheres are connected by a substantial bridge of white matter, the **corpus callosum**, which has a length of about 10 cm (4 in) from its front to its rear end. Surprisingly little is known about the function of this impressive structure. In those rare cases where it has been congenitally absent or has been divided by the surgeon's knife, there has been remarkably little apparent disturbance of function. Sophisticated tests do, however, reveal functional defects in these cases. The corpus callosum conveys an estimated 200 million nerve fibres from one cerebral hemisphere to the other. It therefore provides a structural basis for the transfer of information (communication) between the two hemispheres and for the integration of their separate activities.

Studies of the effects of damage to the corpus callosum have confirmed that there is an *asymmetry of function* in the cerebrum. One hemisphere is *dominant* and possesses the attribute of verbalisation, by which the individual can communicate with other individuals. The dominant hemisphere is usually the left one and associated with right-handedness. The cortical area for motor control of speech (Broca's area) and use of the contralateral hand are close together in the dominant hemisphere. Thus a stroke, due to a left cerebrovascular accident, causing paralysis of the right hand, also usually causes a motor aphasia.

The attributes of the dominant cerebral hemisphere are verbal, linguistic, mathematical, sequential and analytical. It seems to have a more direct link to consciousness than does the non-dominant ('minor') hemisphere. The latter is almost non-verbal but its attributes are musical and geometrical, and concerned with spatial comprehension and temporal synthesis. The two hemispheres are undoubtedly different in their specialization but it should be emphasized that they are united and that their abilities are complementary. The memory stores of each hemisphere appear to be accessible to the other hemisphere, so that total integration of function is possible.

Each cerebral hemisphere has frontal, parietal, temporal and occipital **lobes** (Fig. 17.3). The anterior part of the frontal lobe is called the **frontal pole** and that of the temporal lobe the **temporal pole**, while the posterior portion of the occipital lobe is called the **occipital pole**. The surface of the hemispheres consists of nerve cells or grey matter which is called the **cerebral cortex**. This is arranged in folds or convolutions thereby greatly increasing the total amount of grey matter. The convolutions are also called **gyri** and the furrows between them are termed **sulci** or fissures.

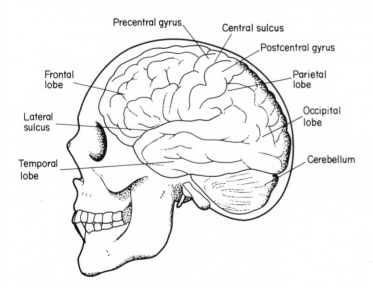

Fig. 17.3 Diagram of the left side of the brain, showing division into lobes and their relation to the skull.

Of the many fissures in the cerebral cortex of each hemisphere, two are of special importance.

(*1*) The **central sulcus** or **fissure of Rolando**, which runs downwards and forwards from the midline, separating the frontal from the parietal lobes.

(*2*) The **lateral sulcus** or **fissure of Sylvius**, which runs backwards and slightly upwards from the temporal pole, separating the frontal and parietal lobes above from the temporal lobe below.

If the margins of the lateral sulcus are separated, a buried portion of cerebral cortex called the **insula** (Fig. 17.6) is exposed. Its fibre connections are still largely obscure. Stimulation of the insula causes such effects as increased salivation, nausea, belching, gastric movements and abdominal sensations.

Localization of cerebral function

It has long been known that the area of frontal lobe cortex immediately in front of the central sulcus was concerned with voluntary movement. It was therefore acceptable to refer to it simply as the motor cortex. Similarly, the area of parietal lobe cortex immediately behind the central sulcus was known as the sensory area, since general sensation was represented there. Other solitary areas of cerebral cortex known to be concerned with special sensation were designated the visual area and the auditory or acoustic area.

The inaccuracies of this oversimplification are no longer acceptable. Firstly, the so-called motor and sensory areas are not exclusively motor or sensory but are of a dual or *sensorimotor* nature. Each has *afferent* fibres conveying information towards it and *efferent* fibres transmitting motor impulses away from it. Not even the visual and acoustic areas are purely sensory; they too are sensorimotor. It is true, nevertheless, that areas previously designated motor or sensory are *predominantly* motor or sensory respectively.

Secondly, the old terms misleadingly implied that in each cerebral hemisphere there was only one area of cortex for each motor and sensory function. The visual area, for example, was the term applied to a specific area in the occipital lobe. It is more accurate to speak of the *first, second and third visual areas*.

The **precentral area** includes the precentral gyrus and the posterior portions of the superior, middle and frontal gyri. The area may be referred to as the *first* or leading **motor area**, whilst bearing in mind

its actual sensorimotor character. The precentral area contains a large number of pyramidal cells of varying size. The fibres projecting from these cells are known as **pyramidal fibres** and they synapse (connect) with the dendrites of cells in the opposite side of the brainstem or spinal cord. The pyramidal cell body together with its long fibre (axon) constitutes the **upper motor neuron**. The cell body in the brainstem or spinal cord, together with its axon constitutes the **lower motor neuron**. Because the axons of the upper motor neurons cross over, a cerebrovascular accident affecting one side of the brain results in *contralateral* paresis (weakness) or paralysis, i.e. hemiparesis of the opposite side of the body. In contrast, damage to a peripheral nerve, resulting in a lower motor neuron lesion, affects the *same* side of the body, causing *ipsilateral* weakness or paralysis. The nature of the paralysis is also different in the two cases, being spastic in the case of upper motor neuron lesions and flaccid when the lesion is of the lower motor neurons. This is because the lower motor neuron is the *final common pathway* of all nervous influences on the skeletal muscle cell. The cell body in the anterior horn of the spinal cord (or in the brainstem) synapses not only with the pyramidal fibres but with many non-pyramidal fibres. Impulses from these fibres can alter the tone of muscle, making it hypertonic or spastic in patients with 'pyramidal' lesions. If the final common pathway is interrupted, however, no nerve impulses can get through and the muscle is not only paralysed but loses its tone, becoming hypotonic or flaccid. If the tendon reflexes are tested, they are exaggerated by upper motor neuron lesions and absent in cases of lower motor neuron lesions.

In the precentral cortex, voluntary movement of the various parts of the body are mediated, the lower limbs and trunk being represented in its upper part, near the longitudinal cerebral fissure separating the two hemispheres, and the head at the lower end of the area, just above the commencement of the lateral sulcus. The cortical areas for movements of the various parts of the body can be represented by superimposing a drawing of a little man (homunculus) on the cerebral cortex. This homunculus is not only upside down but grotesque in appearance, his hands and his lips being disproportionately large (Fig. 17.4). This is because the amount of cortex concerned with any particular movement depends not on the bulk of muscle

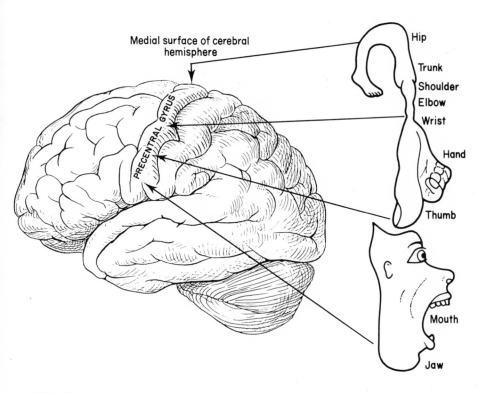

Fig. 17.4 The motor homunculous.

involved but on the skill customarily exercised in performing that movement. It should be remembered that in human beings the mouth is used for speech as well as for feeding.

The leading motor speech area, or Broca's area, is an extension of the precentral motor area and is situated in the inferior frontal gyrus of the dominant hemisphere. Damage to it, or to the fibres emanating from it, results in motor or expressive *aphasia*, i.e. loss of speech. A stroke causing a left hemiplegia frequently causes aphasia also.

Two other motor speech areas are located in the cortex of the dominant hemisphere. One (the second motor speech area of Wernicke) curves around the posterior end of the lateral sulcus and occupies a large part of the temporal lobe and rather less of the parietal lobe. The other is in the frontal lobe in the supplementary motor area.

The **supplementary motor area** is on the medial surface of the upper part of the frontal lobe, immediately anterior to the upper part of the first motor area. Like the latter, it is sensorimotor in nature but predominantly motor.

The first and second sensory areas in the parietal lobe are also sources of 'pyramidal' fibres, indicating that they have a motor function in addition to their main sensory role. The same is probably true of other areas of cerebral cortex.

The primary or **first sensory area** lies in the parietal lobe, immediately behind the central sulcus, and occupies most of the postcentral gyrus. In the latter, as in the precentral gyrus, the various parts of the body are represented in proportion to their skill rather than their size. An inverted and grotesque sensory homunculus can be drawn to illustrate this, just as a motor homunculus was drawn for the pre-

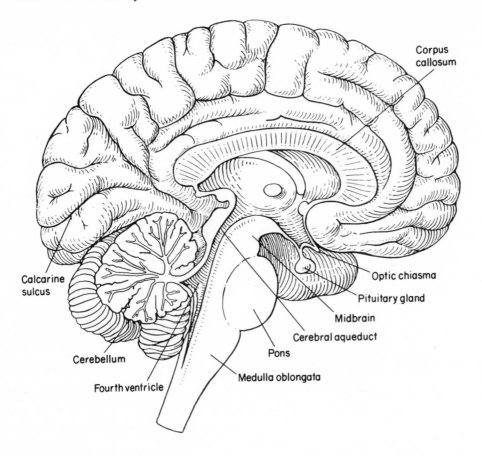

Fig. 17.5 Diagram showing the medial surface of the left cerebral hemisphere.

central gyrus. Sensations aroused in the first sensory area are mostly from the opposite (contralateral) side of the body but some are from the same (ipsilateral) side and some arise bilaterally.

The **second sensory area** is in the lowest part of the postcentral gyrus whilst the **third sensory area** is the previously mentioned supplementary 'motor' area in the medial frontal gyrus. All three sensory areas receive impulses conveying sensory information from the thalamus.

The special sense of *sight* is controlled by the grey matter in the occipital lobe. There are **three visual sensory areas**, linked with one another and with their equivalents in the opposite hemisphere, and with other cortical areas in both hemispheres, by association and commissural fibres. The primary visual cortex (area 17) or **striate cortex**, also known as the first visual area, is sited around the posterior

part of the calcarine sulcus (Fig. 17.5). It receives impulses from the temporal (lateral) half of the retina of the ipsilateral eye and from the nasal (medial) half of the retina of the contralateral eye. The striate area is therefore concerned with vision in the contralateral half of the total (binocular) visual field (Fig. 18.6). A left occipital lesion causes a right *homonymous hemianopia* (half-blindness) in which some or all objects to the right of the nose are not seen. Macular vision, as distinct from peripheral vision, is often spared because macular representation is separate from that of the peripheral visual fields and larger. The impulses from the eye pass first to the **lateral geniculate body** where they are processed and relayed to the visual cortex along pathways collectively named the **geniculate or optic radiation**. There is orderly, point-to-point, representation of the retina in the lateral geniculate body and in the striate cortex. Optic nerve fibres from the upper half of the retina (subserving vision in the lower half of the visual field) terminate in the medial half of the lateral geniculate body. The geniculate fibres from this site terminate in the *superior* lip of the calcarine sulcus. Fibres from the lower retina relay in the lateral half of the geniculate body, the fibres of which terminate in the *inferior* lip of the calcarine sulcus. Thus the inverted image formed on the retina (as on the film in a camera) is transmitted to the first visual cortex. However, the brain perceives it the right way up.

The other visual areas (areas 18 and 19) have been called the *visual association areas* and are concerned with the further elaboration of visual information reaching the first visual area. While area 17 appears to emphasize orientation, area 18 is more concerned with stereoscopic depth and area 19 with colour and movement. Other parts of the brain are involved in the manipulation and interpretation of all these visual stimuli.

The visual cortex is primarily sensory in nature. However, it can initiate eye movements and is therefore more accurately described as *sensorimotor*.

The **temporal lobe** is highly evolved in human beings and is involved in *hearing*, language and perception. Most of the **first acoustic area** is in the anterior transverse temporal gyrus, hidden in the lateral sulcus, but it extends a little into the superior temporal gyrus. A **second acoustic area** is located just inferior to the first acoustic area. Although the acoustic areas are primarily sensory they do give rise to *efferent* fibres which descend to the brainstem. Like the visual areas, therefore, the acoustic areas are *sensorimotor*. Because the auditory connections are bilateral, even an extensive tumour in one temporal lobe has little effect on hearing, although it may interfere with the interpretation of language.

In considering the parietal lobe, reference was made to the **second motor speech area** which occupies an even greater area of the **temporal lobe**, posterior to the acoustic region. The proximity of the acoustic and speech areas to one another is of interest, both being concerned with language.

The temporal lobe is also thought to have a *vestibular area*, concerned with orientation in space and maintenance of balance, but its site is uncertain.

The sense of *smell* (olfaction) is served by a relatively small part of the limbic lobe or *limbic system* (formerly called the rhinencephalon) in the human brain. Together with the hypothalamus, the limbic system is concerned with instincts, emotion (rage and fear), motivation and feeding, social and sexual behaviour, and complex neuroendocrine regulatory functions. The limbic system consists of cortical gyri on the medial aspect of each hemisphere (surrounding the corpus callosum) and associated deep nuclei, together with the olfactory nerves and associated structures.

The sense of *taste* does not have separate representation in the cerebral cortex but is subserved by that part of the postcentral gyrus which is concerned with cutaneous sensation from the face.

Identification of the sensorimotor areas leaves large parts of the cerebral cortex unaccounted for. These so called '*silent areas*' are probably concerned with such functions as the further elaboration, integration and interpretation of sensory information.

The interior of each hemisphere consists mainly of a mass of white matter formed by nerve fibres passing to and from the brain and between various parts of the brain itself. Collections of grey matter found deep in the cerebral hemispheres include the basal ganglia and hypothalamus.

The **basal ganglia** are specialized structures which are concerned in the planning and programming of muscular movements, including their modulation, so that they are not coarse and clumsy. This is achieved by a fine balance of *excitation* (facilitation) and *inhibition*. Prominent among the basal ganglia are the **caudate nucleus** and the **lentiform nucleus** which together comprise the **corpus striatum**. The lentiform nucleus has two distinct parts, a medial **globus pallidus** and a lateral **putamen**, which is of darker hue. There are connections between the individual members of the basal ganglia and between them and the cerebral cortex, the thalamus, various nuclei and the reticular formation of the brainstem.

The basal ganglia form part of a complex known as the **extrapyramidal system**, which is made up of those parts of the central nervous system, other than the pyramidal and cerebellar systems, which are concerned with *movement*. The pyramidal system of upper motor neurones is concerned with the execution of discrete movements. The extrapyramidal system makes these movements precise and harmonious with other muscular movements.

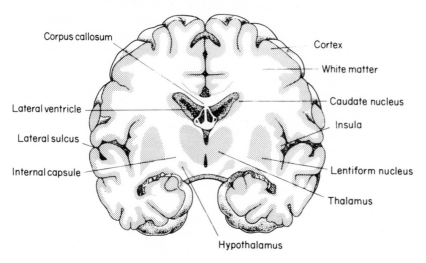

Fig. 17.6   Coronal section of the cerebrum.

Formerly it was believed that the main outflow of nerve fibres from the corpus striatum was downwards to the lower motor neurons, at which level they were thought principally to exert their modulating action. It is now clear that the major outflow is to the thalamus, from which ascending pathways carry the impulses up to the motor cortex. They are thus able to influence the pyramidal cells from which the programmed motor impulses are finally conducted down the corticospinal ('pyramidal') fibres to the lower motor neurons.

Diseases of the pyramidal system cause spastic paralysis. Diseases of the extrapyramidal system cause muscular *rigidity* (which is different from spasticity), *tremor*, and impoverishment of movement, as in *parkinsonism*; or other disorders of movement, especially the excessive (hyperkinetic) and abnormal movements known as chorea and athetosis. *Chorea* consists of rapid, irregular, semi-purposive, often jerky movements and is associated with lesions of the caudate nucleus. *Athetosis* is characterized by slow writhing movements and is due to lesions in the lentiform nucleus.

In many cases of parkinsonism, levels of *dopamine*, a neurotransmitter chemical in the *inhibitory* pathway, are reduced in the striatum (putamen and caudate nucleus). Dopamine given as replacement therapy would not cross the blood-brain barrier but levodopa does and is converted to dopamine in the brain, often resulting in dramatic improvement in parkinsonian patients. The *excitatory* afferent pathways are cholinergic, utilizing

acetylcholine – another neurotransmitter, and some improvement is obtained in parkinsonism if the cholinergic influence is reduced by anticholinergic drugs such as benzhexol.

Occasionally patients may be seen with involuntary movements caused by the large doses of phenothiazines commonly prescribed by psychiatrists. These drugs block dopamine receptors so that the impulses in the inhibitory dopaminergic fibres are rendered ineffective. To counterbalance this somewhat, patients receiving high doses of phenothiazines are, at the same time, routinely prescribed an anticholinergic drug.

The thalamus

The thalamus is predominantly *a sensory relay station*, with incoming fibres from the spinal cord and brainstem and onward fibres to the cerebral cortex. Of the various nuclei in the thalamus, the **non-specific projection nuclei** receive impulses from the reticular activating system of the brainstem (p. 236) and relay them widely to the cerebral cortex which is thereby alerted. Three groups of nuclei in the dorsal thalamus project to specific areas of the cortex. They are:

(1) Specific sensory relay nuclei. Of these, the ventrobasal group relay general sensory information to the postcentral gyrus and the medial and lateral geniculate bodies respectively relay auditory and visual impulses to the auditory and visual areas of the cerebral cortex.

(2) Nuclei concerned with efferent control mechanisms. Several of these are concerned with motor function. They influence the motor cortex in accordance with the information they receive from the basal ganglia and cerebellum. Other nuclei project to the limbic cortex and are concerned with recent memory and emotion.

(3) Nuclei concerned with complex integrative functions. These project to the cortical association areas and are concerned, for example, with language.

Lesions of the thalamus may cause pain, often continuous and of a burning nature, in the opposite side of the body. Accompanying emotional disturbance and lack of response to analgesics may give the false impression that the pain is psychogenic. The threshold to sensory stimuli in the affected half of the body may be raised but effective stimuli may excite sensations of a peculiarly unpleasant character, a form of *hyperpathia*.

The hypothalamus

This is composed of a number of nuclei and areas below the thalamus, at the base of the brain. The hypothalamus has neural connections to the posterior lobe of the pituitary gland (hypophysis) and vascular connections (known as portal hypophyseal vessels) to the anterior lobe of that gland.

The *functions of the hypothalamus* are:

(*a*) *Synthesis of vasopressin (antidiuretic hormone, ADH) and oxytocin* (see p. 204) which are subsequently stored in the posterior lobe of the pituitary gland prior to their release into the bloodstream.

(*b*) *Control of anterior pituitary secretion* by means of chemical agents (hypothalamic hormones) produced by the hypothalamus and transported to the pituitary gland by the portal hypophyseal blood vessels. These hypothalamic hormones either release or inhibit the release of anterior pituitary hormones. For example, growth hormone-releasing factor (GHRF) releases growth hormone (GH) from the pituitary gland and growth hormone release-inhibiting hormone (GHRIH) inhibits its release.

(*c*) *Control of appetite.* The hypothalamus contains both a 'feeding centre' and a 'satiety centre',

the function of the latter being to inhibit the former after ingestion of food. Lesions affecting the satiety centre result in overeating (hyperphagia) and hypothalamic obesity. Simple obesity, which is by far the commonest form of obesity, is not due to disease of the hypothalamus or endocrine glands.

(*d*) *Control of thirst.* Osmoreceptor cells in the hypothalamus are stimulated by an increased osmotic pressure (plasma hyperosmolality) to provoke thirst. Reduction in the extracellular fluid (ECF) volume without osmolality change, as after haemorrhage, induces thirst by a separate mechanism involving stimulation of baroreceptors and activation of the renin-angiotensin system (p. 202).

(*e*) *Regulation of body temperature.* Normal bodily function depends on chemical reactions, the speeds of which vary with temperature, aided by enzymes which function optimally within a narrow temperature range. A relatively constant body temperature is therefore necessary and this is maintained by the integration of reflex thermoregulatory responses in the hypothalamus.

(*f*) *Participation in autonomic (sympathetic and parasympathetic) responses.* The hypothalamus is involved in autonomic activity affecting heart rate, cardiac output, vasomotor tone, ventilation and the motility and secretory activity of the stomach and intestines.

(*g*) *Emotional feeling and expression.* The hypothalamus, limbic system and prefrontal cortex together form a complex system in which subjective feelings and objective physical accompaniments are integrated with changes in the environment (both external and internal) and with other cerebral activities.

In addition, positive and negative reward centres, having broadly opposite actions, are located in the hypothalamus and elsewhere. Stimulation of the positive reward centres causes pleasurable sensations; hence the term 'pleasure centres'.

(*h*) *Sexual behaviour* is influenced by the hypothalamus in concert with the limbic system.

(*i*) *Control of circadian rhythms.* Regular fluctuations occur in many bodily functions, with a periodicity of about 24 hours. Such circadian rhythms include the fluctuations which regularly occur in sleeping and wakefulness, in body temperature and in the secretion of cortisol by the adrenal cortex. In many cases, the hypothalamus exerts overall control over these rhythms. One nucleus in particular, the suprachiasmatic nucleus, has been called a *biological clock*. Fibres from the retina terminate in this nucleus and, by transmitting information about changes in environmental lighting, they could be involved in the photic regulation of circadian rhythms.

The internal capsule

Most of the important nerve fibres passing to and from the cerebral cortex lie in a relatively narrow tract of white matter situated near the basal ganglia, called the **internal capsule**. In horizontal section through the cerebrum (Fig. 17.7) this is seen as a broad bundle of white fibres sandwiched between the lentiform nucleus on its lateral side and the caudate nucleus and thalamus on its medial side.

The internal capsule has anterior and posterior limbs, connected by a genu, and parts below and behind the lentiform nucleus. The anterior limb contains fibres which arise in the frontal cortex and synapse with cells in the pons, the axons of which pass to the cerebellar hemisphere of the opposite side. It also contains fibres of the anterior thalamic radiation which connect the thalamus with the frontal cortex.

The **genu** contains fibres which connect the cerebral cortex with the motor nuclei of the cranial nerves, mostly of the opposite side of the head.

The **posterior limb** contains the corticospinal ('pyramidal') tract of fibres concerned with motor innervation of the limbs and trunk. It also contains fibres of the superior thalamic radiation carrying general sensory impulses from the thalamus to the postcentral gyrus of the cerebral cortex.

The **retrolentiform part** of the internal capsule contains the posterior thalamic radiation, which includes the optic radiation. The fibres of the latter arise in the lateral geniculate body and sweep backwards in close relationship to the lateral ventricle.

The **sublentiform part** of the internal capsule contains the acoustic radiation, the fibres of which sweep forwards and laterally from the medial geniculate body to reach the superior and transverse temporal gyri of the cerebral cortex.

The ventricles of the brain

Inside each hemisphere is a hollow space called the **lateral ventricle**. The left and right lateral ventricles are almost completely separated from each other by the septum pellucidum but each communicates with a midline cavity called the **third ventricle**. The

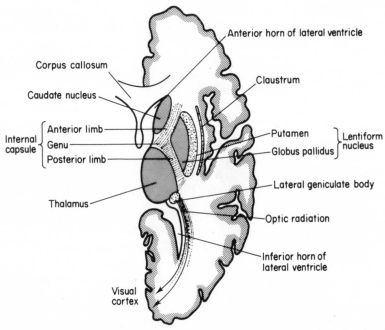

Fig. 17.7   Diagram of a horizontal section of the right cerebral hemisphere showing the internal capsule.

latter in turn communicates posteriorly with the **fourth ventricle**. The posterior surface of the pons and the upper part of the medulla oblongata form the floor of the fourth ventricle and its roof is formed by the undersurface of the cerebellum. It is continuous caudally with the central canal of the medulla oblongata and spinal cord. All of the ventricles contain cerebrospinal fluid (CSF) which is secreted by the choroid plexuses. CSF reaches the subarachnoid space via openings in the roof of the fourth ventricle. Of these, the **median aperture** is always present but either of the two lateral apertures may be absent.

The brainstem

This connects the forebrain and spinal cord. It consists of three parts, the midbrain and two portions of the hindbrain, namely the pons and the medulla oblongata.

The **midbrain** (mesencephalon), which is the shortest segment of the brainstem, joins the forebrain above to the pons and cerebellum below. For descriptive purposes the midbrain is divided into right and left **cerebral peduncles**. The two ventral portions of these are distinctly separate from each other and are called the **crura cerebri**. A layer of grey matter, called the substantia nigra, separates them from the dorsal part of the midbrain which is continuous across the midline and is known as the **tegmentum** (L. covering). The latter is traversed by the **cerebral aqueduct**, which connects the third and fourth ventricles. The part of the tegmentum dorsal to the aqueduct is called the **tectum** (L. roof) and it presents four rounded elevations known as the colliculi. The upper pair, the **superior colliculi**, are visual reflex centres and the lower pair, the **inferior colliculi**, are auditory reflex centres. There are connections between the superior and inferior colliculi which are involved in the integration of visual and auditory function.

The **pons** (L. bridge) is that part of the brainstem which lies between the midbrain and the medulla oblongata. On each side it is connected to the cerebellum by a middle cerebellar peduncle. The dorsal surface of the pons forms the upper part of the floor of the fourth ventricle. The white matter of the pons consists of nerve fibres passing up and down the brainstem and its grey matter comprises the nuclei of some of the cranial nerves.

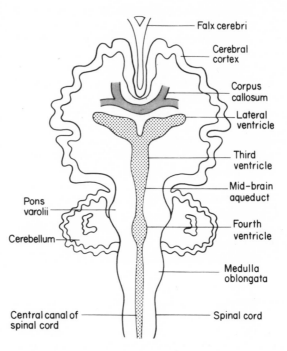

Fig. 17.8 Diagram of the ventricles of the brain (anterior view).

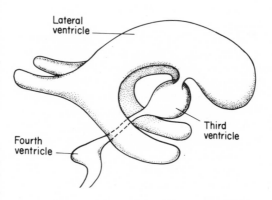

Fig. 17.9 Diagram of the ventricles of the brain (lateral view).

The **medulla oblongata** also consists of white and grey matter. About 3 cm (1.2 in) in length, it is continuous above with the pons and below, through the foramen magnum, with the spinal cord. The upper portion of the posterior surface of the medulla forms the lower part of the floor of the fourth ventricle. Two inferior cerebellar peduncles connect the pons with the cerebellum.

The anterior surface of the medulla is characterized by two longitudinal swellings, one on each side of the midline fissure, caused by the pyramidal tracts and known as the **pyramids**. They contain the fibres which descend from the cerebral cortex to the spinal cord, namely the corticospinal fibres. In the lower part of the medulla most of these fibres cross the median plane, in what is known as the **decussation of the pyramids**, and pass down the opposite side of the spinal cord in the lateral corticospinal tract.

The medulla oblongata contains collections of grey matter known as the **vital centres**, because death usually follows damage to them. They integrate the complex reflexes which regulate heart rate, blood pressure and ventilation of the lungs; hence the terms cardiac centre, vasomotor centre and 'respiratory' centre.

Other autonomic reflex responses which are integrated in the medulla are swallowing, gagging, vomiting, coughing and sneezing.

The reticular formation

The reticular formation, which is a part of the reticular activating system (RAS), is a network of intermingled grey and white matter deep in the brainstem, extending throughout the medulla, pons and midbrain. It consists of numerous islets of grey matter interlaced with nerve fibres (white matter) running in all directions. Information converges on it from all the major parts of the nervous system to alert the individual. Conversely it directly or indirectly projects back to these regions, including the cerebral cortex via the thalamus, to modulate their activities.

It is the activity of the reticular formation which produces arousal, awakening and the conscious alert state with attention and perception. This results from the sensory input received from all parts of the body, including the special sense organs of sight, hearing, taste and smell. In turn, the reticular formation exerts its influence on the activity patterns of skeletal (striated) muscle, partly

via the reticulospinal tracts, cardiac and involuntary (unstriated) muscle, the limbic system, the hypothalamus and pituitary gland, and the pineal gland. Thereby, it contributes to cardiac and respiratory function (hence the concept of controlling 'centres' for these functions), emotional behaviour, learning, memory, and the functioning of the 'biological clocks' responsible for circadian rhythms.

Sleep and wakefulness

The state of consciousness of an individual is reflected in the electrical activity of the brain which is revealed by the electroencephalogram (EEG). When special pads or electrodes are applied to the head and connected to the apparatus, waves or oscillations can be recorded which indicate that there is a continous discharge of energy from the neurons which normally spreads in an orderly manner. When individual parts of the brain are called into activity there is an alteration in the character of these waves.

For example, if records are taken from the region of the occipital cortex which is concerned with sight, the waves alter when the eyes are opened and shut.

An *alpha rhythm* is the dominant rhythm recorded from the scalp of a normal conscious adult resting with closed eyes and allowing the mind to wander. It consists of 8–12 waves per second in a fairly regular pattern.

When the eyes are opened, sensory stimulation is received or mental concentration is attempted, the alpha rhythm is replaced by fast, irregular (desynchronized) low-voltage activity. This *desynchronization* of the resting rhythm is sometimes referred to as the arousal or alerting response.

In sleep, the alpha rhythm is replaced by other waves. Normal sleep is made up of two distinct kinds of sleep which alternate with each other.

Slow-wave sleep is associated with synchronized large *slow* waves interspersed with bursts of alpha-like waves known as *sleep spindles*. The term synchronized as applied to alpha rhythm and the rhythm of slow wave sleep, implies the rhythmic discharge of neurons. Normal sleep commences with Stage 1 of slow-wave sleep. There are four stages, Stage 4 being the deepest.

Rapid eye movement (REM) sleep, or paradoxical sleep, in which dreaming occurs, is characterized by rapid roving movements of the eyes and

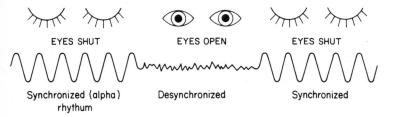

Fig. 17.10 Electroencephalogram — rhythym recorded from occipital cortex.

rapid, irregular (desynchronized) low voltage electrical activity in the EEG, arising from the pontine reticular formation and the occipital visual cortex.

Slow-wave sleep is sometimes called non-REM sleep. It occupies about 75% of sleeping time. Discharge of serotonin-secreting (serotonergic) neurons appears to be involved in the production of slow-wave sleep. No single synaptic transmitter is apparently responsible for REM sleep. Surprisingly, sleepwalking and bed-wetting (nocturnal enuresis) are associated not with dream sleep but with arousal from slow-wave sleep.

Arousal can be initiated either by sensory impulses ascending in the reticular activating system or by impulses descending from the cerebral cortex down corticofugal fibres to the reticular formation.

Normal adult sleep is deepest during the first two hours. Most people spend about one-third of their lives asleep. The need for this is still something of a mystery and active research into sleep continues. It may be said that sleep is a restorative process, during which growth and repair can be given greater priority in the body's economy. The amount of sleep required is highly variable between individuals but young people generally need more than middle-aged and elderly people. Growing children need the most sleep and it is probably relevant that growth hormone is secreted mainly during sleep.

Most physiological functions are reduced to their lowest activity during sleep.

In summary:

(*1*) General metabolism and heat production are reduced.
(*2*) The heart is slowed and blood pressure falls.
(*3*) Respiration is slowed and the thoracic type of breathing predominates over abdominal breathing.

(*4*) Renal excretion is diminished.
(*5*) Skeletal muscle tone is markedly diminished in REM sleep and tendon reflexes are reduced. Sphincter muscles remain contracted, however, and peristalsis continues in the alimentary tract.
(*6*) Changes in the vasomotor system result in a diminished blood supply to the brain while blood vessels in the skin are dilated and perspiration is increased.

Prolonged lack of sleep may affect an individual's work performance but insomnia is generally less important than it is commonly thought to be. The body will tend to get the sleep that it needs and it is helped by the fact that a sleep deficit does not have to be paid off by sleeping for the number of hours lost. Even a cumulative loss may be adequately recompensed by sleeping for an extra hour or two.

Some individuals, anxious about not sleeping, tend to exaggerate their insomnia. The same individuals, or others, may fail to add their afternoon or evening nap into their total daily hours of sleep.

Some people are naturally better sleepers than others but the habit of sleeping soundly can be acquired. Many simple methods are available to help people get off to sleep and others who awake during the night may find they can sleep again after making themselves a warm drink, reading, writing or listening to some pleasant music. Failure to get off to sleep, however, may be due to *anxiety* and awakening in the early morning hours (2–3 a.m.) may be due to *depression*. Improvement in sleeping is often one of the earliest responses to treatment of these conditions. Apart from psychological causes of insomnia, including grief, physical symptoms may prevent or impair sleep. *Pain* may necessitate analgesic therapy and *paroxysmal nocturnal dyspnoea*, another cause of early morning awakening, may indicate the need for diuretics or other cardiac drugs.

The electroencephalogram (EEG) is not only a valuable research tool for monitoring sleep. It is also useful in routine clinical investigation, especially in patients with suspected epilepsy.

The cerebellum

The cerebellum, the largest part of the hindbrain, is situated in the posterior cranial fossa of the skull and lies below the occipital lobes of the cerebrum, from which it is separated by the fold of dura mater called the tentorium cerebelli (p. 224).

Like the cerebrum, the cerebellum has a fissured outer cortex of grey matter and an inner core of white matter containing nuclei of grey matter. It consists of right and left cerebellar hemispheres joined by a median portion called the **vermis**, of which the most anterior part seen on the inferior surface is known as the **nodule**. Closely related to the nodule are partially detached portions of the cerebellum known as the **flocculi**, of which there is one on each side. Each flocculus has a **peduncle** which is a narrow band of afferent (incoming) and efferent (outgoing) fibres. The flocculi, their peduncles and the nodule make up the **flocculo-nodular lobe** which is one of the two fundamental parts of the cerebellum, the other being the **corpus cerebelli**, which is the remainder of the cerebellum.

The cerebellum is connected to the brainstem on each side by superior, middle and inferior peduncles. Stretching between the superior cerebellar peduncles, and forming a part of the roof of the fourth ventricle, is the superior medullary velum, which is continuous with the white matter of the vermis.

The internal structure and functional organization of the cerebellum is complex but the general principles of its function are readily understood. It is primarily concerned with the co-ordination of movement. To perform this task it requires information and it receives this from (*a*) the motor areas of the cerebral cortex and (*b*) sensory receptors, namely vestibular and auditory receptors, visual receptors, proprioreceptors (detecting sensations from within the body, e.g. as to the positions of its parts), cutaneous tactile (touch) receptors, and visceral receptors.

The connections of the flocculonodular lobe are essentially all vestibular. If it is damaged, as in childhood by a malignant cerebellar tumour known as a medulloblastoma, the patient walks on a broad base with a staggering gait and tends to fall. The flocculonodular lobe is also the seat of motion sickness.

Other portions of the cerebellum are stimulated by auditory and visual impulses, proprioceptive impulses from the whole body, tactile and other sensory impulses and impulses from the motor cortex. Homunculi can be plotted on the cerebellar cortex as on the cerebral cortical surface.

The auditory and visual impulses travel to the cerebellum from the inferior and superior colliculi

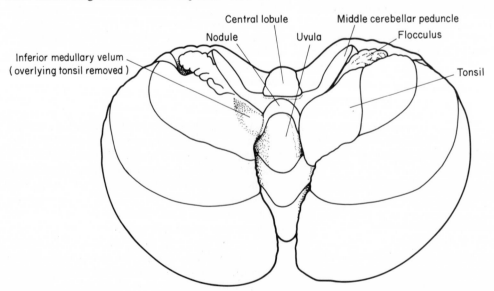

Fig. 17.11 The cerebellum from below.

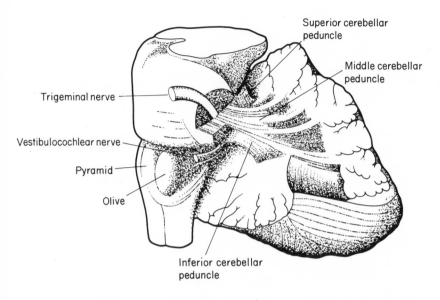

Superior cerebellar
peduncle

Middle cerebellar
peduncle

Trigeminal nerve

Vestibulocochlear nerve

Pyramid

Olive

Inferior cerebellar
peduncle

Fig. 17.12   The dissected left cerebellar hemisphere.

along the tectocerebellar tract. Stimulation of this region of the cerebellum causes the head and eyes to turn to the ipsilateral (same) side.

Cerebellar destruction results in diminished muscle tone or *hypotonia*. At rest, the patient may otherwise seem normal but when movements are made *ataxia* is manifest. This is incoordination due to errors in the rate, range, force and direction of movement. An ataxic gait is a wide-based unsteady gait which may suggest that the patient is drunk. Ataxia of the skilled movements necessary for speech results in a slurred voice or scanning speech. These are forms of *dysarthria*, which is a disorder of articulation, the motor function by means of which words formed in the brain are given sound. Dysarthria has to be distinguished from dysphasia, due to lesions of the speech areas of the cerebral cortex, which is a disorder of the comprehension and expression of meanings by means of words.

Nystagmus is the oscillating movement of eyes which are trying to fix on an object. In this case the ataxia is affecting the external ocular muscles.

The ataxic patient overshoots when making a voluntary movement. An attempt to touch some-

thing with a finger results in *past-pointing*. Excessive correcting action then causes the erring finger to overshoot to the other side. The finger consequently oscillates wildly from side to side, a phenomenon known as *intention tremor*.

Unilateral lesions of the cerebellum cause symptoms and signs on the ipsilateral (same) side of the body, not on the opposite side as in the case of cerebral lesions.

The cerebellum is principally concerned with muscle tone and coordination. It adjusts and smoothes out movement, exerting its regulatory influence both before and during movement. The muscles primarily responsible for a movement are called *agonists* and they are assisted by *synergistic* muscles and by *fixation muscles*, which give anchorage. At the same time, the tone of *antagonistic* muscles must diminish by just the right amount at the right time. A precise interaction of all these muscles underlies both reflex and skilled voluntary movements and it is the function of the cerebellum to organize the coordinated activity of all the muscles. The anatomical basis for the control of movement is represented in Fig. 17.13.

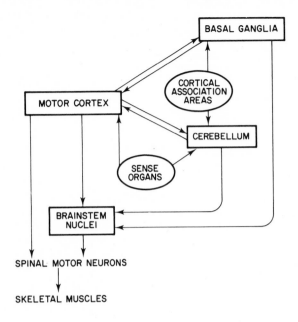

Fig. 17.13   Pathways involved in the regulation of movement.

Summary of functions of the major parts of the brain

Functions of the cerebrum

(*1*) To receive all sensory stimuli and to convey most of them to consciousness.
(*2*) To initiate all voluntary movement.
(*3*) To correlate and retain all impulses received, thereby forming the basis of memory.
(*4*) To formulate and associate ideas giving rise to intelligence.
(*5*) To exercise unconscious control over many functions of the body.
(*6*) To exercise control over the lower parts of the brain.

The symptoms of cerebral disease depend upon which parts of the brain are involved. For instance, a lesion (e.g. thrombosis) in the internal capsule or the motor cortex may cause spastic hemiplegia. Damage to the visual pathway will cause a loss of part of the visual field.

Functions of the cerebellum

The activity of the cerebellum is unconscious and not under the control of the will. It is concerned with posture and movement.

(*1*) It helps to maintain muscular tone.
(*2*) It coordinates muscular movement.
(*3*) It helps to maintain balance and equilibrium. It is able to do this because of the impulses it receives from the semicircular canals, joints and muscles.

Cerebellar disease therefore results in erratic and uncoordinated movements (ataxia) which may be associated with an intention tremor (not present at rest) and a tendency to fall to one side when standing or walking.

Functions of the brainstem

The midbrain conveys impulses to and from the cerebral hemispheres.

The pons conveys impulses to and from the cerebrum and also to and from the cerebellum.

Further, all parts of the brainstem contain collections of grey matter, which are the nuclei of the cranial nerves, and the grey and white matter of the reticular formation.

The medulla oblongata connects the other parts of the brain to the spinal cord. In addition to the nerve fibres passing up and down in its substance and the nuclei of many of the cranial nerves, it contains collections of grey matter known as the vital centres. These are:

(1) The respiratory centre which controls the rate and depth of ventilation.
(2) The vasomotor centres which control the calibre of the blood vessels.
(3) The cardiac centre which influences the heart rate.
(4) Visceral centres such as the swallowing centre, the vomiting centre, and centres for movements of the stomach and the secretion of saliva and gastric juice.

The brainstem centres are similar in function to the reflex centres in the spinal cord. They receive afferent impulses from the periphery and send out motor impulses in response to these stimuli. They may also be influenced by the higher centres of the brain. Thus thoughts may induce nausea or a quickening of the heart.

With so many vital structures contained in the small space of the medulla, together with the fact that all impulses to and from the brain pass through it, it is clear that any disease, injury or pressure on the medulla is very serious and often fatal.

Poliomyelitis is one of the diseases which may affect the brainstem, causing bulbar paralysis affecting breathing and swallowing and necessitating artificial ventilation.

Neurotransmitters

The receiving processes extending from neurons are called dendrites and the efferent processes are called axons. The axon, of which there is normally one per cell, branches near its termination into several fine **axon terminals**. These end in close proximity to other cells. When the other cell is also a neuron, the junction is known as a **synapse**. When the other cell is a muscle cell, a **neuromuscular junction** is formed. The axon terminals may also end in glands or other structures.

Transmission of information from one neuron to another at a synapse is in most cases neurochemical, although there are also electrically acting synapses similar to the electrical junctions in cardiac and unstriated muscle. *Neurochemical transmission* depends on the ability of axon terminals to synthesize, store and release the chemical transmitter. In some cases the axons also reabsorb the transmitter, a process called reuptake, and store it for subsequent use.

Many substances have been identified as neurotransmitters. Among the major ones are *acetylcholine*, the catecholamines *noradrenalin* and *dopamine, serotonin* and the amino acid *gamma aminobutyric acid (GABA)*. The transmitter substance binds with special protein molecules called *receptors* on the surface of the cell to which information is being transmitted. The effect on that cell may be either *facilitatory* (excitatory) or *inhibitory* according to whether the transmitter raises or lowers the resting potential of the cell. The effect is required only for a limited time and this limitation is achieved either by reuptake of the transmitter into the axon terminals, as in the case of the catecholamines, or by enzymatic destruction of the transmitter. An example of the latter is the destruction of acetylcholine by the action of *cholinesterase*, which is present in the synaptic gap. In the muscle disease known as *myasthenia gravis* anticholinesterase drugs, such as prostigmine, are used to inhibit the enzyme and thereby enhance the action of acetylcholine.

Nerve fibres which secrete acetylcholine are called *cholinergic* whilst those which secrete noradrenalin, dopamine and serotonin are respectively termed *noradrenergic, dopaminergic* and *serotonergic*.

Acetylcholine is the transmitter at synapses in the autonomic ganglia (p. 256) and at the terminals of parasympathetic and somatic motor nerve fibres, i.e. in the latter case, at neuromuscular junctions. Acetylcholine also acts as a transmitter in virtually all parts of the brain including the corpus striatum. Depletion of dopamine at this site in cases of Parkinson's disease may cause an unbalanced state in which the action of acetylcholine is largely unopposed. Anticholinergic drugs, such as benztropine, are therefore useful in parkinsonism.

Histochemical techniques are used to map out the distribution of groups of neurons which secrete a particular neurotransmitter substance. Thus cholinergic, noradrenergic, dopaminergic and serotonergic neuron groups have all been identified in the reticular formation.

Apart from its presence in parts of the central nervous system, noradrenalin is the transmitter at peripheral (post-synaptic) sympathetic nerve endings. Its action depends upon the type of receptor present in the innervated cell surface. Alpha-adrenergic (α) receptors are stimulated to cause contraction of vascular smooth muscle and hence vasoconstriction. An alpha-adrenergic blocking drug such as phentolamine will prevent this constriction of blood vessels. Beta-1 (β_1) receptors in cardiac muscle respond to adrenalin and noradrenalin by increasing the rate and force of contraction of the heart; effects which can be prevented by administering a beta-adrenergic blocking drug such as propranolol. Beta-2 (β_2) receptors in the bronchi show an inhibitory response to adrenalin and noradrenalin and bronchodilation results from the relaxation of bronchial smooth muscle. In this case, a non-selective beta-adrenergic blocking agent like propranolol will encourage bronchoconstriction. Propranolol is therefore contraindicated in asthmatic patients.

Another class of neurotransmitters is characterized by having polypeptide chains of five or more amino acid residues. These include *substance P* and the opioid peptides, of which the *enkephalins, dynorphins* and *endorphins* seem to be the most important. The opioid peptides bind to opiate receptors and function as either short-acting neurotransmitters or long-acting neurohormones. They are found mainly in the sensory and autonomic systems, limbic system and neuroendocrine systems. They are still being investigated but because they have analgesic activity, the endorphins have been described as the body's own naturally occurring morphine-like substances.

The cranial nerves

There are 12 pairs of cranial nerves, which arise from the brain and brainstem. Some of them are *sensory*, or afferent, bringing impulses to the brain; others are *motor*, or efferent, carrying impulses from the brain to the periphery, while a few are mixed and contain both motor and sensory fibres.

The cranial nerves are arranged from above downwards as follows:

I	Olfactory
II	Optic
III	Oculomotor
IV	Trochlear
V	Trigeminal
VI	Abducent
VII	Facial
VIII	Vestibulocochlear
IX	Glossopharyngeal
X	Vagal
XI	Accessory
XII	Hypoglossal

I The olfactory nerves

These are sensory and serve the sense of smell. Their fibres arise from olfactory cells in the mucous membrane of the upper part of the nose. These fibres ascend through foramina in the cribriform plate of the ethmoid bone and synapse with cells in the **olfactory bulb**. In turn, the axons of these cells pass back in the olfactory tract to the primary olfactory cortex in the undersurface of the temporal lobe.

Loss of the sense of smell, *anosmia*, may result from conditions such as the common cold affecting the nasal mucosa, frontal lobe tumours and head injuries. In addition to anosmia, fractures involving the anterior cranial fossa may cause cerebrospinal fluid to leak into the nose, from which it may drip, a condition known as CSF rhinorrhoea.

II The optic nerve

This is the nerve of sight and is entirely sensory. Its fibres commence in the retina of the eye as the axons of ganglion cells. These axons converge from all parts of the retina to form the optic disc, which may be viewed in living subjects using an instrument called an ophthalmoscope. Behind the optic disc, the fibres pierce the sclera (outer coat of the eye) to form the optic nerve.

The optic nerve (and retina) is an extension of the brain and is sheathed with extensions of the meninges. Raised intracranial pressure is therefore transmitted along the extension of the subarachnoid space around the optic nerve, resulting in swelling of the nerve 'head', or optic disc, a condition which is known as *papilloedema*. This may be detected by ophthalmoscopy and may provide a

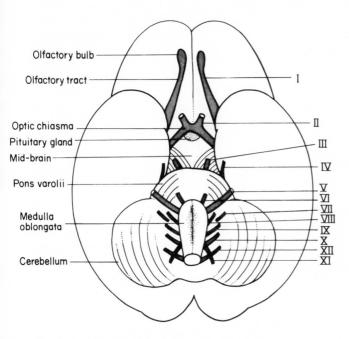

Olfactory bulb

Olfactory tract

I

Optic chiasma

II

Pituitary gland

III

Mid-brain

IV

Pons varolii

V
VI
VII
VIII
IX
X
XII
XI

Medulla oblongata

Cerebellum

Fig. 17.14   Diagram illustrating the origin of cranial nerves from the base of the brain.

valuable clue to the presence of a space-occupying lesion such as a tumour.

A lesion affecting one optic nerve causes unilateral loss of vision. Inflammation of the nerve (optic neuritis) may be responsible; if it involves the optic disc (papillitis) it may be detected by ophthalmoscopy but often the disc is spared and the nerve is affected behind the eye (retrobulbar neuritis).

The optic nerve passes backwards through the optic foramen and meets its fellow nerve on the undersurface of the brain at the **optic chiasma** (Fig. 17.14) where the medial fibres of each nerve cross to the optic tract of the opposite side. These are the fibres from the medial half of the retina, concerned with the temporal half of the visual field. A large adenoma of the pituitary gland, or a suprasellar tumour (above the sella turcica) may press on the optic chiasma and cause loss of vision in both temporal fields, i.e. a *bitemporal hemianopia*.

The fibres from the lateral half of the retina, serving the nasal portion of the visual field, pass backwards in the optic tract of the same side, together with the fibres which have crossed from the other side. Most of them relay in the lateral geniculate body, from which fibres pass to the visual

cortex in the occipital lobe. Because the optic tract is composed of crossed and uncrossed fibres, lesions cause visual loss in the temporal half of one visual field and in the nasal half of the other, resulting in *homonymous hemianopia*, i.e. blindness to one side, the side opposite to that of the lesion.

III (Oculomotor), IV (trochlear) and VI (abducent)

The oculomotor, trochlear and abducent nerves may be considered together because they are all motor nerves of the eyeball. They supply the extrinsic muscles which move the eye and injury to any one of them results in double vision (diplopia) and a squint (strabismus). The **third cranial nerve** also conveys parasympathetic fibres which synapse in the ciliary ganglion, a small ganglion situated in the apex of the orbit. The efferent (post-ganglionic) fibres from this ganglion supply the ciliary body, which effects accommodation of the lens, and the iris muscles which constrict the pupil. A complete third nerve lesion therefore causes not only a diplopia and a divergent squint but also dilatation of the pupil, due to the unopposed action of sympathetic dilator pupillae nerve fibres. Pupillary reflexes (accommodation-convergence and reaction to light) are lost because of paralysis of the constrictor pupillae fibres. As the third nerve also supplies the muscle which suspends the upper eyelid, a complete third nerve lesion also causes drooping of the lid, or *ptosis*.

The **trochlear nerve** supplies only one eye muscle, the superior oblique. A lesion of the fourth cranial nerve results in paralysis of this muscle and diplopia so that the patient is unable to turn the eye downwards and laterally. Attempts to do so cause the eye to be rotated medially, resulting in diplopia.

The **abducent nerve** supplies only the lateral rectus muscle. Paralysis of this nerve results in diplopia and convergent squint. The sixth cranial nerve has a long intracranial course and is vulnerable to injury at the same time as fractures of the base of the skull are sustained. Lateral squint may also be a *false localizing sign* of an intracranial lesion as a result of the abducent nerve, in its long course, being stretched in patients with raised intracranial pressure. The reader will note that two possible signs of raised intracranial tension have now been mentioned, the first being papilloedema.

V The trigeminal nerve

This is the largest cranial nerve and is the sensory nerve of the face and anterior part of the scalp. It also has a small motor portion which supplies the muscles of mastication and is therefore a nerve of mixed type. As its name suggests, it has three divisions:

   (1) the **ophthalmic division**, which supplies the forehead, a large part of the scalp, the upper eyelid, the conjunctiva of the eye and most of the nose;

   (2) the **maxillary division**, which supplies the cheek and upper jaw, including its teeth, the nasal septum, the palate, nasopharynx, uvula and tonsils;

   (3) the **mandibular division** distributed to the temple, parts of the ear and lower face, the lower jaw and its teeth, the anterior two-thirds of the tongue and the floor of the mouth. The mandibular nerve also conveys the small motor root to the muscles of mastication and secreto-motor fibres to the parotid salivary gland.

The **trigeminal ganglion**, which lies near the apex of the petrous temporal bone, is the first cell station for the sensory fibres of the fifth cranial nerve. It is the equivalent of the dorsal sensory ganglion of a

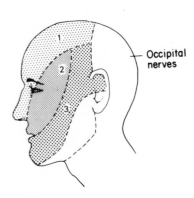

Fig. 17.15 Distribution of the branches of the trigeminal nerve. **1** Opthalmic branch. **2** Maxillary branch. **3** Mandibular branch.

spinal nerve. Controlled heat-coagulation of the trigeminal ganglion is one of the surgical procedures used in cases of trigeminal neuralgia in which drug therapy has proved ineffective.

VII The facial nerve

The seventh cranial nerve is predominantly a motor nerve supplying the muscles of facial expression. It also conveys secreto-motor fibres from the parasympathetic system to the sublingual and submandibular salivary glands and the lacrimal gland. In addition, it carries sensory fibres of taste from the anterior two-thirds of the tongue, other types of sensation being conveyed by the mandibular branch of the fifth cranial nerve. The motor fibres originate in the pons and the sensory fibres originate in cells in the genicular ganglion, within the temporal bone.

A lesion of the seventh nerve or its nucleus causes paresis (weakness) or paralysis of all the muscles of the face on that side. The commonest cause is Bell's palsy which is thought to be due to a viral infection; the swollen nerve is compressed within the facial canal of the temporal bone and corticosteroids may be used to reduce the swelling and decompress the nerve. The patient with Bell's palsy may at first be thought to have had a 'small stroke'. However, upper facial movements are preserved after a stroke (an upper motor neuron lesion) because they are represented in the cortex of both cerebral hemispheres. Consequently, the patient can still close the eye on the affected side as well as on the unaffected side. The patient with Bell's palsy (a lower motor neuron lesion) cannot do this because it is the final common pathway, in the seventh nerve trunk, which is affected.

VIII The vestibulocochlear nerve

The eighth cranial nerve is a sensory nerve consisting of two sets of fibres which convey impulses from the internal ear to the brain. One set of fibres forms the vestibular nerve, which is concerned with equilibrium (p. 264). The other set forms the cochlear nerve, which is the nerve of hearing.

The **vestibular nerve** is composed of fibres arising from the cells of the **vestibular ganglion** which is situated in the internal acoustic meatus. Impulses from the semicircular ducts are conveyed along the nerve, which lies in the internal acoustic meatus, to the vestibular nuclei in the medulla. From these nuclei, efferent fibres pass via the inferior cerebellar peduncle to the cerebellum, mostly to the flocculus and nodule. The inferior cerebellar peduncle also conveys some vestibular nerve fibres which bypass the vestibular nuclei and pass directly to the cerebellum. Other fibres in the inferior cerebellar peduncle convey impulses in the opposite direction, i.e. from cerebellum to vestibular nuclei. Other connections of the vestibular nuclei are to the motor nuclei of the cranial nerves (III, IV and VI) supplying the eye muscles and to motor nuclei of muscles of the neck. Efferent fibres from the lateral vestibular nucleus descend to form the vestibulospinal tract. The connections of the vestibular system enable it to influence the muscles of the eyes, neck, trunk and limbs so as to preserve balance.

The fibres of the **cochlear nerve**, or nerve of hearing, originate in the spinal ganglion of the cochlea. They pass along the internal acoustic meatus to the medulla and terminate in the ventral and dorsal cochlear nuclei.

The vestibulocochlear and facial nerves are near to each other as they leave the brain stem and they enter the internal acoustic meatus together. An acoustic neuroma in the cerebello-pontine angle therefore soon compresses the facial nerve and causes unilateral facial paralysis of lower motor neuron type. Fractures of the base of the skull may cause loss of taste in the anterior two-thirds of the tongue (VIIth nerve) and loss of hearing (VIIIth nerve) on the same side.

IX The glossopharyngeal nerve

The ninth cranial nerve is a mixed nerve. It contains sensory fibres which supply the pharynx, tonsil and posterior one-third of the tongue, including its taste buds. A branch named the carotid nerve serves the carotid sinus and carotid body. It provides the afferent pathways for the baroreceptor and chemoreceptor reflexes associated with these structures.

The motor fibres of the glossopharyngeal nerve supply the stylopharyngeus muscle which helps to elevate the pharynx in swallowing and speaking. Secretomotor fibres are supplied to the parotid salivary gland.

X The vagus nerve

The vagus nerve is also mixed. It has the most extensive course and distribution of all the cranial nerves. It arises from the medulla oblongata, passes through the jugular foramen in the base of the skull and continues downwards in the neck to the thorax, where it accompanies the oesophagus before passing through the diaphragm and into the abdomen. It is distributed to the pharynx, larynx, trachea, bronchi, lungs, heart, oesophagus, stomach, upper intestine and kidneys.

The recurrent laryngeal branch of the vagus nerve has a different origin and course on the two sides. On the right side it arises in front of the subclavian artery behind which it passes and ascends obliquely to the side of the trachea. It is intimately related to the inferior thyroid artery. The left recurrent laryngeal branch arises as the vagus nerve crosses the arch of the aorta. It winds below the ligamentum anteriosum, under the arch, and passes upwards in the groove between the trachea and oesophagus. Both recurrent laryngeal nerves give branches to the muscles and mucous membrane of the larynx.

Damage to one of the recurrent laryngeal nerves may result in paralysis of the ipsilateral vocal cord and hoarseness of the voice. The nerve may be damaged in the neck by malignant lymph nodes or a carcinoma of the thyroid gland. The left recurrent laryngeal nerve may be affected in the thorax by a bronchial or oesophageal carcinoma, a mass of enlarged mediastinal lymph nodes, or an aneurysm of the aortic arch. Rarely, an enlarged left atrium, in a patient with mitral stenosis, causes a recurrent laryngeal nerve palsy by pushing the left pulmonary artery upwards so that it compresses the nerve against the aortic arch.

The superior and inferior **cardiac branches** of the vagus nerve are inhibitory to the heart, which is slowed when they are stimulated. The vagus nerve may be stimulated reflexly by applying external pressure to the carotid sinus in the neck. This is sometimes done to stop a fast abnormal cardiac rhythm but is potentially dangerous and should be performed only by a physician. In a minority of people with a very sensitive carotid sinus, mild stimulation, such as the pressure of a shirt collar when the neck is turned, may be sufficient to cause loss of consciousness (carotid sinus syncope) due to vagal inhibition of the heart.

A little above the diaphragm the vagus nerves form **anterior and posterior vagal trunks** in front of the oesophagus and behind it, respectively. These enter the abdomen with the oesophagus, through the oesophageal opening in the diaphragm, and supply **gastric branches** to the stomach. The surgeon divides the two vagal trunks in the operation of truncal vagotomy in order to reduce the acid-pepsin secretion of the stomach in patients with a duodenal ulcer. However, total *truncal vagotomy* denervates not only the stomach but also the liver, gallbladder, pancreas and most of the intestines. As this widespread denervation is a possible cause of diarrhoea and other postvagotomy complications, some surgeons divide only the gastric branches of the vagus nerve (*selective vagotomy*) or only those supplying the parietal cell mass (*proximal gastric* or *highly selective vagotomy*).

Amongst the other branches of the vagus nerve the **auricular branch** is of interest when there is a need to syringe wax out of ears. The distribution of this branch includes the posterior wall and floor of the external acoustic meatus and stimulation of this area may reflexly cause coughing, vomiting or cardiac syncope.

XI The accessory nerve

This is a motor nerve possessing cranial and spinal roots. The cranial root is distributed in branches of the vagus nerve to the muscles of the palate, pharynx and larynx. Paresis of these muscles results in dysphagia and dysphonia.

The spinal root of the accessory nerve supplies the sternomastoid and trapezius muscles in the neck. Central irritation causes clonic spasm of these muscles, a condition known as spasmodic torticollis.

XII The hypoglossal nerve

The twelfth cranial nerve is the motor nerve of the tongue. A lesion of this nerve, or its nucleus in the medulla, causes paralysis, or weakness (paresis), and wasting of the muscles of the tongue on the same side and deviation of the protruded tongue towards that side. In contrast to this lower motor neurone lesion, an upper motor neurone lesion (e.g. as part of a stroke) causes paresis without any wasting of the tongue muscles.

The spinal cord

The spinal cord is a part of the central nervous system which lies in the vertebral canal. It commences above at the level of the foramen magnum where it is continuous with the medulla oblongata. In the adult, it is about 45 cm (18 in) long and tapers into the **conus medullaris** to end at the lower level of the first lumbar vertebra; it therefore does not completely fill the lower part of the vertebral canal. In childhood, the cord extends to a lower level than the first lumbar vertebra. Surrounding the spinal cord are the meninges. The dura mater and arachnoid mater form a sac which encloses the cord and extends downwards to the lowest part of the vertebral canal, while the pia mater closely covers the surface of the cord in the same way as it adheres to the surface of the brain. A prolongation of the pia mater, the **filum terminale**, descends from the conus medullaris to be attached to the back of the coccyx. Cerebrospinal fluid fills the subarachnoid space between the arachnoid mater and pia mater.

The spinal cord which is oval in cross-section (Fig. 17.17), is increased in circumference in two areas, namely the **cervical and lumbar enlargements**. From these enlargements the important nerves to the arms and legs arise (see Fig. 17.18).

The spinal nerves arise from the sides of the spinal cord by two **roots**, anterior and posterior, which unite as they leave the vertebral column to form the **spinal nerve** trunk.

There are, in all, thirty-one pairs of spinal nerves corresponding to the segments of the vertebral column, namely:

8 cervical,
12 thoracic or dorsal,
5 lumbar,
5 sacral,
1 coccygeal.

The nerve roots from the lumbar, sacral and coccygeal regions occupy the space in the lower part of the vertebral canal below the termination of the spinal cord, before passing between their corresponding vertebrae. This leash of nerve roots is called the **cauda equina** (like a horse's tail).

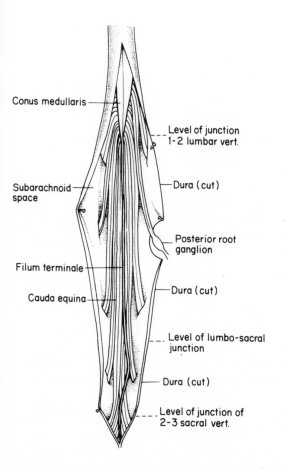

Conus medullaris

Level of junction 1-2 lumbar vert.

Subarachnoid space

Dura (cut)

Posterior root ganglion

Filum terminale

Cauda equina

Dura (cut)

Level of lumbo-sacral junction

Dura (cut)

Level of junction of 2-3 sacral vert.

Fig. 17.16 Lower end of the dura and arachnoid laid open to show the cauda equina.

Structure

The structure of the spinal cord is best understood by study of transverse cross-sections. It has been noted that the grey matter of the brain is situated on the surface and the white matter in the interior. In the spinal cord this arrangement is reversed, the **white matter** being on the surface while the **grey matter** is arranged in an H-shaped manner in its interior.

It will be seen (Fig. 17.17) that the cord is oval in transverse cross section and that the H-shaped grey matter divides the white matter into ventral (anterior), lateral and posterior columns or funiculi. The ventral columns of the two sides are separated from one another by the anterior median fissure. In the centre of the spinal cord is a minute canal, the **central canal**, which is continuous above with the fourth ventricle in the medulla oblongata and contains cerebrospinal fluid. Most of the latter fluid is, however, contained in the subarachnoid space.

The grey matter of the spinal cord

The H-shaped grey matter is commonly described as having anterior and posterior horns, which represent the **anterior and posterior grey columns**. There is also a **lateral grey column** on each side, appearing as a lateral horn in cross-sections of the thoracic and upper lumbar segments of the cord. The lateral grey columns contain the cells from which the preganglionic fibres of the sympathetic nervous system originate.

The ventral or anterior grey columns contain the motor cells (anterior horn cells) from which arise the nerve fibres forming the ventral or anterior roots of the spinal nerves.

The posterior horns are capped by the **substantia gelatinosa**, a crescentic mass of small nerve cells and fibres intimately concerned with the connections of incoming afferent nerve fibres. The dorsal or posterior root of the spinal nerve, its sensory root, enters the posterior horn of grey matter.

The white matter of the spinal cord

The white matter of the spinal cord is white because of the high proportion of myelinated fibres in it. Sensory fibres ascend in the white matter to the brain and motor fibres descend from the brain. The motor fibres are situated in the lateral columns of

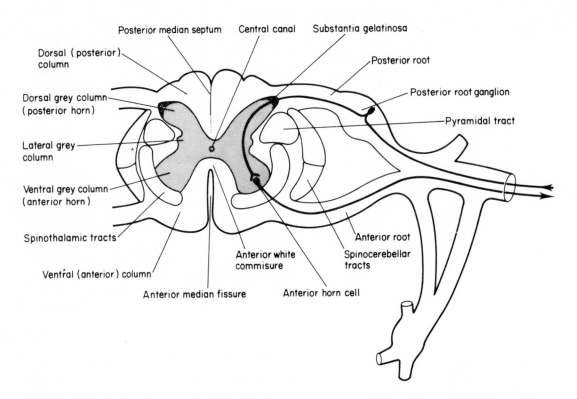

Fig. 17.17   Diagram of transverse cross-section of spinal cord and nerve roots, showing the relative positions of the grey matter and various tracts of white matter.

white matter in a special tract, the **lateral corticospinal tract** or **pyramidal tract**. Its fibres are the axons of the pyramidal cells (upper motor neurons) of the motor cortex. They cross over (decussate) in the medulla and descend in the lateral corticospinal tract of the contralateral side of the cord. The fibres terminate in the anterior columns of grey matter, where they synapse with motor cells (anterior horn cells, lower motor neurons) in successive spinal segments.

Injury or disease causing complete transection of the lower part of the spinal cord results in paralysis of both lower limbs – *paraplegia*. If the lesion is higher in the cord, all four limbs may be affected – *quadriplegia*. These conditions present many nursing and social problems, such as care of the bladder, prevention of bed sores and provision of a suitable invalid carriage.

The sensory fibres enter the cord in the posterior roots of the spinal nerves. Fibres conveying sensations of fine touch (tactile discrimination), pressure, vibration and *proprioception* (position

and movement) ascend in the **posterior tracts** (dorsal columns) of white matter to the medulla oblongata. There they synapse with neurons in the **gracile and cuneate nuclei**. From these, most of the axons of the second set of neurons cross the midline and ascend, in a band known as the medial lemniscus, to the thalamus. From this relay centre, fibres convey sensory impulses to the cerebral cortex (postcentral gyrus). Some of the axons of the second neurons, in the gracile and cuneate nuclei, pass to the cerebellum, which also receives the fibres of the spinocerebellar tracts. Unconscious adjustments of muscular tone can therefore be made in response to proprioceptive information received.

Fibres conveying *pain and temperature sensation* synapse with neurons in the posterior grey columns (posterior horns). Most of the axons from the secondary neurons cross the midline, decussating with the corresponding fibres from the opposite side, to ascend in the anterolateral portion of the spinal cord. These ascending pathways are known as the **anterior and lateral spinothalamic tracts**.

There are many other ascending and descending tracts in the spinal cord. Fibres also ascend and descend to and from the brainstem reticular system and these are known as spinoreticular and reticulospinal fibres.

The spinal nerves

It has been seen that there are thirty-one pairs of spinal nerves, each arising from the spinal cord by two roots, ventral (anterior) and dorsal (posterior). The latter unite to form the main nerve trunk as it leaves the vertebral column. On each dorsal nerve root there is a collection of nerve cells called the **spinal ganglion** (posterior root ganglion).

It has also been pointed out that all sensory nerve fibres reach the spinal cord by the posterior root and all motor nerve fibres leave by the anterior root. It follows that the peripheral nerve trunk is a

mixed nerve containing both sensory and motor fibres.

The individual nerve trunks arising from certain regions of the spinal cord join up together to form what is called a plexus from which they emerge rearranged as the individual peripheral nerves. Two plexuses are formed by the cervical nerves and two by the lumbar and sacral nerves.

Cervical plexus

The upper four cervical nerve trunks unite to form a plexus lying deeply in the upper part of the neck from which peripheral nerves are distributed mainly to the skin and muscles of the head and neck.

The most important branch of the cervical plexus is the **phrenic nerve** which passes down in the neck to enter the thorax, where it is closely related to the side of the pericardium. It terminates in the diaphragm of which it is the motor nerve. Paralysis of the diaphragm may result from a lesion in the neck or

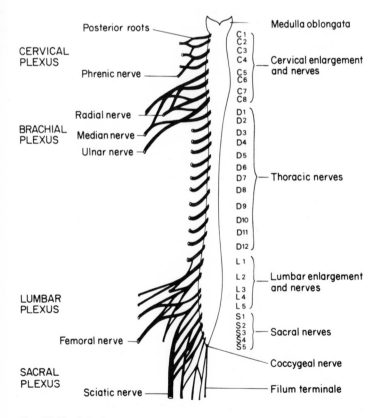

Fig. 17.18 Spinal nerves.

thorax. Thus it may be due to entrapment of the phrenic nerve in the mediastinum by a bronchial carcinoma. The affected side of the diaphragm is then displaced upwards and moves paradoxically; it is seen to move upwards instead of downwards, on fluoroscopy, when the patient takes a vigorous sniff.

Brachial plexus

This is formed from the lower four cervical and the first thoracic nerves, and is situated deeply in the lower part of the neck behind the clavicle. From this plexus the important nerves of the arm are derived. These nerves enter the axilla where they are closely related to the axillary artery.

Among the more important branches of the brachial plexus are:

The **axillary (circumflex) nerve**, which is related to the surgical neck of the humerus and supplies the deltoid muscle. In this region it may be injured by a fracture of the bone or in dislocations of the shoulder joint.

The radial nerve which, after leaving the axilla, winds round the posterior aspect of the shaft of the humerus in the spiral groove and supplies branches to the skin, the triceps muscle and the extensor muscles in the back of the forearm. It is liable to injury in the upper part of its course and when paralysed produces the characteristic deformity known as '*wrist drop*', in which the patient is unable to extend the wrist or fingers. The nerve is liable to crutch pressure in the axilla.

The **ulnar nerve** which is at first related to the axillary artery and then accompanies the brachial artery in the arm. It passes down behind the medial epicondyle of the humerus to reach the ulnar side of the forearm. In addition to supplying muscles in the forearm, it gives branches to the skin of the ring and little fingers and small muscles in the hand.

It is the close relationship with the medial epicondyle of the humerus, where the nerve is relatively superficial, which has given rise to the popular name '*funny bone*' for this part of the elbow. A severe knock in this region stimulates the ulnar nerve and produces pain together with a tingling sensation in the ring and little fingers, which clearly demonstrates the distribution of the nerve. The nerve is most commonly injured at this site, behind the medial epicondyle, and the resulting paralysis is

characteristic. The hand assumes a claw shape, in which the metacarpophalangeal joints, especially of the ring and little fingers, are extended and the interphalangeal joints are flexed, owing to the unopposed action of opposing muscles.

The **median nerve** which also lies close to the brachial artery in the arm. Thereafter it passes down in the mid-line of the front of the forearm to the hand. It supplies muscles in the forearm and hand and gives cutaneous branches to the thenar eminence and central part of the palm and to the first three and one-half digits.

Injury to the median nerve usually occurs just above the wrist, resulting in inability to oppose the thumb and loss of sensation on the palmar surfaces of the thumb, index and middle fingers and the radial half of the ring finger. If the nerve is damaged higher in the forearm, there will be many other effects, including weakness in the action of pronation.

At the wrist, the median nerve lies in a restricted space, between the flexor retinaculum and the carpal bones, called the carpal tunnel. If this space

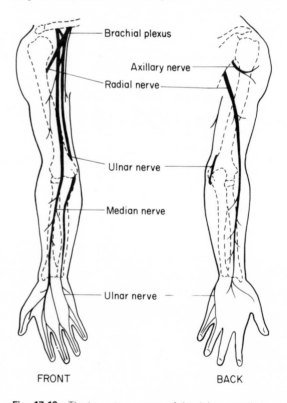

Fig. 17.19 The important nerves of the right upper limb.

is compromised by bony injury or arthritis, or by soft tissue swelling in some patients with pregnancy, myxoedema or acromegaly, the *carpal tunnel syndrome* results. In many cases the compression is 'spontaneous', meaning that it is incompletely understood. There is weakness and wasting of the outer thenar muscles (short muscles of the thumb). Pain and tingling are felt and sensation is impaired in the digits supplied by the nerve but not in the palm because this is supplied by the palmar cutaneous branch of the nerve, which arises above and lies superficial to the retinaculum.

The thoracic nerves

These are twelve in number and one passes to each intercostal space. They run forwards between the ribs, supply branches to the intercostal muscles and muscles of the anterior abdominal wall, and are finally distributed to the skin of the thorax and front of the abdomen.

The lumbar plexus

This plexus originates from the first four lumbar nerves and lies within the psoas major muscle in front of the transverse processes of the lumbar vertebrae. Its principal branches are the femoral and obturator nerves.

The **femoral nerve** is the largest branch of the lumbar plexus. It passes behind the inguinal (Poupart's) ligament to enter the femoral (Scarpa's) triangle in the front of the thigh. It supplies the muscles of the front of the thigh (principally quadriceps), the hip and knee joints (sensory fibres) and skin on the front of the thigh and medial side of the leg, ankle and foot.

The skin of the lateral aspect of the thigh is supplied not by branches of the femoral nerve but by the **lateral cutaneous nerve of the thigh**, which arises directly from the lumbar plexus. It enters the thigh after passing behind or through the inguinal ligament, usually about a centimetre medial to the anterior superior iliac spine. Entrapment of this

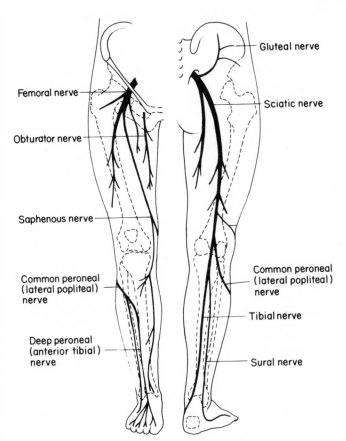

Fig. 17.20 The important nerves of the right lower limb. **(a)** Front of limb; **(b)** back.

nerve results in the condition known as *meralgia paraesthetica* (from the Greek meros, thigh, and algos, pain), with pain, numbness and paraesthesiae referred to the lateral aspect of the thigh. The condition may remit spontaneously; if not, it is speedily relieved by surgical division of the fibrous tissue compressing the nerve.

The **obturator nerve** runs downwards and forwards to leave the pelvis through the obturator foramen in the innominate bone, thereupon entering the thigh. It supplies the adductor muscles of the thigh, the hip and knee joints, and an area of skin over the medial aspect of the thigh. In cases of spastic paraplegia, spasm of the adductor muscles can be relieved by division of the obturator nerve.

The sacral plexus

This originates from the fourth and fifth lumbar and first four sacral nerves. Its branches supply the muscles of the pelvis and hip and the skin of the buttock and back of the thigh. The plexus terminates as the pudendal nerve and the much larger sciatic nerve.

The **pudendal nerve** provides the principle motor and sensory innervation of the perineum. Pudendal nerve block, with a local anaesthetic injected through a long needle, has been used in Obstetrics prior to forceps delivery, to relax the pelvic floor muscles and abolish sensation in the lower vagina and vulva.

The **sciatic nerve** is the largest nerve in the body. It passes out of the pelvis through the greater sciatic foramen of the innominate bone into the gluteal region. Thence it descends down the back of the thigh, the hamstring muscles of which it supplies, to about its lower one-third, where it terminates by branching into the tibial and common peroneal nerves. The sciatic nerve can be represented on the back of the thigh by a broad line. First, a line is drawn between the ischial tuberosity and the apex of the greater trochanter, both of which bony landmarks can easily be felt through the skin. From just medial to the midpoint of this line, another line is drawn down to meet the popliteal fossa just medially to its apex. A knowledge of this anatomy will prevent intramuscular injections being given too medially and misplaced into the sciatic nerve, causing severe pain and weakness of dorsiflexion and plantar flexion of the foot.

In the buttock, the sciatic nerve may also be damaged by penetrating injuries and by posterior dislocations or fracture dislocations of the hip joint. Higher up, in the pelvis, the sciatic nerve may be compressed in the later stages of pregnancy or damaged by the direct extension of a pelvic neoplasm.

A complete sciatic nerve lesion causes paralysis of the hamstrings and of all the muscles below the knee, with a foot drop deformity. All cutaneous sensibility below the knee is also lost except for the area, medially, innervated by the saphenous nerve, which is a branch of the femoral nerve.

The term '*sciatica*' has been used for a syndrome, characterized especially by pain, due to compression of lumbar spinal nerves by one or more protruding intervertebral discs. However, it must be realised that sciatica is a symptom (pain in the lower limb), and not a disease, and that it may occasionally be due to serious intrapelvic disease.

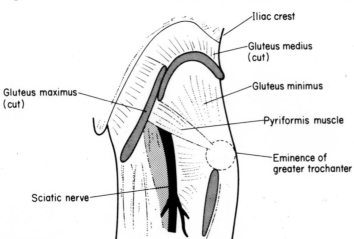

Fig. 17.21 Diagram showing the position of the sciatic nerve.

Sensation and the sensory path

Two types of sensation may be considered, special sensation and general sensation (see also Chapter 18).

Special sensations are those which can only be detected by specialized organs, i.e. smell, sight, hearing and taste. Sensory impulses received from these organs are conveyed to the brain by the appropriate cranial nerve (see p. 242). *General sensations* are the feelings which are appreciated by all parts of the body. They include the superficial sensations detected by the skin and the deep sensations felt in the muscles, joints and other organs.

Superficial sensation

The important superficial sensations are appreciated by the sensory nerve endings in the skin and include:

pain;
touch; and
temperature (heat and cold).

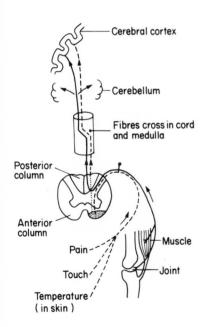

Posterior column

Anterior column

Pain
Touch
Temperature (in skin)

Cerebral cortex

Cerebellum

Fibres cross in cord and medulla

Muscle

Joint

Fig. 17.22 The sensory path.

Deep sensation

In addition to pressure and deep pain, the most important deep sensation is that of the position of muscles and joints. We are always aware of the exact position in space of any limb or part of a limb. This may be demonstrated by closing the eyes and bringing the tip of the index finger to the end of the nose. This can only be done accurately because we know exactly where the respective parts are by muscle and joint sense, without the aid of vision. Vibration is also a deep sensation.

The sensory path

Both superficial and deep sensations travel in the peripheral nerve from the skin, joint or muscle towards the spinal cord. After reaching the trunk of the nerve, the sensory fibres enter the posterior horn of grey matter in the spinal cord via the posterior nerve root. The two types of nerve fibre then take separate courses.

The fibres conveying sensations of position, pressure, vibration and light touch pass upwards in the posterior columns. Those taking the superficial sensations of pain and temperature go to the anterior columns.

Both sets of fibres cross to the opposite side of the cord either before or when they reach the medulla oblongata. Thence they are conveyed via the brainstem, thalamus and white matter of the brain to the parietal and other sensory areas of the cerebral cortex. The sensory impulses of one side of the body are, therefore, like the motor impulses, dealt with by the opposite side of the brain.

Movement and the motor path

Movement may be voluntary or involuntary. Involuntary movement is considered in the section on reflex action.

Voluntary movement commences with an impulse sent out by the pyramidal cells of the motor cortex situated in front of the central sulcus (fissure of Rolando). The axons of these cells pass through the white matter of the brain and brainstem. In the medulla oblongata they cross to the opposite side and travel down in the lateral column of the spinal cord as the lateral corticospinal or pyramidal tract. At the appropriate level the fibres leave the pyramidal tract and end around the cells in the anterior horn of grey matter.

The pyramidal nerve cell in the motor cortex and its axon extending as far as the anterior horn cell is called the **upper motor neuron**.

The motor impulse is then relayed through the anterior horn cell whose fibre passes via the anterior nerve root to form, with the incoming fibres of the posterior root, the main nerve trunk. The motor nerve reaches its destination in the muscle via the peripheral nerve.

The anterior horn cell and its axon passing in the peripheral nerve to the muscle is called the **lower motor neuron**.

Because the upper motor neurons cross over in the medulla (at the decussation of the pyramids), one side of the brain controls the muscles on the opposite side of the body.

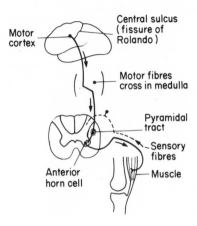

Fig. 17.23   The motor path.

Reflex action

Reflex action may be defined as the automatic motor response to a sensory stimulus, and is therefore independent of the will. The structures concerned in the production of a reflex constitute 'the reflex arc'.

They consist of:

(*1*) a sense organ such as the skin or the nerve endings in a muscle, tendon or other organ;

(*2*) an afferent or sensory nerve passing from the sense organ via the peripheral nerve and posterior nerve root to the spinal cord;

(*3*) the spinal cord; and

(*4*) an efferent or motor nerve commencing in the anterior horn cells of the cord and passing via the peripheral nerve to the motor organ, for example muscle or gland.

It is clear that if this route is taken, the time elapsing between the application of the sensory stimulus and the motor response will be much less than if the impulse had to pass the whole length of the spinal cord to the sensory area of the brain and thence to the motor cortex, which in turn would send out a voluntary impulse down the motor path in the spinal cord before it could reach the peripheral nerve and motor end-organ.

It follows, therefore, that many reflexes are protective in character and designed to obtain the quickest possible motor response. For example, the finger is withdrawn from a hot object before we have time to think about it.

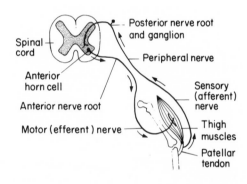

Fig. 17.24   Diagram illustrating reflex action.

Other reflexes are concerned with automatic control of functions which do not require the supervision of consciousness, such as the secretion of gastric juice when food enters the stomach.

Common examples of reflex action include:

Withdrawal of a part of the body from any painful stimulus such as a pinprick or excessive heat.

The knee jerk or patellar reflex. If the patellar tendon is tapped sharply, when the muscles of the leg are relaxed, the leg is jerked suddenly forward (extended) at the knee.

Stroking the lateral border of the sole of the foot results in the downward movement (flexion) of the big toe.

A crumb in the larynx causes coughing.

In some cases reflex action may be operated via the organs of special sense, for example the sight or smell of appetizing food causes a reflex flow of saliva ('makes the mouth water').

Inhibition
It has been seen that in infancy the control of the bladder and rectum is automatic or reflex in character. As age increases, voluntary control over these organs is developed. This means that the higher centres of the brain are able to control or inhibit certain reflex actions. When the controlling influence of the brain is removed by injury or disease, the uninhibited reflex mechanism may again operate and, in the case of the bladder and rectum, the patient will become incontinent.

The autonomic nervous system

In addition to those parts of the central and peripheral nervous system already described, with somatic sensory and motor functions, there is a second system of nerve cells and nerve fibres which have special and very important visceral functions. This visceral or autonomic system innervates the viscera (e.g. heart and stomach), glands, blood vessels and unstriped muscle generally. The word autonomic means 'self-controlling'.

Although the involuntary system can exercise its functions independently of the somatic part of the central nervous system, they are both closely associated and are able to influence each other.

Whereas the somatic part of the central nervous system is largely concerned with the interpretation of sensory impulses and in sending motor impulses to the voluntary or striped muscle in the body, the involuntary system is concerned mainly with the control of all the involuntary, unstriped, smooth muscle in the body such as is found in the blood vessels, abdominal viscera and bladder. Many glands, for example the salivary and sweat glands, also come under its influence. In general, the somatic part of the nervous system is concerned with responses of the body to the external environment, whilst the autonomic system is concerned with control of the internal environment.

The essential feature of the involuntary system is a series of **ganglia** or masses of nerve cells forming synapses outside the brain and spinal cord. These ganglia may be classified into three groups.

(*1*) the cranial component;
(*2*) the thoracolumbar component;
(*3*) the sacral component.

The cranial autonomic and sacral autonomic portions belong to the parasympathetic system. The thoracolumbar component belongs to the sympathetic system.

It will be seen later that, in many instances, the involuntary system is capable of sending impulses of directly opposite character to the same organ. Thus there are involuntary fibres which increase the rate of the heart-beat and fibres which decrease the rate. In such instances the opposite impulses are conveyed by the sympathetic and parasympathetic respectively. In fact almost all the unstriped involuntary muscles have both sympathetic and parasympathetic supplies.

1. The cranial autonomic system (parasympathetic)

This consists of the ganglia and fibres which are associated with the cranial nerves. The most important are:

(*a*) fibres which control the calibre of the pupil of the eye and are associated with the IIIrd or oculo-motor nerve;
(*b*) fibres which run with the Xth or vagus nerve to the heart, bronchi and alimentary tract.

These fibres convey impulses opposite in action to those distributed by the sympathetic system to the same organs.

2. The sympathetic system

This consists of two chains of ganglia situated one on either side of the front of the vertebral column. Each of these ganglia has connecting fibres which pass to the corresponding nerve roots of the spinal nerves. Instead of separate ganglia for each segment of the spinal cord, however, in certain segments a number are collected together to form a single large ganglion (e.g. the cervical ganglion in the neck and the stellate ganglion in the upper thoracic region).

From certain of these ganglia nerve fibres are collected into special groups of plexuses from which efferent fibres pass to the organs, from which afferent fibres may be received. The most important plexuses are:

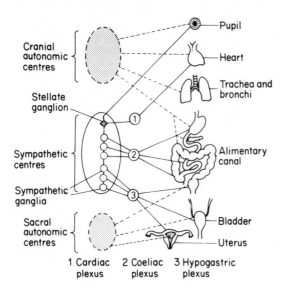

Fig. 17.25 Diagram showing the distribution of the autonomic nervous system.

(a) The **cardiac plexus** which is connected via the stellate ganglion to the lower cervical and upper thoracic (dorsal) spinal nerves and supplies the heart and lungs.

(b) The **coeliac plexus** (solar plexus) connected with the lower thoracic (dorsal) spinal nerves. This lies in the epigastrium behind the stomach and close to the coeliac axis artery. It supplies sympathetic fibres to the stomach, intestine, suprarenal glands and other viscera in the abdominal cavity.

3. The sacral autonomic system

This supplies parasympathetic fibres to the structures in the pelvis including the rectum, bladder, uterus and other reproductive organs.

It is important to remember that reflex action can take place through the afferent and efferent fibres and the ganglia of the involuntary system just as in the central nervous system.

In mentioning the opposite actions of the sympathetic and parasympathetic systems it was indicated that both sets of fibres were conveyed to the heart. The parasympathetic fibres reaching the heart from the vagus carry impulses which tend to slow its rate (inhibitors), while those from the sympathetic (cardiac plexus) tend to increase its rate (accelerators).

It is also true to say that the structures innervated by the involuntary nervous system are very largely influenced by the secretions of ductless glands. In particular, noradrenalin from the suprarenal gland stimulates the smooth muscle in the walls of small arteries and acts as a vasoconstrictor.

Chemical transmitters

Messages are transmitted from one neuron to another and from neurons to muscle or gland cells by chemical transmitters. The transmitter at all somatic and parasympathetic nerve terminals is acetylcholine. This is also the transmitter at the terminals of preganglionic sympathetic nerve fibres. All of these fibres are therefore referred to as *cholinergic*. Adrenalin and noradrenalin are the transmitters at the terminals of postganglionic sympathetic nerve fibres which are therefore termed *adrenergic*. The only exception is the postganglionic sympathetic fibres to sweat glands, which are cholinergic. The term preganglionic refers to those fibres entering the ganglia. They synapse with cells having fibres (postganglionic) which leave the ganglia to supply the organs.

Adrenergic receptors

The sympathetic nervous system exerts its effects through special receptors which are found in certain tissues. There are two distinct types of receptor, known as the alpha (α)-receptor and the beta (β)-receptor. α-receptors are generally excitatory in function; their stimulation causes, for example, contraction of smooth muscle in the walls of blood vessels, resulting in vasoconstriction. β-receptors are generally inhibitory in function; their activation causes relaxation of smooth muscle in the walls of blood vessels and bronchi, resulting in vasodilatation and bronchodilatation. The β-receptors in cardiac muscle are exceptional in being excitatory; their stimulation causes the heart to beat more rapidly and more forcibly. These are called beta-1 receptors whereas those in the bronchi are beta-2 receptors.

These facts have important applications in Pharmocology and Therapeutics where, for example β-receptor stimulants (e.g. salbutamol) are used in asthma and β-receptor blocking agents (e.g. propranolol) are used in the treatment of certain diseases of the heart.

Questions

1. Describe what you know of the meninges.
2. Which structures secrete the cerebrospinal fluid and where does it flow? Describe a lumbar puncture and how the nurse can help during and after the procedure.
3. What are the main parts of the brain?
4. Which parts make up the brainstem and what is meant by the term 'brain death'?
5. Describe the main features of the cerebrum.
6. What are the basal ganglia and what physical signs may result from disease of these structures?
7. Write brief notes on the thalamus.
8. What are the functions of the hypothalamus?
9. What is the reticular formation?
10. Write notes on sleep and the physiological changes which occur during it.
11. Describe the cerebellum and its functions.
12. What are neurotransmitters? Name two of them.
13. List the cranial nerves. Name the functions with which any six of them are concerned.
14. Give an account of the spinal cord. What is meant by the terms paraplegia and quadriplegia?
15. Write notes on the sciatic nerve and its importance in Medicine and Nursing.
16. What is your understanding of reflex action? Give examples. Give also an example of inhibition of a reflex action.
17. Write notes on the autonomic nervous system.

18
The Sense Organs

The afferent or sensory nerves reaching the central nervous system have their beginnings (although these are sometimes referred to as nerve endings) in various peripheral structures such as the skin, muscles, joints and special organs such as the eye and ear.

Sensations are the conscious results of processes taking place in the brain following the arrival of impulses derived from the sensory nerves.

The structures concerned in the production of a sensation are:

(*1*) An end-organ or **sensory receptor**, situated in the periphery at the terminations of the sensory nerves;

(*2*) The afferent nerve fibres in the peripheral nerve and spinal cord;

(*3*) The thalamus, which is a cell station relaying sensations to the cerebral cortex;

(*4*) The sensory reception areas in the brain which are connected with various psychic areas where the impulse is interpreted and may be stored as memory.

It is important to remember that, although the brain receives and appreciates the sensation, it projects it back to the site or end-organ at which it was received and it is actually felt by the individual as being in the peripheral region.

The end-organs for each sense are specially adapted in structure to respond to the particular stimulus for that sense. Thus, waves of light are described as being the '*adequate stimulus*' for the nerve endings in the retina of the eye. Sound waves are the adequate stimulus for the endings of the cochlear nerve in the ear, and have no effect on the eye, nose or skin.

Sensation may be classified in various ways. In describing the nervous system it was sufficient to make the simple subdivisions of (*a*) special senses, i.e. sight, hearing, taste and smell, (*b*) general sensations, including all the others. Numerous other classifications of the senses have been made, however, and the reader may encounter some of the terms used to describe the various receptors. These include:

(a) **telereceptors** or 'distance receivers' (e.g. for sound and smell);

(b) **exteroceptors**, concerned with elements of the immediate external environment (e.g. for taste);

(c) **interoceptors**, concerned with the internal environment (e.g. hunger and thirst);

(d) **proprioceptors**, concerned with the position of the body or its parts in space (e.g. receptors in the joints and semicircular canals).

Sensation in the skin

The sensations from the skin are those of pain, temperature (hot and cold) and light touch. Situated in the skin is a mosaic of minute sensory areas, which correspond with the nerve endings. Specific endings are present for each of the varieties of sensation. Further, the endings for one type of sensation may be more numerous in one area of skin than in another.

The sense of smell

The mechanism of smell is dependent on:

(*1*) the receptor or end-organ, i.e. the endings of the 1st cranial or olfactory nerve in the mucous membrane of the upper part of the nasal cavity;

(*2*) the olfactory bulb and tract which convey the impulses to the brain;

(*3*) the limbic lobe situated in the medial surface of the cerebral hemisphere.

The appropriate or adequate stimulus for the sense of smell must be either in the form of gas or minute particles which are soluble in the secretions of the nasal mucous membrane. These gases or particles are conveyed to the nasal cavities by the air and the quantity reaching the uppear part of the nasal cavities may be increased by the process of 'sniffing.'

Smell is a sense characterized by its extreme delicacy and the ease with which it becomes fatigued. This is shown by the fact that persons sitting in a closed room for some time may cease to be aware of an odour which is very apparent to someone entering from the fresh air.

The sense of smell is also diminished by inflammation or excess of secretion in the nasal mucous membrane which occurs in the common cold.

Odours are most simply classified into pleasant and unpleasant. It is important to remember that the sense of smell is closely associated with the sense of taste and that the majority of flavours are actually appreciated by the olfactory organ. This is clearly shown by the loss or change of taste which accompanies a severe head cold. Loss of the sense of smell is called anosmia.

The sense of taste

The true sense of taste is localized in the tongue. There are four basic tastes, namely bitter, sweet, sour and salt. These only are appreciated by the tongue. All other flavours are appreciated by the sense of smell as already stated.

The tongue is a very mobile organ, consisting of muscles some of which arise from the hyoid bone and the lower jaw. It is covered by mucous membrane. The roughness of its upper surface is due to numerous minute elevations called papillae. The end-organs for the sense of taste are called **taste buds**, which are situated most densely at the tip, sides and base of the tongue. They consist of collections of receptor cells and supporting cells together with the sensory nerve endings which are wrapped intimately around the receptor cells. The taste bud has a pore which opens on to the epithelial surface of the tongue. A number of hairs projects from each receptor cell into this taste pore. The nerve fibres conveying impulses to the brain from the taste buds of the *anterior two-thirds* of the tongue travel in the **facial nerves** while those from the *posterior third* of the tongue travel in the **glossopharyngeal nerves**. The sensations of pain and touch are conveyed by the trigeminal nerves. There is no area of the cerebral cortex concerned solely with taste. This sense is represented in that portion of the postcentral gyrus which subserves cutaneous sensation from the face. The motor nerve to the tongue is the XIIth cranial or hypoglossal nerve.

The functions of the tongue are:
(*1*) Motor: (*a*) mastication and the act of swallowing
           (*b*) speech
(*2*) Sensory: (*a*) taste (bitter, sweet, sour and salt)
           (*b*) touch

As in the case of smell, in order that substances may have taste they must be soluble in the watery secretion of the mouth and salivary glands.

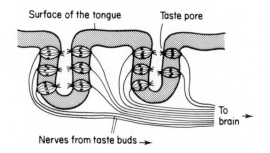

Fig. 18.1 Taste buds and their nerve supply. NB Only one nerve fibre is shown for each taste bud.

Fig. 18.2 Diagram illustrating the parts of the tongue in which the various sensations of taste are appreciated.

The sense of hearing

The auditory apparatus consists of (*1*) the external ear, (*2*) the middle ear, (*3*) the cochlea of the internal ear, (*4*) the cochlear nerve and acoustic areas in the temporal lobe of the brain.

The external ear

This consists of:
- (*a*) the auricle or pinna
- (*b*) the external acoustic meatus
  - (*i*) cartilaginous portion
  - (*ii*) bony portion

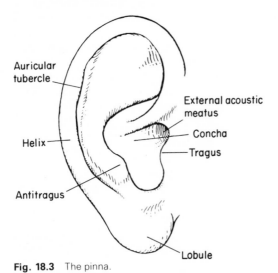

Fig. 18.3   The pinna.

The **auricle** or **pinna** is attached to the side of the head about midway between the forehead and the occiput. It has a deep shell-like cavity called the **concha**. Beneath the skin, the auricle is composed of yellow elastic cartilage, apart from the **lobule** or ear lobe which is soft because it is composed of fibrous and adipose tissues.

The function of the auricle is to collect sound waves and conduct them to the external acoustic meatus. This function is more marked in many animals than in man and in consequence their auricles are relatively larger and more mobile. Although there are several small muscles attached to the human ear, in only a few people is a limited amount of movement possible.

The **external acoustic meatus** leads from the pinna to the tympanic membrane or ear-drum. It is a tubular passage about 2.5 cm (1 in) long. Its course is not straight but shows a slight double or S-shaped bend, being directed at first medially, forwards, and slightly upwards, and then medially and slightly backwards. In order to bring the meatus into a straight line the pinna should be lifted upwards and backwards in adults, but in children owing to a slightly different course the pinna should be pulled downwards and backwards. This is of importance when it is desired to examine the tympanic membrane.

Structurally the external acoustic meatus consists of two parts: the outer *cartilaginous* portion (lateral one-third) which is continuous with the cartilage of the pinna; and the inner *bony* part in the temporal bone.

The meatus is lined by skin which is continuous with that covering the pinna, but is characterized by special glands, the **ceruminous glands**, which secrete a yellow greasy substance called cerumen or wax. They are modified sweat glands and their secretion helps to prevent the entry into the meatus of foreign bodies, especially insects. A few hairs are present which also assist in this function. If there is excessive production or retention of wax, the meatus may become blocked and require syringing to dislodge and wash out the wax and restore normal hearing.

The middle ear

The middle ear or tympanic cavity is a small irregular cavity situated in the petrous portion of the temporal bone. It contains a chain of small bones or **ossicles** by which the sound waves are transmitted from the **tympanic membrane** to the internal ear. Roughly speaking, it is a narrow oblong box having anterior, posterior, medial and lateral walls with a roof and a floor.

Its walls are both bony and membranous in structure. The outer or lateral wall is formed by the tympanic membrane. Its medial wall, though mainly consisting of bone, has two openings which are covered by membrane, namely the **fenestra vestibuli** (f. ovalis) or oval window of the vestibule above, and the **fenestra cochleae** (f. rotunda) or round window of the cochlea below.

In addition to these membranous defects in its outer and inner walls, both the anterior and posterior walls have openings. Entering the middle ear in its anterior wall is the outer or lateral end of the **auditory (Eustachian) tube** which communicates with the nasopharynx. Posteriorly the middle ear

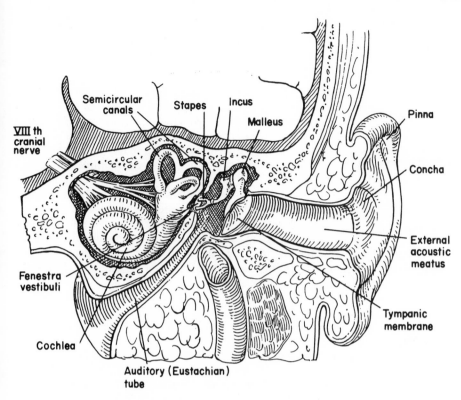

Fig. 18.4 The structure of the ear.

communicates with the mastoid antrum and the mastoid air cells which occupy the mastoid process of the temporal bone.

The whole of the cavity of the middle ear is lined by mucous membrane which, therefore, forms the inner lining of the tympanic membrane and is continuous with the mucous lining of the Eustachian tube and with that of the mastoid antrum and cells.

The auditory tube connects the nasopharynx with the middle ear and is less than 4 cm ($1\frac{1}{2}$ in) in length. There is thus a connection between the middle ear and the outer air so that the pressure of air on each side of the tympanic membrane is equalized.

The pharyngeal opening of the Eustachian tube is normally closed by the approximation of its walls, but is opened by the action of muscles in swallowing and yawning. In catarrhal conditions of the nasopharynx, as in a severe head cold, the opening of the tube may be defective and a degree of deafness is not uncommon from this cause.

The effect of atmospheric pressure is also demonstrated in going to the top of a high hill rapidly in a car, or going up in an aeroplane. Until air escapes from the middle ear by the Eustachian tube the atmospheric pressure on the outer surface of the tympanic membrane, being diminished by reason of the height, is less than the pressure on its inner side. An uncomfortable sensation in the ear and slight temporary deafness is produced but disappears with the adjustment of pressure, which is often accompanied by clicking sounds and is aided by the act of swallowing.

The main function of the Eustachian tube is, therefore, to equalize the pressure in the middle ear with the atmospheric pressure outside.

The ossicles

The ossicles or small bones of the middle ear are three in number, the **malleus**, the **incus** and **stapes**. They stretch from the tympanic membrane to the fenestra vestibuli or oval window of the vestibule.

The **malleus** or hammer bone consists of a head which articulates with the incus and a handle which is attached to the tympanic membrane. The **incus** or anvil is the middle of the three bones and consists of a body and two short legs, one of which articu-

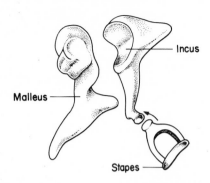

Fig. 18.5   The bones of the middle ear.

lates with the roof of the middle ear, the other with the stapes. The **stapes** or stirrup bone is the smallest of the three. Its head articulates with the incus while its base or foot-plate is attached to the membrane covering the fenestra vestibuli.

These three bones act as a series of levers transmitting the movements or vibrations of the tympanic membrane, caused by sound waves impinging upon it, to the membrane covering the fenestra vestibuli. It will be seen later that from the fenestra vestibuli the vibrations are passed on to the internal ear.

The movements of the ossicles are controlled to some extent by two tiny muscles, the **tensor tympani** inserted into the handle of the malleus, and the **stapedius** muscle inserted into the neck of the stapes. These muscles act as dampers to prevent excessive movement of the ossicles in response to loud noise. They also attenuate low frequency components of sound so that weaker high frequency components are not masked. This improves the perception of sounds obscured by background noise.

The tympanic membrane

The **tympanic membrane** (tympanum, or eardrum) is situated at the deepest part of the external acoustic meatus which it separates from the middle ear. It lies obliquely so that its upper part is nearer the exterior than its lower part. In structure, its outer surface consists of epithelium continuous with the skin lining the external acoustic meatus, while its inner lining is mucous membrane continuous with that of the middle ear. Between these two layers is a small amount of fibrous tissue. Firmly attached to its inner wall (and passing downwards and slightly backwards from its upper edge to a point just below and behind its centre) is the handle of the malleus.

It appears as an almost circular structure tightly stretched between the walls of the bony meatus except for a small area in its upper part (known as the flaccid membrane of Shrapnell). It is easily seen with the aid of an auriscope and it appears red and sometimes bulging in patients with acute middle ear infections (acute otitis media). In some cases it perforates and pus escapes through the perforation into the external acoustic meatus.

The internal ear

The internal ear or labyrinth consists of a series of irregular cavities situated in the petrous portion of the temporal bone. These cavities constitute the bony or **osseous labyrinth**. Within these bony walls is a membranous structure, which more or less follows the shape of the bony labyrinth and is called the **membranous labyrinth**.

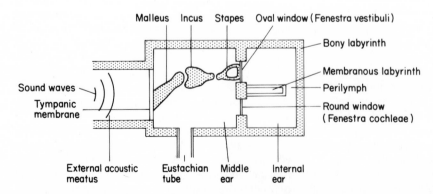

Fig. 18.6   Diagram illustrating the anatomy of the auditory apparatus.

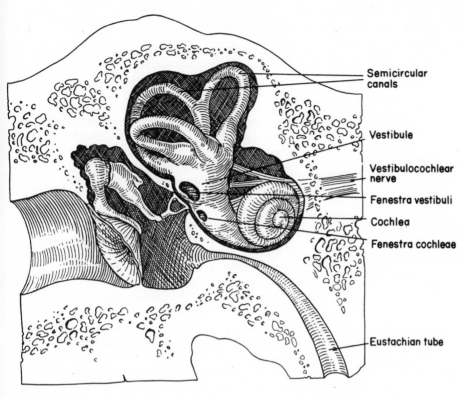

Semicircular
canals

Vestibule

Vestibulocochlear
nerve

Fenestra vestibuli

Cochlea

Fenestra cochleae

Eustachian tube

Fig. 18.7   Diagram showing the osseous labyrinth.

Between the bony walls and the membranous part of the labyrinth is a clear fluid called **peril-ymph**, while the membranous labyrinth itself is a sac filled with a similar fluid called the **endolymph**.

The osseous labyrinth (Fig. 18.7) consists of the following parts:

(*1*) the vestibule or entrance which communicates with

(*2*) the cochlea (little shell) or organ of hearing in front and

(*3*) the semicircular canals behind. These are concerned with equilibrium and the sense of position.

The **vestibule** is closely connected to the middle ear, from which it is separated by the membrane covering the fenestra vestibuli (the oval window of the vestibule) with the attached foot-plate of the stapes.

The **cochlea** or anterior portion of the labyrinth contains the organ of hearing. In some respects it resembles the shell of a small snail and, in section, looks something like a spiral staircase with a central

bony structure called the **modiolus** from which ridges project to the outer wall. The membranous portion of the cochlea, called the duct of the cochlea, is contained within these bony walls and in it is the most important part of the structure, the organ of Corti, which is the true end-organ of hearing.

The **organ of Corti**, is a spiral structure contained in the membranous labyrinth, which follows the shape of the cochlea and is bathed in endolymph. It consists of special epithelial cells, called rods of Corti, on either side of which are layers of cells (hair cells) having hair-like processes on their free surface rather like ciliated epithelium in appearance. It is around these cells that the fibres of the cochlear or auditory division of the VIIIth cranial nerve commence. The nerve cell bodies are located in the spiral ganglion within the modiolus.

Mechanism of hearing

Sound is due to waves or vibrations in the air and has three main qualities.

(*1*) **Pitch**, which depends on the *frequency* of the vibrations. The more rapid the frequency the higher the pitch of the note produced.

(*2*) **Intensity** or loudness, which depends on the *amplitude* of the vibrations.

(*3*) The **quality**, which is due to the combination of various vibrations. These may blend to produce harmony or music, or fail to unite giving rise to a discord or noise.

Sound waves in the air are collected by the pinna and directed along the external acoustic meatus to the tympanic membrane which they cause to vibrate. These vibrations are transmitted across the middle ear by the movements of the malleus, incus and stapes to the membrane covering the fenestra vestibuli. The inner surface of this membrane is in contact with the perilymph in the vestibule which picks up the vibrations and, in turn, passes them on to the endolymph by means of which they reach the organ of Corti.

The stimulus thus reaching the organ of Corti is conveyed by the cochlear portion of the VIIIth cranial or vestibulocochlear nerve, which leaves the petrous portion of the temporal bone by a foramen (the internal acoustic meatus) to reach the brain-stem. The fibres are then carried to the acoustic areas of the brain situated in the temporal lobe of the opposite side.

Summary of the sense of hearing

Sound waves → pinna → external acoustic meatus → tympanic membrane → malleus → incus → stapes → fenestra vestibuli → vestibule → cochlea (perilymph → endolymph) → organ of Corti → VIIIth cranial nerve → temporal lobe of brain (of opposite side).

Hearing loss may be classified as conductive or sensorineural. In *conductive deafness*, sound is not transmitted to the cochlea because of a lesion affecting the external acoustic meatus or middle ear.

Sensorineural deafness may be of *sensory* type, resulting from a lesion within the cochlea, or *neural* due to a lesion affecting the VIIIth cranial nerve or any part of the pathway from the cochlea to the temporal lobe.

Audiometry, with an electronic audiometer, is used to measure hearing acuity and to localize the site of a lesion causing loss of hearing.

The semicircular canals

These are three canals situated in the posterior part of the bony labyrinth and set at right angles to each other. They are anterior (superior), posterior and lateral canals. Within the bony walls are the membranous canals or ducts surrounded by perilymph and containing endolymph. Each canal is enlarged at one end into an **ampulla** where the special nerve fibres end around cells which have fine hair-like processes projecting from them. Movements of the head and alteration in its position cause movement of the endolymph in the semicircular canals. This movement of fluid acts as a stimulus to the nerve endings in the ampullae and the impulses are conveyed to the brain (both the cerebrum and cerebellum) by the vestibular division of the VIIIth cranial nerve. Lesions of this portion of the nerve cause vertigo (dizziness).

The ampullae of the semicircular ducts open into a sac called the **utricle**. A thickened part of the floor and lateral wall of this sac is known as the **macula** or otolithic organ. It contains hair cells innervated by fibres which join those from the ampullae to form the vestibular nerve. Surmounting the hair cells and their supporting cells is a membrane containing the **otoliths** ('ear dust') which are crystals of calcium carbonate.

The semicircular canals provide the brain with information on head movement. The otolith organs give information on the position of the head.

It will be remembered that the sense of equilibrium and position is also dependent on impulses received from the eyes and from the muscles and joints which contain receptor organs described as proprioceptive.

The sense of sight − vision

The **eye** or organ of sight is situated in the orbital cavity of the skull and is well protected by its bony walls except on its anterior aspect.

In addition to the essential organs of the visual apparatus, namely, the eyeball, the optic nerve and the visual centres in the brain, there are certain accessory organs which are necessary for the protection and functioning of the eye. These include: (*1*) the eyebrows; (*2*) the eyelids; (*3*) the conjunctiva; (*4*) the lacrimal apparatus; and (*5*) the muscles of the eye.

Accessory organs of the eye

The eyebrows

These are formed by the skin covering the orbital process of the frontal bones, which in its natural state is plentifully supplied with short thick hairs. Their main function is protective and by their shape prevent the sweat of the brow from pouring into the eye.

The eyelids

These are two movable folds, upper and lower, which form the anterior protection for the eye. The upper is the larger and more mobile of the two and is provided with a muscle which elevates it. The eyelids are covered externally with skin and their inner lining consists of mucous membrane, the conjunctiva. Between these layers is a dense plate of fibrous tissue called the **tarsus**. Into the tarsal plate of the upper lid the muscle which raises it is inserted (levator palpebrae superioris). Surrounding both lids is a circular sphincter muscle (orbicularis oculi) which closes them and, when fully contracted, 'screws them up.'

The space between the two lids is the **palpebral fissure**. Its **lateral angle** is also called the lateral canthus and its **medial angle** the medial canthus.

The eyelids blink every few seconds. This movement keeps the front of the eye free from dust and helps to move the tears across the conjunctival sac. In addition to protecting the eyes from the entrance of foreign bodies the eyelids also prevent the entry of excessive light.

The eyelashes

A row of short thick hairs project from the free margin of each eyelid. Arranged immediately behind the eyelashes are the openings of the **tarsal (Meibomian) glands,** which are modified sebaceous glands and are sometimes the site of small cysts. Infection of the hair follicles of the eyelashes results in the common condition known as stye.

The conjunctiva

This is a delicate mucous membrane which lines the inner surface of the eyelids and is then reflected on to the outer surface of the eyeball. The space between the two layers is called the conjunctival sac.

The lacrimal apparatus

This is concerned with the formation of tears and consists of the following structures: (*1*) the lacrimal glands; (*2*) the lacrimal canaliculi; and (*3*) the lacrimal sac and nasolacrimal duct.

The lacrimal glands

These are situated in the orbital cavity immediately above the lateral angle of the eye. Each is almond-shaped and lies in a depression in the orbital plate of the frontal bone. A number of small canals leads from it to the lateral angle of the conjunctival sac.

The lacrimal canaliculi

If the medial end of each eyelid is carefully examined the orifice of a minute duct can be seen. From this opening the lacrimal canaliculus passes inwards to enter the lacrimal sac.

The lacrimal sac and nasolacrimal duct

The lacrimal sac may be regarded as the upper expanded portion of the nasolacrimal duct which passes downwards inside the bony wall of the nasal cavity to open into the inferior meatus of the nose.

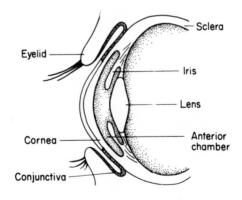

Fig. 18.8 Diagram of the conjunctiva.

Tears are a slightly alkaline watery fluid containing a small amount of sodium chloride which give them a salty taste. Normally there is a constant secretion from the lacrimal glands just sufficient to keep the interior of the conjunctival sac moist and free from dust. By the frequent movement of the eyelids the tears pass across the front of the eye from the lateral to the medial side, where they pass through the openings of the lacrimal canaliculi and are drained into the nose by the nasolacrimal duct to mix with the secretions of the nose.

Having observed the minute openings of the lacrimal canaliculi at the medial corners of the eyelids it is clear that any excess of tears must overflow from the conjunctival sac and run down the cheek. Some of the excess, however, does pass down the ducts and accounts for the excess of watery secretion from the nose after crying which, if severe, requires the use of a handkerchief, although in these circumstances its place is frequently taken by 'sniffing.'

The secretion of tears is increased by the presence of foreign bodies and inflammation caused by bacteria or irritating vapours. Irritation of the nasal mucous membrane and very bright light provoke reflex lacrimation, while emotional states and pain also result in the flowing of tears.

The *functions of the tears* may be summarized as:
  (*1*) keeping the eyes moist, thereby allowing free movement of the lids;
  (*2*) removal of dust and foreign bodies, including bacteria;
  (*3*) acting as a mild antiseptic;
  (*4*) expression of emotion or pain.

The extraocular muscles of the eye

Each eyeball is moved by muscles which arise from the posterior wall of the bony orbit close to the entrance of the optic nerve and are inserted into the outer fibrous coat (**sclera**) of the eye. There are four straight and two oblique muscles in addition to the muscle elevating the upper lid (levator palpebrae superioris).

The straight muscles are the **superior rectus, inferior rectus, medial rectus** and **lateral rectus**. Their position in relation to the eyeball is indicated by their names. The action of these muscles is not too difficult to follow if two facts are remembered. **(1)** If one eye alone is considered, contraction of the superior rectus turning the eye upwards will be associated with relaxation of the inferior rectus, and vice versa. The medial and lateral recti move the eye to one side or the other respectively and work together in the same way. This action is similar to the opposing action of the flexor and extensor muscles of the forearm. When one set contracts the opposite set relaxes. **(2)** In normal vision both eyes move together (*conjugate deviation*). Therefore, if the eyes are turned upwards both right and left superior recti will contract and both inferior recti will relax. On the other hand, if both eyes are turned to the right it follows that the right lateral rectus and the left medial rectus will

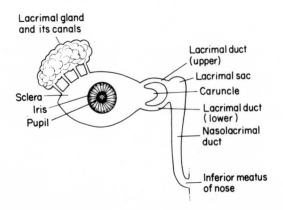

Fig. 18.9 The lacrimal apparatus.

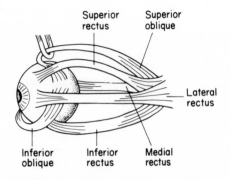

Fig. 18.10 Diagram of eye muscles. Left orbit (lateral aspect).

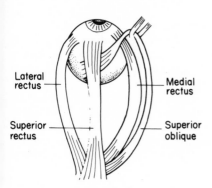

Fig. 18.11 Diagram of eye muscles. Left orbit (from above).

contract while the right medial rectus and left lateral rectus will relax. The converse is true if the eyes are turned to the left. This is quite simple, and with a little thought the individual can work it out for herself or himself.

Of the two oblique muscles, the **superior oblique** is so arranged as to direct the eye downwards and outwards. The **inferior oblique** muscle turns the eye upwards and outwards.

The muscles of the eyes are supplied by the cranial nerves (IIIrd, IVth and VIth). The lateral rectus is supplied by the VIth or abducens nerve. The superior oblique is supplied by the IVth or trochlear nerve. All the others (levator palpebrae superioris, medial rectus, inferior rectus, inferior oblique and superior rectus) are supplied by the

IIIrd or oculomotor nerve. These extraocular muscles are sometimes referred to as the extrinsic muscles of the eye. The internal or intrinsic muscles are mentioned later.

The common condition of squint is usually due to imperfect balance between opposing muscles and may be treated by operations designed to shorten or lengthen the appropriate muscles.

The eyeball or bulb of the eye

The eyeball is situated in the anterior part of the orbital cavity and is almost spherical in shape. It is surrounded by a pad of fat.

The structure of the eye may be considered in the following way.

A *Three tunics or layers*
  (1) fibrous
     (a) sclera
     (b) cornea
  (2) vascular
     (a) choroid
     (b) ciliary body
     (c) iris
  (3) nervous
     retina

B *The light-transmitting mechanism*
  (1) aqueous humor
  (2) the lens
  (3) the vitreous body

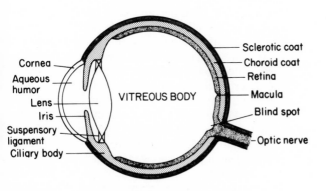

Fig. 18.12 Section through eyeball viewed from above.

The fibrous coats

(a) The **sclera** (or sclerotic coat). The posterior five-sixths of the outer coat of the eyeball consists of strong, opaque fibrous tissue and is called the sclera. It is protective in function and helps to maintain the shape of the eyeball. When viewed from the front it is that portion which is referred to as 'the white of the eye' and, in this position, is covered by the conjunctiva. Posteriorly, the optic nerve passes through it to reach the retina inside the eye, and in the orbit the nerve is protected by a sheath of fibrous tissue continuous with the sclera.

(b) The **cornea** occupies the anterior one-sixth of the external surface of the eyeball and, being transparent, allows light to enter the interior of the eye. The cornea is sometimes described as the 'window of the eye' and its anterior surface can be seen to be slightly curved or convex. Over the cornea·the conjunctiva becomes very thin and is only represented by a few layers of epithelial cells. The cornea has no blood vessels but derives its nourishment from the aqueous humor.

The vascular coat

This is the middle layer of the eye. It contains many blood vessels and capillaries, which are derived from the ophthalmic branch of the internal carotid artery, and is pigmented. The choroid, ciliary body and iris together form the uveal tract. Hence the term uveitis for inflammation of the iris (iritis) and the structures in continuity with it.

(a) The **choroid**. This is a thin pigmented membrane, dark brown in colour, which lines the posterior compartment of the eye. It is situated between the inner surface of the sclera and the retina.

(b) The **ciliary body**. This is a circular structure, continuous with the anterior part of the choroid, which surrounds the periphery of the iris immediately behind the outer margin of the cornea where it joins the sclera (Fig. 18.12). It contains muscle fibres (the ciliary muscle) and to it is attached the ligament which helps to suspend the lens in position.

(c) The **iris**. This is the pigmented membrane which surrounds the pupil of the eye. It arises from the margin of the ciliary body and forms a diaphragm with a black central opening (the pupil) immediately in front of the lens. The colour of the eye is dependent on the pigment in the iris. In dark eyes the pigment is plentiful, but in blue eyes it is scanty.

The iris contains two sets of muscle fibres. Those comprising the **sphincter pupillae** encircle the pupil. The fibres of the other muscle, the **dilator pupillae**, pass in a radial direction from the outer margin of the iris to the edge of the pupil. It will be clear that the circular muscle acting as a sphincter will reduce the size of the pupil when it contracts. Contraction of the radial fibres, on the other hand, increase its size and they are, therefore, dilators. These, with the ciliary muscle, are the *intrinsic muscles* of the eye.

The function of the iris is to regulate the amount of light entering the posterior part of the eye. Thus, when a bright light shines on the retina the pupil contracts.

The iris is under the control of the autonomic nervous system, the effect of sympathetic stimulation being pupillary dilatation (mydriasis) and that of parasympathetic stimulation being pupillary constriction (miosis). Parasympathetic blockade with anticholinergic drugs such as atropine therefore dilates the pupils. Drugs which dilate the pupils are called mydriatics and one which is frequently used as a topical solution (drops) for this purpose is the anticholinergic agent cyclopentolate. Drugs which act like or prolong the action of acetylcholine (parasympathomimetic drugs) constrict the pupil. An example is physostigmine (Eserine) which blocks the enzyme cholinesterase, which normally terminates the action of acetylcholine. Physostigmine is therefore an example of an anticholinesterase.

The retina

The retina is the innermost coat of the eye. It lines the posterior chamber and ends anteriorly at the margin of the ciliary body. It is a delicate membrane consisting of neurons, that is, nerve cells and nerve fibres, together with a layer of special structures called **rods** and **cones**. These are situated on the outer or choroidal surface of the retina, while the nerve fibres are on the inner surface facing the chamber of the eye. The rods and cones are the actual receptors of sight and light reaching them sets up the impulses which are transmitted to the nerves. The impulses are generated by the action of light on photosensitive pigments in the rods and cones. In the rods, the pigment is known as *visual purple* or *rhodopsin*. The cones are responsible for

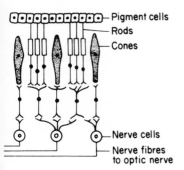

Pigment cells
Rods
Cones
Nerve cells
Nerve fibres
to optic nerve

Fig. 18.13   Diagram illustrating the layers of the retina.

the choroid, particularly in people with severe myopia (short sightedness).

The light-transmitting mechanism

1. The aqueous humor

Situated between the cornea in front and the iris and ciliary body behind is the anterior chamber of the eye which contains a clear watery fluid called the aqueous humor.

2. The lens

This is a firm transparent structure, convex in shape, which is suspended in its capsule by a ligament attached to the ciliary body. It is placed immediately behind the iris and pupil of the eye. Its function is to focus rays of light entering the eye through the pupil on to the retina.

3. The vitreous body

This is a colourless, transparent jelly-like substance which fills the posterior four-fifths of the eye. It helps to preserve the spherical shape of the eyeball and to support the retina.

colour vision. The rods give only monochromatic (black and white) vision but their much greater sensitivity enables us to see in very poor lighting conditions.

The retina is the nervous portion of the eye and is therefore the true end-organ of vision. The fibres of the optic nerve commence in the cells of the retina and are collected together at a point just medial to the most posterior part of the eyeball where they pierce the choroid and sclera and pass backwards as the optic nerve through the orbit to the optic chiasma and brain.

The point at which the optic nerve fibres all converge contains no nerve cells and no rods and cones. It is, therefore, insensitive to light and is called the *blind spot*.

The retina is supplied with blood by a branch of the ophthalmic artery which enters the eye with the optic nerve and is called the **central artery of the retina**.

The **macula** is situated just to the lateral side of the exit of the optic nerve, i.e. at the very centre of the posterior part of the eye. It is a small area of the retina of great sensitivity on which the images seen by direct or near vision are focused. No rods are present in this area but cones are especially numerous.

The retina may become partially detached from

The presence of fluid and semi-fluid material or gel in the interior of the eye maintains its shape by keeping up a constant pressure on its walls. This is referred to as the *intraocular tension*.

In certain conditions drainage of the fluid may be impaired and there will be a consequent increase in the intraocular tension, a serious effect which may disturb the nutrition of the retina and lead to blindness. This is known as glaucoma. It may be treated medically and surgically

(*1*) by instilling drops which cause the pupil to contract (miotics e.g. physostigmine), thereby helping to open up the canals situated at the point of attachment of the iris to the ciliary body where the excess of fluid is normally drained off into the circulation;

(*2*) by making a small hole (trephine) through the sclera into the anterior chamber of the eye so that fluid can drain under the conjunctiva and so relieve the tension within the eyeball.

Diminution in intraocular tension is seen in cases of severe shock and marked fluid loss from the body. Complete loss of tension is observed after death.

The mechanism of sight

From a structural point of view the eye may be compared with a camera. The eyelids act as a shutter and there is an entrance window for light — the cornea; a diaphragm to regulate the aperture and therefore the amount of light entering — the iris; a lens to focus the image; a darkened interior formed by the choroid, and a light-sensitive plate which receives the image — the retina.

The optic nerve and its connections convey the details of the image to the occipital region of the cerebral cortex where they are processed before reaching consciousness.

In order to understand the mechanism of vision it is necessary to know something about light and the action of lenses. Light consists of electromagnetic waves which travel faster than anything else known in the universe, at the rate of 186 000 miles per second.

Some objects, such as the sun, electric light or candle, emit light rays and are self-luminous sources of light. Other objects, such as the things we normally see, merely reflect light received from other sources. If there is no source of light, complete darkness exists and no object can be seen.

Rays of light ordinarily travel in straight lines. A lens, which may be roughly defined as a curved transparent structure, has the power of bending or refracting rays of light. A lens which is thicker at

the centre than at the periphery is described as convex. One which is thinner in the middle is called concave.

A convex lens has the power of bending rays of light so that they converge and meet at a point of focus behind the lens. The stronger the lens, i.e. the greater the degree of curvature of its surfaces, the nearer is the focal point. A concave lens, on the other hand, bends the light rays so that they diverge and do not focus behind the lens. The lens of the eye is convex and focuses the rays of light passing through it on to the retina.

Actually the image reaching the retina is inverted but this is turned the right way up by the visual cortex in the brain.

Accommodation

Rays of light from distant objects are for all practical purposes parallel and therefore strike the vertical axis of the lens at right angles. The eye is so adjusted that such rays are bent by the lens to focus exactly on the retina, forming a sharp image. Rays of light from a near object (say 25 cm or 10 in) are divergent and strike the lens obliquely. In order that such rays may be accurately focused on the retina the lens must be made more powerful (of greater focusing power) by increasing its curvature i.e. making it more convex. This is accomplished by the action of the ciliary muscles. At the same time the clearness of the image is increased by cutting down the number of rays entering the eye by contraction of the iris. The process of altering the shape of the lens is called *accommodation* and operates every time a near object is looked at. The nearest point at which an object can be brought clearly into focus by accommodation is called the *near point*. For a normal child of 10 years it is about 9 cm (3.5 in) but it recedes throughout life and by the age of about 45 may become so distant that reading is difficult without spectacles. This defect of accommodation with advancing age is due to decreasing elasticity of the lens and is called *presbyopia*.

When an object is placed near the eyes, in order to obtain a clearly focused picture on both retinae the eyes turn slightly inwards towards each other. This is called *convergence*. The extreme of convergence is illustrated by 'squinting down the end of the nose'. The triple response of accommodation, convergence and pupillary constriction is called the *near response*.

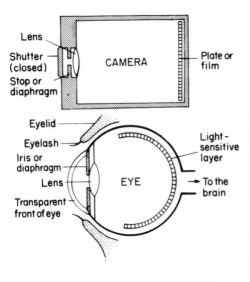

Fig. 18.14   The eye compared with a camera.

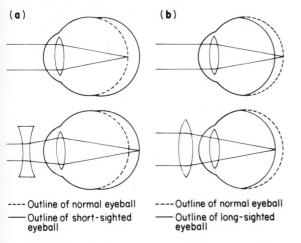

(a)   (b)

---- Outline of normal eyeball
—— Outline of short-sighted
        eyeball

----Outline of normal eyeball
——Outline of long-sighted
       eyeball

Fig. 18.15 **(a)** Myopia (short sight) and its correction. **(b)** Hypermetropia (long sight) and its correction.

It is of interest at this point to note some of the common defects of vision requiring the use of spectacles. Whereas the normal eye is practically spherical, in some people it tends to be slightly elongated and in others flattened. In other words, in the former case the distance from the lens to the retina is increased and in the latter it is decreased. It follows, therefore, that the lens will not naturally focus the image accurately on the retina in these conditions. In the former, elongated or myopic eye (short sighted), the image will tend to fall in front of the retina, while in the shortened or hypermetropic eye (long sighted) it will fall behind the retina. In both instances the objects seen will be blurred and out of focus.

These defects can be compensated by using additional lenses in the form of spectacles. By placing a concave glass lens in front of a myopic eye the rays of light will become divergent before reaching the lens, so that its point of focus is shifted back on to the retina. A convex lens in front of a hypermetropic eye will bring the image nearer the front of the eye by increasing the convergence of the rays so that they are focused on the retina.

Astigmatism is due to unequal curvature of the surfaces of the cornea, i.e. it may be curved more vetically than horizontally. This also may require correction with spectacles (cylindrical lenses).

Vision

Rays of light reflected from visible objects fall on the retina and, in near vision, are focused on the macula. The light falling on the retina produces certain chemical changes there which stimulate the endings of the optic nerve. These stimuli are conveyed by the optic nerve to the optic chiasma where some of the fibres cross and are then carried back to the cortex of the occipital lobe to be interpreted into consciousness.

Binocular vision

In considering the sense of sight it must be remembered that although we can see with each eye separately, normal stereoscopic vision is obtained by the simultaneous use of both eyes.

Rays of light strike the retina from all directions. Those coming from the left-hand side of the body will fall on the nasal side of the retina of the left eye and the temporal side of the retina of the right eye. These images produced by objects on the left side of the body are eventually received by the occipital lobe on the right side of the brain. This explains the crossing of certain fibres of the optic nerve in the optic chiasma and is part of the principle that one side of the body is controlled by the cerebral hemisphere of the opposite side. These facts are easily appreciated from Fig. 18.16.

Summary of the sense of sight

Light waves → cornea → aqueous humor → lens → vitreous body → retina → optic nerve → optic chiasma → occipital lobe of brain.

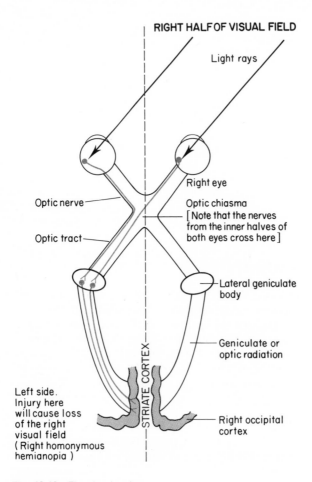

RIGHT HALF OF VISUAL FIELD

Light rays

Right eye

Optic nerve

Optic chiasma
[Note that the nerves
from the inner halves of
both eyes cross here]

Optic tract

Lateral geniculate
body

Geniculate or
optic radiation

STRIATE CORTEX

Left side.
Injury here
will cause loss
of the right
visual field
(Right homonymous
hemianopia)

Right occipital
cortex

Fig. 18.16   The visual pathways.

Questions

1. Which are the 'special senses'? Give an account of one of these.
2. Write brief notes on the sense of taste.
3. What are the main features of the external ear?
4. Describe the middle ear.
5. Write notes on the structure and functions of the internal ear.
6. What is the mechanism by which we see? Which defects are correctable with spectacle lenses?

19

The Reproductive System

The power of reproduction is one of the essential characteristics of life. It is illustrated in its primitive form by the action of the single-celled amoeba in dividing into two by the process of mitosis (p. 12). Most of the cells of the human body have the same power of division by virtue of which growth and repair are possible.

Reproduction of the species in humans and the higher animals is a complicated process involving the existence of two sexes, each contributing a **gamete** (mature sex cell) for the formation of the new individual.

The nucleus of every human cell contains **forty-six chromosomes** (44 somatic and 2 sex chromosomes, p. 10). Sexual reproduction results from the fusion of an **ovum** and a **sperm**, and if each of these contained 46 chromosomes the new individual would have ninety-two chromosomes. In order to maintain the number of chromosomes characteristic of the human species each gamete undergoes a special type of cell division called meiosis.

Meiosis occurs in two stages and reduces the number of chromosomes in the gametes by half so that the resulting ovum or sperm contains 23 chromosomes (22 somatic and 1 sex chromosome). Thus the fusion of an ovum and a sperm results in a new individual with forty-six chromosomes.

During the meiotic divisions the genetic material contained in the chromosomes becomes mixed between homologous chromosomes. This results in new combinations of genetic material and accounts for the uniqueness of each human being.

The reproductive organs of the male and female differ in anatomical structure and arrangement, each being adapted to the functional activities they are required to perform.

The function of the male organs is to form spermatozoa and implant them within the female so that they can meet the ova. The female organs are adapted to form ova which, if fertilized by the spermatozoa, remain in the cavity of the uterus. Here an **embryo** is formed which will grow into a **fetus**, and is retained and nourished until the new individual is capable of a separate and independent existence.

The male reproductive system

The scrotum

The scrotum is a pouch of pigmented skin situated below the root of the penis, and is continuous with the skin of the perineum and groin. It is divided into two by a midline fibrous septum, which is marked on the surface by a ridge, the **scrotal raphe**. Each half of the scrotum contains a testis, an epididymis and the lower end of the spermatic cord.

The subcutaneous tissue of the scrotum contains smooth muscle fibres, constituting the **dartos muscle**. This muscle contracts in response to cold or exercise to hold the testes closer to the body, causing the scrotum to become smaller and wrinkled. The muscle relaxes in response to warmth and thus helps to maintain an optimal temperature for spermatogenesis.

The testes

The testes (singular: testis) are small ovoid glands suspended in the scrotum. They are the reproductive glands, or gonads of the male.

In the embryo the testes develop within the upper abdominal cavity and during the seventh month of fetal life they migrate down the posterior abdominal wall and leave the abdominal cavity by passing through the inguinal canals into the scrotum, drawing with them the blood vessels and ducts which form the spermatic cords. As it passes into the scrotum, each testis carries with it a coat of peritoneum which normally becomes separated from the rest of the abdominal peritoneum. If this separation does not take place a channel remains

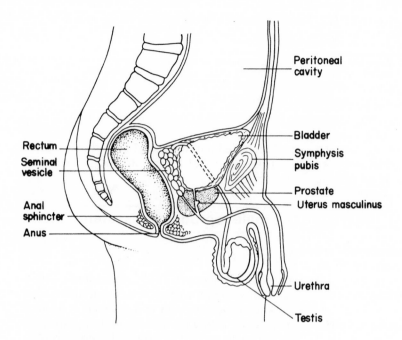

Fig. 19.1   Section through the male pelvis.

between the abdominal cavity and the scrotum, and thus abdominal viscera may protrude into the scrotum, giving rise to a hernia or rupture.

Structure of the testis

The separated coat of peritoneum forms the serous outer covering of the testis, the **tunica vaginalis**. The testis and the tunica vaginalis are attached to the lower part of the scrotum by fibrous tissue.

Each testis is surrounded by a dense white fibrous capsule, the **tunica albuginea**, which projects into the substance of the testis to divide it into 200 or more cone-shaped **lobules**. Lining the tunica albuginea is a delicate layer of connective tissue which supports a network of blood capillaries called the **tunica vasculosa**.

The lobules of the testis contain the **seminiferous tubules**. Each seminiferous tubule is highly convoluted, and if unravelled would measure about 70 cm (27.5 in) in length. The tubules are coiled in such a way that both ends join a series of **straight tubules** which converge to form a network called the **rete testis**.

The seminiferous tubules are lined by **germinal epithelial cells** (spermatogonia) resting on a base-

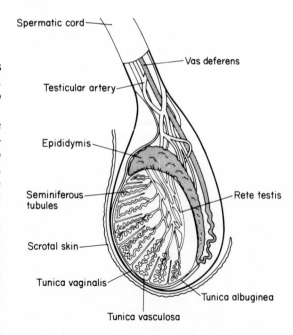

Fig. 19.2   Structure of testis and epididymis.

ment membrane. The germinal epithelial cells form the **spermatozoa**. Lying between the germinal cells of the tubule are the **sustenacular cells** (of Sertoli). Immature spermatocytes become attached to these cells and appear to be nourished by them at an early stage of their development.

The seminiferous tubules are embedded in loose connective tissue containing blood vessels, nerves and groups of **interstitial cells** (the cells of Leydig), which secrete the male hormones.

The rete testis is continuous with a series of efferent ductules which drain into the epididymis.

Functions of the testes

(1) Spermatogenesis

Spermatogenesis is the process by which primitive male gametes (spermatogonia) become mature sperm.

The spermatogonia are the germinal epithelial cells lining the seminiferous tubules; they are present at birth but do not produce spermatozoa until puberty. At puberty, spermatogonia begin to actively divide under the influence of *follicle stimulating hormone* (FSH) from the anterior pituitary gland, and will continue to do so throughout life, although there may be a decrease in the numbers of spermatozoa produced in the later years due to a decrease in testosterone secretion.

Some of the spermatogonia begin to enlarge and undergo mitotic division to form **primary spermatocytes**, each containing 44 somatic and 2 sex chromosomes (XY). The primary spermatocytes move away from the basement membrane and undergo their first meiotic division, to produce a pair of **secondary spermatocytes** each containing 22 autosomes and one sex chromosome, either X or Y. Thus half the spermatozoa contain the Y chromosome and become male spermatozoa, while the other half contain the X chromosome and become female spermatozoa.

Each of the secondary spermatocytes rapidly divides again by mitosis to form a pair of **spermatids** (each containing 23 chromosomes). The spermatids become attached to the Sertoli cells and elongate to develop a head and a tail. The **head** contains the **pronucleus**, and is covered by the **acrosome** which is thought to contain proteolytic enzymes. The **tail**, comprising the neck, middle piece, main piece and end piece, is the organ of motility. The **neck** contains the centriole and the

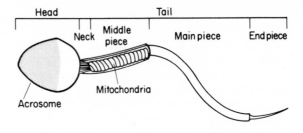

Fig. 19.3 Diagram of spermatozoon. The head is almost entirely composed by the nucleus.

middle piece contains mitochondria, arranged around an axial sheath, which provide energy for motility. The axial sheath extends into the **main piece** which is covered by a tough fibrous coat that terminates at the end piece. The **end piece** consists of the exposed axial sheath which forms a fine filament.

The spermatozoa pass from the seminiferous tubules of the testis into the epididymis. They are non-motile and are transported by contraction of the smooth muscle of the epididymis. It is here that the spermatozoa become mature and capable of fertilizing an ovum. The secretion of testosterone is necessary for the complete maturation of the spermatozoa.

(2) Secretion of hormones

Testosterone

The interstitial cells of Leydig, which lie in compact groups in the connective tissue surrounding the seminiferous tubules, secrete the male sex hormones collectively known as *androgens*. Testosterone, a steroid compound, is the principal hormone secreted; dehydroepiandrosterone and androstenedione are produced in lesser amounts.

Very little testosterone is secreted in childhood, but at puberty the secretion of *luteinizing hormone* (LH) by the anterior pituitary gland stimulates the interstitial cells to produce testosterone. (In the male LH is often referred to as *interstitial cell-stimulating hormone* or ICSH.)

Testosterone secretion promotes maturation of the male reproductive organs and causes the appearance of secondary sexual characteristics. It causes growth of hair over the chest, abdomen and pubis, and on the face. It causes enlargement of the larynx which results in the voice 'breaking' and gradually changing to the deeper base voice of the adult male.

Although FSH initiates spermatogenesis, testosterone must be secreted simultaneously for development of the spermatozoa beyond the primary spermatocyte stage. Testosterone also has a powerful anabolic action, stimulating protein synthesis and growth of bones.

Oestrogen

The testes secrete small amounts of oestrogen, but its exact function in the male is unknown.

The epididymis and vas deferens

The **epididymis** is a tightly coiled tubule surrounded by connective tissue. It is described as having a **head** which is connected to the testes by the efferent ductules, a **body** and a **tail** which is continuous with the vas deferens. The convoluted tubule is lined with pseudostratified columnar epithelium. The epithelial cells have long cellular processes on their surfaces through which cellular secretions enter the lumen of the duct. The secretions contain hormones, enzymes and nutrients and may be important in the maturation of spermatozoa. The duct is surrounded by a circular layer of smooth muscle fibres which contract to aid the passage of spermatozoa along the duct.

At its termination, the duct of the epididymis straightens and becomes continuous with the **vas deferens**. This is a thick-walled fibromuscular tube approximately 45 cm (17.7 in) long which passes upwards behind the testis and through the spermatic cord and inguinal canal to enter the pelvic cavity. It passes backwards to the base of the bladder and joins the seminal vesicle on its own side to form the ejaculatory duct. At its termination the vas deferens becomes dilated and tortuous to form an **ampulla**. The thick muscular walls of the duct move spermatozoa along the lumen by peristaltic contraction towards the ampulla. The pseudostratified epithelial lining of the duct is thrown into longitudinal folds. The walls of the ampulla are thinner and the lining is folded to form pocket-like recesses.

Mature spermatozoa are stored in the vas deferens and the ampulla. Vasectomy is the surgical interruption of the vas deferens via an incision through the scrotum. It renders a man sterile since it prevents spermatozoa leaving the epididymis.

The spermatic cord

The spermatic cords extend through the inguinal canals and then pass in front of the pubis to reach the scrotum and testes. Each spermatic cord contains the testicular artery, the pampiniform plexus of veins, lymphatic vessels, nerves and the vas deferens.

Blood supply

The **testicular artery** arises from the aorta immediately below the renal artery. It passes downwards through the spermatic cord, supplying branches to the vas deferens and the epididymis before reaching the testis.

Veins emerge from the posterior surface of the testis to form the **pampiniform plexus**, which passes upwards through the inguinal canal in the spermatic cord to form the **testicular vein**. The right testicular vein drains into the inferior vena cava, while the left vein drains into the left renal vein.

Lymphatic drainage

The lymphatic vessels of the testis and epididymis accompany the veins and drain into the lateral and preaortic nodes.

Nerve supply

The testis receives its nerve supply from the 10th and 11th thoracic nerve segments of the spinal cord.

Seminal vesicles

The seminal vesicles are two small convoluted pouches situated behind the bladder and above the prostate gland. Each vesicle is about 5 cm (2 in) long and consists of three coats.

(*1*) An outer coat of connective tissue containing elastic fibres.

(*2*) A middle coat of smooth muscle fibres consisting of an outer layer of longitudinal fibres and an inner layer of circular fibres.

(*3*) An inner coat of secretory columnar epithelium which is folded to form numerous pockets. Under the influence of testosterone the epithelium secretes a viscid liquid containing fructose and other nutrients, which is stored within the seminal

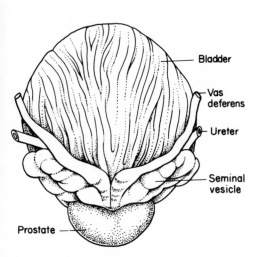

Bladder

Vas deferens

Ureter

Seminal vesicle

Prostate

Fig. 19.4 The structures at the base of the bladder (from behind).

The prostate gland

The prostate gland is about the size of a chestnut and is situated below the base of the bladder surrounding the urethra. The prostate is enclosed by a sheath of fibro-elastic tissue which is called the **false capsule**. It contains an extensive plexus of veins, the vesicoprostatic plexus. Beneath this lies the **true capsule** consisting of dense fibrous tissue.

The glandular tissue of the prostate is formed by secretory alveoli and tubules of very irregular size and shape, composed of columnar epithelial cells. These are surrounded by a fibro-elastic stroma containing smooth muscle fibres. The prostate secretes a thin, milky, slightly alkaline fluid which contains calcium, acid phosphate and citric acid. During emission the prostatic capsule contracts and expels the fluid into the urethra through numerous small ducts.

Blood supply

The arterial blood supply of the prostate gland is from the inferior vesical, middle rectal and internal pudendal arteries.

The veins form a plexus around the gland (which also receives venous blood from the deep dorsal vein of the penis) and drain into the internal iliac veins.

Nerve supply

The prostate receives both sympathetic and parasympathetic nerve fibres from the inferior hypogastric (pelvic) plexus.

In many elderly men the prostate gland may become enlarged and by pressing on the urethra can obstruct the flow of urine from the bladder, thus causing retention of urine. Surgical removal of the gland (prostatectomy) is usually required.

Bulbourethral glands

The bulbourethral glands (of Cowper) lie, one on each side, in the connective tissue behind the membranous urethra just below the prostate gland. Each gland is about the size of a pea and consists of compound tubulo-alveolar tissue and has a duct which enters the penile portion of the urethra. The bulbourethral glands produce a clear, viscid alkaline secretion.

vesicles until ejaculation occurs. Just prior to ejaculation, the seminal vesicles empty their contents into the ejaculatory ducts to join the spermatozoa from the vas deferens.

Blood supply

The seminal vesicles are supplied with arterial blood by the inferior vesical and middle rectal arteries.

The veins drain into the vesico-prostatic venous plexus.

Nerve supply

The seminal vesicles are innervated by nerve fibres from the pelvic plexuses.

The ejaculatory ducts

Each ejaculatory duct is formed by the union of the vas deferens with the duct of the seminal vesicle. It is about 2 cm (0.8 in) in length and passes through the prostate gland to enter the prostatic urethra.

The penis

The penis is composed mainly of three cylindrical columns of **erectile tissue** (i.e. tissue which becomes firm and rigid when congested with blood). The two larger dorsal columns are the **corpora cavernosa penis,** and the single inferior column is the **corpus spongiosum**. Numerous trabeculae divide each column of tissue into cavernous spaces (sinuses) giving the entire structure a spongy appearance.

The penis has a fixed **root** arising from the perineum and a free **shaft**. The corpora cavernosa originate separately but unite beneath the pubic arch and run forward together. Each corpus cavernosum is surrounded by a thick fibrous sheath. Between the two columns lies a fibrous septum through which the cavernous spaces of both sides communicate.

The corpus spongiosum is surrounded by a thin fibrous sheath containing elastic and smooth muscle fibres. It encloses the penile portion of the urethra. The corpus spongiosum arises at the root of the penis and passes forwards inferiorly in the deep groove formed between the corpora cavernosa. Its tip is expanded to form a cap, the **glans penis,** overlapping the terminal ends of the corpora cavernosa.

The three columns of erectile tissue are surrounded by a small amount of subcutaneous tissue containing numerous smooth muscle fibres; this is covered by thin delicate skin, which forms a double fold over the glans penis called the **prepuce** (foreskin).

The urethra serves as a common outlet for urine and for semen.

If the prepuce covers the glans penis too tightly a condition known as phimosis is caused which results in difficulty in micturition. This is not uncommon in young children and may require circumcision.

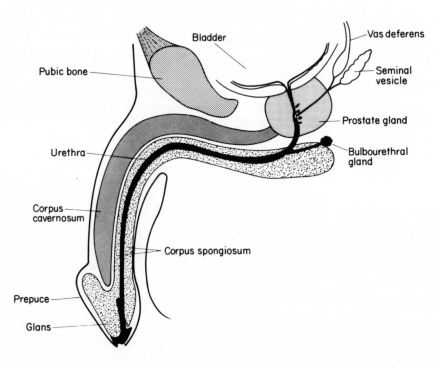

Fig. 19.5 Structure of the penis and related organs.

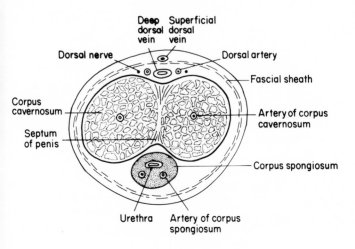

Fig. 19.6 Transverse section through the body of the penis.

Blood supply

Arterial blood is supplied to the penis by branches of the internal and external **pudendal arteries**. The deep arteries of the penis and branches of the dorsal arteries of the penis supply the cavernous spaces, some branches dividing to form convoluted vessels opening directly into the sinuses of the erectile tissue.

The veins draining the cavernous spaces converge on the dorsum of the penis to form the **deep dorsal vein** which empties into the prostatic venous plexus.

Nerve supply

The penis is innervated by the 2nd, 3rd and 4th sacral spinal nerves through the pudendal nerve and the pelvic plexuses.

Semen

Semen (seminal fluid) is a mucoid, milky white fluid consisting of the spermatozoa suspended in the mixed secretions from the vas deferens, seminal vesicles, prostate gland and the bulbourethral glands. The average pH of the semen is approximately 7.5.

Functions of the male reproductive system

The male reproductive system is concerned with spermatogenesis and the introduction of spermatozoa into the female vagina during sexual intercourse (coitus).

Erection of the penis is necessary for penetration of the vagina. Parasympathetic stimulation causes dilation of the penile arterioles and constriction of the veins, this results in engorgement of the erectile tissue, and the penis becomes elongated and rigid.

Emission is the reflex movement of spermatozoa and secretions from the epididymis, vas deferens and seminal vesicles into the urethra.

Ejaculation is caused by rhythmic contractions of the prostate and the bulbocavernous muscles which result in wave-like increases in pressure on the erectile tissue causing propulsion (*ejaculation*) of the semen out of the urethra.

The Female Reproductive System

The external organs of generation

The vulva

The female external genitalia are known collectively as the **vulva**, and are enclosed in an area bounded in front by the mons veneris and on either side by the labia majora.

The **mons veneris** is a pad of fat lying over the symphysis pubis.

The **labia majora** (singular: labium majus) are two folds of skin extending backwards from the mons veneris to unite posteriorly in the skin of the perineum. They are composed of fibrous and fatty tissue and contain numerous sebaceous and apocrine glands.

The **labia minora** (singular: labium minus) are two delicate folds of modified skin lying between the labia majora. They are highly vascular and contain numerous sebaceous glands. The labia minora enclose a triangular area called the **vestibule**, within which are the openings of the urethra and vagina.

Posteriorly the labia minora unite in a small fold, the **fourchette**. Anteriorly they unite to surround the clitoris, forming the **prepuce** above and the **frenulum** below the clitoris.

During puberty the mons veneris and the labia majora become covered by coarse hair.

The **clitoris** is a small structure which corresponds developmentally with the penis in the male. It is composed of erectile tissue and is richly supplied with nerves. The clitoris is the most sensitive part of the vulva.

The **hymen** is a thin membranous diaphragm guarding the vaginal entrance. It is normally perforated, which permits the exit of menstrual blood. In rare instances the hymen is imperforate, but this condition is not discovered until after puberty when the menstrual flow is unable to escape.

The **greater vestibular glands** (Bartholin's glands) lie one on either side of the vaginal orifice. Each gland has a long duct which opens into the space between the labium minus and the hymen. The glands produce a colourless mucoid secretion

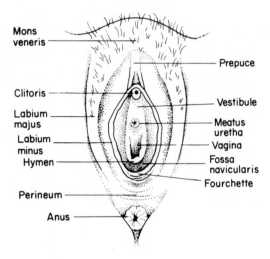

Fig. 19.7   The external genital organs.

which lubricates the vulva. Large amounts of the secretion are produced in response to sexual excitement.

The lesser vestibular glands (Skene's glands) are a group of minute mucous glands which open via two small ducts on the posterior wall of the urethral meatus.

The perineum

The perineum is the area between the fourchette and the anus. It consists of a triangular mass of connective tissue, muscle and fat which contains the perineal body. The perineal body provides a central attachment for the muscles and fascia of the pelvic floor.

The perineum stretches to a remarkable extent during labour, but in spite of this elasticity is frequently torn and requires surgical repair. When this rupture appears inevitable a deliberate surgical cut may be made (episiotomy) immediately before the baby is born. This is easily sutured and minimizes the damage to the pelvic floor.

The internal organs of generation

The vagina

The vagina is a fibromuscular canal which passes upwards and backwards from the vulva. The posterior wall of the vagina is 7.5–8 cm (about 3 in) in length, while the anterior wall is shorter because the cervix of the uterus protrudes into its upper end. The upper blind end of the vagina is called the **vault,** and is divided into four areas or **fornices** by the protrusion of the cervix (the anterior, posterior and lateral fornices).

The walls of the vagina consist of *three coats.*

(*1*) An *inner* lining of thick stratified squamous epithelium, which is thrown into prominent transverse folds or rugae. The epithelium does not contain glands, but is kept moist by the secretion of the mucous glands present in the cervix. The cells of the vaginal epithelium contain glycogen. Desquamation of these cells forms part of the normal vaginal discharge and leads to liberation of the glycogen. Doderlein's bacillus, a normal inhabitant of the vagina, acts upon the glycogen to produce lactic acid and thus maintain an acid environment within the vagina.

(*2*) A *middle* muscle coat of smooth muscle fibres arranged in circular and longitudinal bundles.

(*3*) An *outer* coat of dense connective tissue.

Anteriorly the vagina is closely related to the base of the bladder and the urethra is embedded within the lower half of the anterior vaginal wall.

Posteriorly the upper third of the vagina is in contact with the peritoneum of the rectovaginal pouch (of Douglas). The middle third is in apposition with the rectum, while the lower third is related to the perineal body.

Laterally the vagina is in relation to the levator ani muscles of the pelvic floor.

Superiorly the ureters and uterine arteries pass just above the lateral fornices.

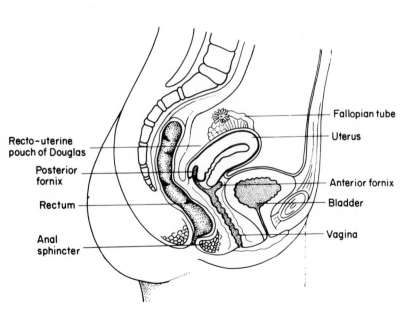

Fig. 19.8   Section through the female pelvis.

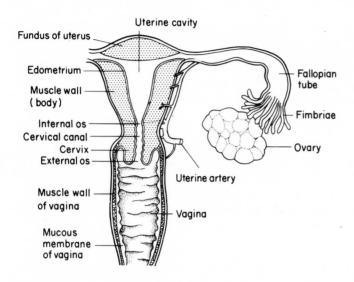

Fig. 19.9   Diagram of vagina, cervix, uterus, Fallopian tube and ovary.

Blood supply

The vagina and surrounding tissues are very vascular and receive arterial blood from branches of the internal iliac and uterine arteries; the lower portion of the vagina is supplied by the middle and inferior rectal arteries and branches of the internal pudendal artery.

The veins form a plexus around the vagina which connects with the veins around the bladder and rectum, ultimately draining into the internal iliac veins.

Lymphatic drainage

The upper two thirds of the vagina drain lymph into the external and internal iliac nodes, the lower third drains lymph into the inguinal and femoral nodes.

Nerve supply

The vagina is innervated by sympathetic and parasympathetic fibres from the autonomic nervous system. In addition the *lower* vagina is supplied with sensory nerve fibres from the **pudendal nerve.**

The uterus

The uterus is a hollow, pear-shaped organ situated in the pelvic cavity. It has thick muscular walls and a small central cavity. In the adult it measures about 7.5 cm (3 in) in length, 5 cm (2 in) in width and 2.5 cm (1 in) in thickness.

It consists of the **fundus**, the **body** and the **cervix**. The fundus is the upper part of the uterus situated between the two uterine tubes.

The body or corpus forms the greater part of the organ and gradually tapers downwards towards the cervix. The cervix or neck is the lowest portion, part of which projects like an inverted cone into the vault of the vagina. It is traversed by a canal opening above into the cavity of the uterus by an orifice called the **internal os**, and below into the vagina by the **external os.**

The cavity of the uterus is a mere slit when viewed from the side, but is flat and triangular when seen from the front. The uterine tubes open into the cavity at the upper outer angles of the fundus. The area of insertion of each tube is called the **cornu.**

The *walls of the uterus* consist of three layers.

(*1*) The **perimetrium**, an outer coat of peritoneum which covers the uterus except at the sides and is closely adherent to the underlying muscle layer. The peritoneum on the anterior surface of the uterus is reflected forwards on to the superior surface of the bladder forming a shallow **uterovesical pouch**. The peritoneum covering the posterior surface continues downwards to cover the upper part of the vagina before being reflected on to the rectum. This space between the uterus and the rectum is called the **rectouterine pouch (of Douglas)**.

The peritoneum passing laterally from the uterus extends to the side walls of the pelvis being continuous with the folds of the broad ligament.

(*2*) The **myometrium** is a thick middle coat of smooth muscle fibres arranged in three layers, an inner layer of circular fibres, a middle layer of oblique fibres and an outer layer of longitudinal fibres. The bundles of muscle fibres are interlaced with elastic and fibrous tissue.

(*3*) The **endometrium** is a specialized form of mucous membrane which varies in thickness according to the phase of the menstrual cycle. It is covered by a layer of partially ciliated columnar epithelium which contains glands that dip down into the basal layer.

Position of the uterus

Normally the uterus is bent forwards on itself in a position of *anteflexion* so that the fundus rests on the bladder. When a woman is standing the uterus lies in a position which is almost horizontal with the cervix inclined forwards at an angle of 90° with the long axis of the vagina. This position is called *anteversion*.

In some women the uterus is angled backwards in a position of *retroversion*.

The uterus is held in place by *four pairs of supporting ligaments* and indirectly by the muscles of the pelvic floor.

The *broad ligaments*. Each broad ligament consists of a double fold of peritoneum continuous with the perimetrium and extending from the uterus to the side walls of the pelvis. The uterine tube is enclosed within the upper margin.

The *round ligaments* are fibromuscular cords extending from the cornua of the uterus through the inguinal canals to be inserted in the labia majora. They help to maintain the uterus in a position of anteversion and anteflexion.

The *transverse cervical ligaments* (cardinal ligaments). The lower border of each broad ligament is thickened and strengthened by fibrous tissue, fascia and some smooth muscle to form the transverse cervical ligament. These fan out from the cervix and upper vagina to the side walls of the pelvis. They are important in preventing the uterus from prolapsing into the vagina.

The *uterosacral ligaments* are continuous with the transverse cervical ligaments and extend backwards around the rectum to the sacrum. By pulling the cervix backwards they help to maintain the uterus in a position of anteversion.

The *pelvic floor* consists of muscles, fascia and connective tissue which fills the irregular shape of the pelvic outlet (p. 100).

Blood supply

The uterus receives arterial blood from the **uterine arteries** which are branches of the internal iliac arteries. Each uterine artery passes forwards in the base of the broad ligament and reaches the uterus at the level of the cervix. It divides into branches which supply the cervix and upper vagina, then turns upwards in a tortuous fashion to supply the body of the uterus.

In addition, branches of the **ovarian artery** pass from the ovary and uterine tube to supply the fundus of the uterus and anastomose with the uterine artery.

Venous drainage is by the uterine and ovarian veins which accompany the arteries.

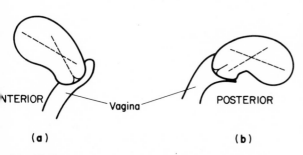

Fig. 19.10 Position of the uterus. **(a)** Normal position of anteflexion and anteversion. **(b)** Retroversion.

Lymphatic drainage

The uterus has a wide distribution of lymphatic vessels; those of the fundus drain with the ovarian vessels into the aortic nodes, while those of the lower part of the body and of the cervix drain into the inguinal, external and internal iliac nodes.

Nerve supply

The uterus receives sympathetic and parasympathetic fibres from the autonomic nervous system.

Functions of the uterus

During the reproductive years the female experiences regular cyclical changes (the menstrual cycle). Each cycle prepares the uterus to receive the fertilized ovum, and to retain and nourish the developing fetus throughout the duration of pregnancy. At the end of pregnancy the muscular walls of the uterus contract to expel the fetus.

The uterine tubes (Fallopian tubes)

Each uterine tube is about 10 cm (4 in) long and extends from the cornu of the uterus to curve round the ovary. The lumen of the tube communicates with the uterine cavity and the outer end of the tube is expanded to open into the peritoneal cavity.

It lies enclosed in the upper margin of the broad ligament, thus being surrounded by peritoneum.

The uterine tube is described in *four parts*.

(*1*) The **interstitial (intramural) part** passes obliquely through the uterine wall and has a very narrow lumen.

(*2*) The **isthmus** is the straight and narrow portion extending from the uterus.

(*3*) The **ampulla** has thinner walls and is wide and tortuous.

(*4*) The **infundibulum** is the expanded funnel-shaped portion which opens into the peritoneal cavity. Its margins are surrounded by a number of fringe-like processes, the **fimbriae**. One of the longer fimbriae lies in contact with the ovary and is called the fimbria ovarica.

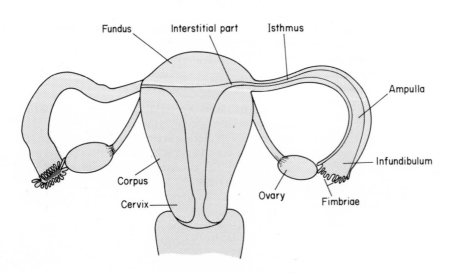

Fig. 19.11   Uterus, uterine tubes and ovaries.

Structure

The outer surface is covered with peritoneum.

The uterine tube has a muscular wall continuous with that of the uterus, but which becomes thinner from the isthmus to the ampulla, and consists of an outer layer of longitudinal muscle fibres and an inner layer of circular fibres.

It has an inner lining of mucous membrane of ciliated columnar epithelium which is thrown into longitudinal folds.

Functions

The uterine tubes serve as ducts to convey the ovum from the ovary to the uterus. *Peristaltic contractions* of the smooth muscle and movement of the cilia propel the ovum towards the uterus.

Fertilization of an ovum by a sperm usually takes place in the uterine tube and occasionally a tubal pregnancy occurs. In this case, the fertilized ovum implants in the tube, but because the tube is unable to distend and accommodate the pregnancy it rarely proceeds beyond six weeks of gestation. The products of conception may be completely absorbed, may be aborted or may result in rupture of the tube. The consequent haemorrhage requires urgent surgical intervention.

The fact that the uterine tube opens into the peritoneal cavity means that there is a potentially open canal leading from the exterior of the body to the abdominal cavity. Infections of the vagina and uterus can spread to the tubes causing salpingitis (which may result in fibrosis of the tubes and infertility) and may lead to peritonitis.

The ovary

The two ovaries are small ovoid structures, measuring 3–4 cm (1.2–1.6 in) in length, 2 cm (0.8 in) in width and 1 cm (0.4 in) in thickness, lying one on either side of the uterus. They are attached to the posterior surface of the broad ligament by the **mesovarium**, a fold of peritoneum. Blood vessels, lymphatics and nerves enter the ovary through the mesovarium at the **hilum** of the ovary. Each ovary is suspended from below the cornu of the uterus by an ovarian ligament.

Structure

The ovary consists of a medulla and a cortex which merge together and are not clearly defined.

The **medulla** is composed of loose connective tissue containing numerous blood vessels, lymphatics and nerves. Close to the hilum and mesovarium it contains small groups of **hilus cells** which are thought to be homologous to the interstitial cells of the testis.

The **cortex** consists of a compact connective tissue stroma containing **ovarian follicles** in all stages of development. Surrounding the cortex is a layer of dense connective tissue called the **tunica albuginea**. The outer surface of the ovary is covered by a single layer of simple cuboidal epithelium called the **germinal epithelium**.

Functions of the ovaries

1. Oogenesis

Oogenesis is the process by which primitive female gametes become mature ova.

Before birth, primitive female sex cells (oogonia) reproduce in the ovaries by mitosis to form **primary oocytes** (immature ova). At birth the ovaries contain about 2 million **primordial follicles**, each containing a primary oocyte surrounded by epithelial cells. During childhood many of these follicles degenerate, so that at puberty only about 400 000 remain. During the reproductive years, some 500 of the follicles will mature and expel their ova; the remainder degenerate and by the end of the reproductive period (menopause) only a few primordial follicles are left.

Very little development takes place in the ovaries between childhood and the onset of puberty. During puberty the internal organs of the reproductive system reach maturity, become active and menstruation begins (menarche).

The mature ovary has a cycle of activity which occupies approximately 28 days (although it may be as short as 21 days or as long as 35 days).

The *ovarian cycle* begins on the first day of menstruation. The secretion of *follicle stimulating hormone* (FSH) by the anterior pituitary gland stimulates several primordial follicles, which begin to grow and develop, although only one will reach maturity.

The primary oocyte within the follicle enlarges and the epithelial cells of the follicle (**membrana granulosa**) proliferate and becomes separated from the oocyte by a membrane called the **zona pellucida**. The stromal cells of the ovary form a capsule

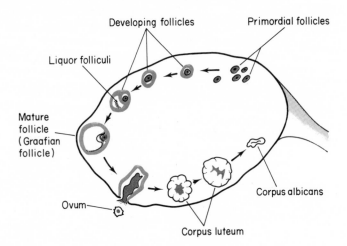

Fig. 19.12   Ovary showing stages in development of an ovum from primordial follicle to corpus albicans.

around the follicle. The capsule consists of two layers, an inner vascular layer, the **theca interna**, and the **theca externa**, an outer fibrous layer. The granulosa cells begin to secrete fluid, the *liquor folliculi* which causes the ovum to be displaced to one side of the follicle, where it becomes surrounded by a mass of granulosa cells called the **cumulus oöphorus**. By the time ovulation occurs the oocyte undergoes the first phase of meiosis.

During development the follicle is known as a **Graafian follicle**, and as it ripens the follicle secretes the hormone *oestrogen* in increasing amounts until production reaches a peak just before ovulation.

Ovulation

During development of the follicle the anterior pituitary also secretes *luteinizing hormone* (LH), which assists FSH to promote final follicular growth and ovulation. A few hours after oestrogen production reaches a maximum there is a marked increase in the secretion of LH. The follicle swells with fluid and ruptures through the surface of the ovary, expelling the ovum into the peritoneal cavity. About the time of ovulation the outer end of the uterine tube moves closer to the ovary to facilitate the entry of the ovum into the tube.

After ovulation the follicle collapses and the granulosa cells enlarge and proliferate. This process is called *luteinization*, and is controlled by LH.

The resulting glandular structure is called the **corpus luteum** (yellow body). The corpus luteum grows for about 7–8 days, secreting increasing amounts of *progesterone* and *oestrogen*, which initiate a feedback mechanism to cause a decrease in the production of FSH and LH by the anterior pituitary.

If the ovum is not fertilized the corpus luteum begins to degenerate and its production of hormones ceases. Connective tissue invades the corpus luteum and it is gradually transformed into a white scar, the **corpus albicans**.

The anterior pituitary, which is no longer inhibited by the secretion of progesterone and oestrogen, begins to secrete increasing amounts of FSH. Menstruation occurs and a new ovarian cycle begins.

If the ovum is fertilized and implants in the uterus the corpus luteum continues to grow and produces large amounts of oestrogen and progesterone during the early months of pregnancy (p. 292).

Ovulation occurs approximately 14 days before the next ovarian cycle commences; thus in a 28 day cycle ovulation occurs about day 14. However in cycles of different lengths the preovulatory phase is variable; for example in a 21 day cycle ovulation would occur about day 7.

Although normally only a single ovum is released from the ovaries in each cycle, multiple ovulation sometimes occurs.

2. Secretion of hormones

Ovarian endocrine activity is mainly concerned with the secretion of oestrogen and progesterone, but in addition the ovaries synthesize androgens.

Oestrogen is a collective name for a group of steroid compounds (oestradiol, oestriol and oestrone) which are of similar structure. Oestrogen is responsible for the development of secondary sexual characteristics at puberty, and for the growth and development of the female reproductive tract and mammary glands.

Progesterone is a steroid compound which can only affect tissues that have already been influenced by oestrogen. Its principal function is to initiate secretory changes in the endometrium in preparation for pregnancy. It also acts in conjunction with oestrogen to promote the proliferation and enlargement of the alveolar cells of the breast.

Androgens are secreted in small amounts by the ovaries, and are thought to be synthesized by the stromal cells and hilus cells of the ovarian medulla.

Control of ovarian functions

The production of ovarian hormones, and thus the ovarian cycle, are controlled by the *gonadotrophic hormones* released by the pituitary gland (p. 204).

Puberty and the menarche

Puberty means being functionally capable of procreation and is characterized by sexual maturation. It is the beginning of *adolescence*, during which mental and emotional maturation occurs and physical growth becomes complete. The *menarche* is the onset of menstruation.

Menstruation

Throughout the reproductive years, from the menarche until the menopause, the endometrium undergoes cyclical changes. The endometrial or menstrual cycle is closely related to the ovarian cycle and takes place over approximately 28 days. It can be described in three phases, each phase passing gradually into the next.

The proliferative phase

During menstruation most of the endometrial lining of the uterus is shed, leaving only a thin basal layer of the endometrium. The proliferative phase begins at the end of menstruation. Oestrogen, secreted in increasing quantities by the ovaries during the first part of the ovarian cycle, causes rapid proliferation of the epithelium. The endometrial blood vessels and glands grow longer and more coiled. By the end of the proliferative phase, (about 14 days from the onset of menstruation) the endometrium is from 2–3 mm in thickness.

This phase ends when ovulation occurs.

The secretory phase

Following ovulation the endometrium continues to hypertrophy under the influence of *progesterone* secreted by the corpus luteum. The endometrial glands become larger and more tortuous, store large amounts of glycogen, and begin secretory activity. The epithelium becomes increasingly vascular and oedematous.

Towards the end of this phase, which lasts approximately 14 days, the endometrium has a

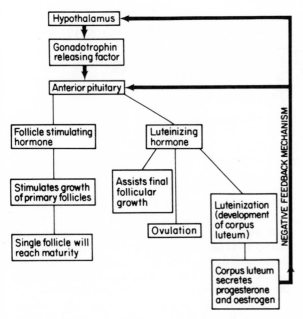

Fig. 19.13   Hormonal control of the ovarian cycle.

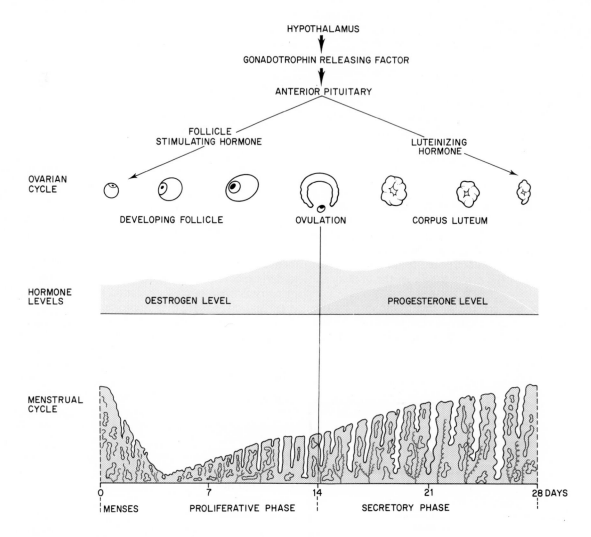

Fig. 19.14 Diagram showing the relationship of ovarian and menstrual cycles.

thickness of 4–6 mm. This thick, soft vascular surface is ready to receive a fertilized ovum. If implantation of a fertilized ovum takes place the endometrium continues to grow and becomes the decidua of pregnancy. In this case the next phase of the menstrual cycle does not occur.

The menstrual phase

If the ovum is not fertilized, the corpus luteum degenerates and the secretion of oestrogen and progesterone decreases rapidly. The endometrium is infiltrated by leucocytes and the blood vessels constrict causing shrinking and ischaemia. The necrotic surface layers of the endometrium are shed as the menstrual flow. The average duration of the menses is about 5 days, although 1 to 10 days may be considered within the normal range.

Normally the blood loss during menstruation causes a slight fall in the haemoglobin level; hence women need a greater intake of iron in their diets (p. 192).

The female climacteric and menopause

The *female climacteric* is a transitional phase which occurs at the end of the reproductive years. Over a period of one to five years ovarian function gradually declines as the supply of ova becomes exhausted. The resulting fall in oestrogen secretion causes many physiological changes. The menstrual cycle and menstruation become irregular and infrequent. Complete cessation of menstruation is known as the *menopause*. Vasomotor instability occurs, resulting in a tendency to sudden flushing (hot flushes), night sweats and palpitations. These effects are often precipitated by anxiety, warm environments and alcohol. The severity of these symptoms varies greatly in each individual. The breasts and the organs of the genital tract atrophy.

Psychological symptoms may occur such as anxiety, irritability, fatigue and loss of concentration.

The climacteric may be induced by surgical removal of the ovaries, or by pelvic irradiation in a woman of any age.

The breasts (mammary glands)

The breasts are accessory glands of the female reproductive system. In childhood and in the male they are present in a rudimentary form only. In the female the breasts begin to develop at puberty due to the influence of the ovarian hormones. Oestrogen stimulates growth of the ducts, while progesterone stimulates development of the alveoli.

Following puberty the breasts continue to enlarge due to deposition of fat and connective tissue. However the breasts remain incompletely developed until pregnancy occurs, when there is further growth of both ducts and alveoli.

Each breast lies over the pectoralis muscles, extending from the second rib downwards to the sixth rib, and horizontally from the margin of the sternum to the mid-axillary line.

The size and shape of the breasts of mature women vary considerably.

Structure

The breast consists of 15 to 20 **lobes** separated by fibrous tissue, which also acts as a supporting framework by forming suspensory ligaments. Each lobe is divided into numerous **lobules** by delicate connective tissue containing fat cells. Embedded in the lobules are clusters of **alveoli**, the secretory cells of the gland. The alveoli are drained by minute ducts, which unite to form one **lactiferous duct** for each lobe. The lactiferous ducts pass towards the nipple, and close to their termination widen to form ampullae or **lactiferous sinuses**.

The **nipple** is composed of *erectile tissue* covered by pigmented epithelium containing smooth muscle fibres, which when contracted harden and elevate the nipple. The lactiferous ducts open onto the surface of the nipple.

Surrounding the nipple is a pigmented area of skin, the **areola**, which contains a number of specialized sebaceous glands (the glands of Montgomery).

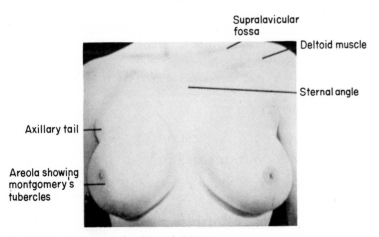

Supralavicular fossa

Deltoid muscle

Sternal angle

Axillary tail

Areola showing montgomery's tubercles

Fig. 19.15 The breasts of a girl aged eighteen years.

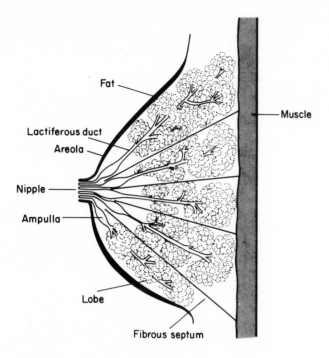

Fig. 19.16   Section of the breast.

Blood supply

Arterial blood is supplied by branches of the thoracic arteries and the anterior intercostal arteries.

The veins of the breast drain into a venous plexus encircling the nipple, which in turn drains into the axillary, internal thoracic and intercostal veins.

Lymph drainage

The mammary glands are drained by superficial and deep lymphatic vessels.

The *superficial* vessels drain the skin and subcutaneous tissues of the breast, and converge to form a diffuse lymphatic plexus with the lymphatic vessels draining the nipple and areola. The *deep* lymphatic vessels are found within the lobes of the breast draining the alveoli and ducts.

The lymphatic plexuses of the superficial vessels join the deep lymphatics. Over 85% of the lymph from the breasts drains into the **axillary nodes**; the remainder drains into the lymph nodes accompanying the internal thoracic artery. In addition there are communicating channels between the lymphatic plexuses of the two breasts.

The extensive lymphatic drainage of the breasts provides an important route for the rapid spread of cancerous cells throughout the body.

Nerve supply

The breast receives sympathetic nerve fibres from branches of the second to sixth thoracic nerves, which are accompanied by sensory nerve fibres. The breast tissue contains numerous free and encapsulated nerve endings, particularly around the nipple.

Functions of the breast

The mammary glands are inactive until stimulated by pregnancy to secrete milk to nourish a newborn infant.

Profound changes occur in the breast during pregnancy in preparation for *lactation*. After conception there is a general enlargement and hardening of the breasts; the veins on the surface become dilated and dark brown pigment is deposited in the areola. These changes are initially due to the increased production of ovarian hormones, but later are due to placental hormones. Oestrogen stimulates growth of the intralobular ducts, and progesterone, acting on the oestrogen-primed tissues, stimulates growth of the alveoli and development of the secretory cells. Other hormones, including growth hormone, prolactin, thyroxine and adrenocorticoids, are also important in the development of the mammary glands.

Towards the end of pregnancy, and for a few days after delivery, a watery fluid called *colostrum* is secreted by the breast. After childbirth there is a fall in the blood levels of oestrogen, which stimulates the anterior pituitary to release prolactin. *Prolactin* stimulates the process of lactation. Milk production commences 3−4 days after childbirth. The posterior pituitary hormone *oxytocin* causes the expulsion of milk from the alveoli to the ducts. The secretion of oxytocin is stimulated by the infant suckling at the breast. Regular suckling of the infant is necessary to maintain lactation.

Pregnancy

Following expulsion of the ovum from the ovary it enters one of the uterine tubes. The ovum remains viable for approximately 12 to 24 hours and is still surrounded by a mass of granulosa cells, the cumulus oophorus, which radiates out from the oocyte and is known as the **corona radiata**. During sexual intercourse sperm enter the female vagina with each ejaculation of semen. Passage of the sperm to the uterine tubes is accomplished partly by sperm motility and partly by contractions of the uterus. Fertilization usually takes place in the uterine tube. Only one sperm is required for fertilization of the ovum. The head of the sperm enters the ovum by penetrating the corona radiata and the underlying zona pellucida. Penetration of the sperm into the ovum is facilitated by proteolytic enzymes. Changes occur in the zona pellucida which prevent penetration of the ovum by other sperm.

When the sperm enters the ovum it loses its tail and body, and the head of the sperm begins to swell forming a male pronucleus. The pronucleus of the ovum and the male pronucleus do not merge but form a new nucleus, each contributing 23 chromosomes to produce a total of 46.

The fertilized ovum begins a series of cell divisions by mitosis, and by the time it has reached the uterine cavity 3−4 days later it forms a mass of cells called a **morula**. During the next few days the morula develops a central fluid filled cavity and is now known as a **blastocyst**. The cells of the blastocyst become differentiated into a double-layered wall called, the **trophoblast** and an **inner cell mass** from which the fetus will be formed.

Following fertilization of the ovum, the endometrial lining of the uterus continues to develop under the influence of progesterone from the corpus luteum, and is called the **decidua**. The trophoblast secretes proteolytic enzymes which digest

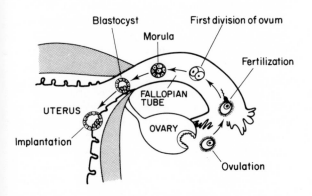

Fig. 19.17 From ovulation to implantation.

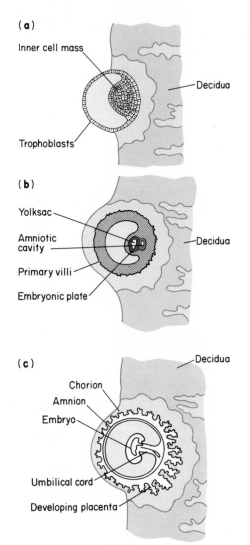

(a)

Inner cell mass

Decidua

Trophoblasts

(b)

Yolksac

Amniotic cavity

Decidua

Primary villi

Embryonic plate

(c)

Decidua

Chorion

Amnion

Embryo

Umbilical cord

Developing placenta

Fig. 19.18 Development of the embryo and placenta: **(a)** at implantation; **(b)** differentiation of inner cell mass; **(c)** development of the umbilical cord and placenta.

the amniotic cavity is the **ectoderm** and that lining the yolk sac is the **endoderm**. A third layer of cells, the **mesoderm** develops between these two, and grows out to form the **umbilical cord** through which the embryonic circulation will extend into the placenta. Blood capillaries grow into the villi of the trophoblast through the umbilical cord from the vascular system of the embryo, and maternal blood sinuses form in the endometrium surrounding the villi. *Oxygen* and *nutrients* pass from the maternal blood sinuses to the blood capillaries of the embryo by diffusion, while *waste products* from the embryo diffuse in the opposite direction to the mother for excretion. At no time during pregnancy do the two circulations come into direct contact, the exchange of nutrients taking place across a membrane called the **chorion**.

As the blastocyst increases in size it bulges into the uterine cavity and the villi on the exposed surface atrophy. The remaining villi become restricted to one area and form the **placenta**. From the time of implantation the trophoblast begins to secrete small amounts of *human chorionic gonadotrophin (HCG)*; this prevents the corpus luteum from degenerating and causes it to continue to secrete progesterone and oestrogen, which maintain the decidual endometrium to provide for the early development of the placenta and embryo.

The **corpus luteum** reaches its peak activity during the first eight weeks of pregnancy, and then begins to wane. By the sixteenth week it has ceased to be active and the **placenta** becomes totally responsible for the secretion of oestrogen and progesterone. The placenta continues to produce HCG in increasing amounts until about the eighth week; from this time the HCG production falls to a low constant level for the remainder of the pregnancy. The presence of HCG forms the basis of pregnancy tests, since it is excreted in the maternal urine soon after the first missed menstrual period.

In addition, from the fifth week of pregnancy the placenta begins to secrete a hormone called *human placental lactogen (HPL)* which has actions similar to those of growth hormone, and probably plays an important role in the growth and development of the fetus.

By the sixth week of pregnancy the ectoderm, mesoderm and endoderm will have formed all the essential structures of the body, and by the eighth week the embryo is recognizable as a human baby. It now becomes known as a **fetus**. By the sixteenth

the cells of the decidua to obtain nourishment, and to embed the blastocyst into the deeper layers of the endometrium. The trophoblast develops projections which invade the endometrium and which will become the **placental villi**.

As the cells of the inner cell mass continue to divide, two cavities appear; one is the **amniotic cavity** and the other forms the **yolk sac**. The two cavities are separated by a double layer of cells, the **embryonic plate**. The cells of the embryonic plate are destined to become the embryo; the layer lining

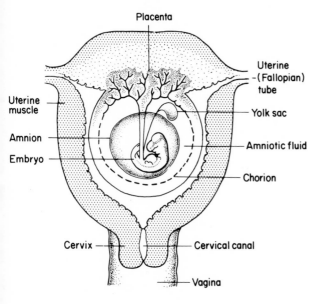

Fig. 19.19   Diagram of uterus and embryo in early pregnancy.

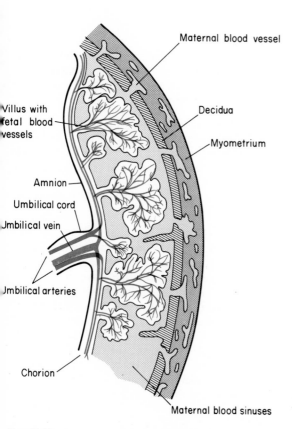

Fig. 19.20   Structure of fully developed placenta.

week it is possible to determine the sex of the fetus and further development is mainly growth.

During the development of the embryo the trophoblasts produce an inner layer of cells which form a membrane called the **chorion**; this covers the fetal surface of the placenta and umbilical cord. The amniotic cavity grows rapidly until it fills the original blastocyst and its cells come into close contact with the chorion. This membrane is the **amnion**; it lines the chorion and covers the umbilical cord, being continuous with the skin of the fetus at the umbilicus. The amnion produces a fluid, **liquor amnii,** in which the fetus is suspended.

The chorion and amnion form the **fetal sac**, which encloses and protects the fetus until it ruptures during labour and allows expulsion of the baby.

Changes in maternal physiology in pregnancy

During pregnancy there are widespread physiological changes in the mother, involving most of the systems of the body, which result in alteration of metabolic, chemical and endocrine balance.

(1) The **uterus** gradually enlarges, mainly due to hypertrophy of individual muscle fibres and partly by the formation of new fibres. The *uterine growth rate* is usually regular and provides a means of estimating the period of gestation by measurement of the *fundal height* (Fig. 19.22).

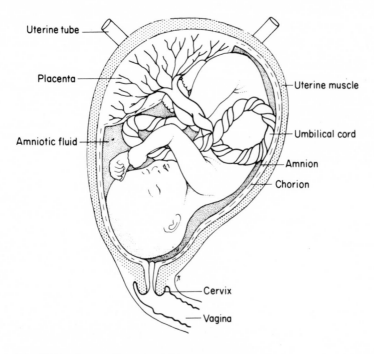

Uterine tube

Placenta

Amniotic fluid

Uterine muscle

Umbilical cord

Amnion

Chorion

Cervix

Vagina

Fig. 19.21   Diagram illustrating full-term pregnancy.

The **cervix** becomes softer and the cervical glands secrete a tenacious mucus which forms a plug that fills the cervical canal. Expulsion of this mucoid plug as the 'show' frequently occurs a few hours before the onset of labour.

(2) The **breasts** begin to enlarge early in pregnancy with increased brown pigmentation of the areola. The sebaceous glands of the areola become raised, appearing as small nodules, and are known as **Montgomery's tubercles**. A small amount of fluid (colustrum) can be expressed from the nipples. In mid-pregnancy, patchy pigmentation develops around the areola and forms the secondary areola.

(3) The **skin** of the abdominal wall becomes stretched and pink striations sometimes develop (striae gravidarum). Increased pigmentation of the skin may also occur on the forehead and cheeks (known as the *chloasma* or mask of pregnancy), and a dark line (the linea nigra) may be seen extending from the umbilicus to the pubis.

(4) *Metabolism* In addition to placental hormone production in pregnancy, there is increased secretion of many hormones including thyroxine, adrenocorticoids and sex hormones. Carbohydrate metabolism is altered, and there may be a lowered renal threshold for glucose resulting in glycosuria. As a result of the increased endocrine functions the basal metabolic rate rises by 15–20% during the latter part of pregnancy.

(5) *Weight gain* in the entire pregnancy averages about 24 lbs (11 kg). The fetus, placenta and liquor amnii account for approximately 10 lbs (4.5 kg). The increase in uterine and breast size accounts for about 4 lbs (1.8 kg). The remainder is due to increased blood volume, fat deposition and fluid retention.

(6) *Nutrition* Early in pregnancy nausea and vomiting commonly occurs ('morning sickness') and is thought to be due to hormonal changes, but usually disappears after the first three months.

During pregnancy the mother's diet must supply the energy requirements for the additional maternal tissues and the growing fetus as well as for her own health. If the diet is inadequate, the fetus will obtain its nourishment at the expense of the mother's tissues. The mother's diet therefore should contain adequate quantities of protein, carbohydrate and fat as well as vitamins, calcium and iron.

(7) *The cardiovascular system* The volume of the

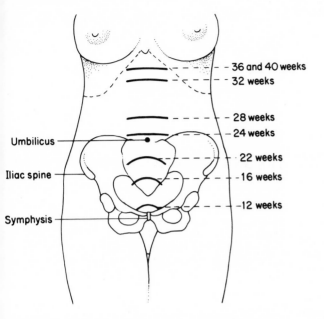

Fig. 19.22 Height of the fundus of the uterus during weeks of pregnancy.

circulating plasma is increased by approximately thirty per cent, mainly during the second half of pregnancy. There is also a small increase in red cells, but the total haemoglobin concentration falls to about 12 g/100 ml of blood. This must not be interpreted as anaemia, although this may occur if the dietary intake of iron and folate is inadequate for the combined needs of the mother and fetus.

As a result of the increased blood volume and the rise in general metabolic demands the cardiac output increases by 25–30%.

(8) Respiration increases in depth rather than rate, to supply the extra oxygen required and to excrete the carbon dioxide produced by the metabolism of the fetus.

Questions

1. Describe either the male or female reproductive system.
2. Discuss the structure and function of the testes.
3. Give an account of the production of spermatozoa.
4. Describe the structure and functions of the ovaries.
5. Describe the relationship between the ovarian and menstrual cycles and the changes which occur in the endometrium.
6. Describe the sequence of events that occurs following fertilization of an ovum.
7. Discuss the physiological changes which occur in a woman during pregnancy.

20
Ageing

Ageing is not something that happens suddenly after 40, 50, 60 or some other determinate number of years. Old age (senescence) is, with certain exceptions, the culmination of many years of slowly declining function and degenerating structure. All physiological functions decline as ageing increases and ageing results in increased susceptibility to stress and disease. Disease and trauma may accelerate the effects of ageing. Healing and recovery are much slower in the elderly. Normal elderly people have a physiology that is adequate for resting conditions but, compared to young people, they show *slower adjustment* to environmental change and a *diminished reserve capacity*.

It is therefore particularly important for doctors, nurses and others to know what changes ageing causes, as these must be taken into account in the treatment of disease.

Changes due to ageing

The skin

With ageing the skin becomes wrinkled because there is a loss of subcutaneous fat and the dermis shrinks, *losing its elasticity*. With less support from the dermis and subcutaneous tissues the epidermis develops folds or wrinkles. Blood flow to the skin is reduced, the epidermis becomes thinner and surface blood capillaries become more fragile. There is an overall decrease in the number of functioning melanocytes, resulting in grey hair and changes in skin colour. However some melanocytes increase in size producing brown 'age spots'.

Sebaceous glands atrophy, skin and hair become dry and brittle and, because it breaks easily, the skin is more susceptible to infection. Sweat glands also atrophy, a fact which contributes to the reduced ability of the elderly to adjust to increased environmental temperatures.

Axillary and pubic hair decrease and nails thicken.

The musculoskeletal system

Body *stature diminishes* with age. While the length of the long bones remains constant there is thinning of the intervertebral discs resulting in decreased height. *Bone mass is reduced* as the bones tend to lose calcium and become thinner and more brittle with ageing. The risk of injury resulting in fractures, especially of the femur, are greater.

Joint capsules and cartilage becomes calcified resulting in stiff joints which limit the elderly person's mobility.

Muscle fibres atrophy and are replaced by fibrous tissue, resulting in decrease in muscle power.

The cardiovascular system

Myocardial function diminishes and cardiac output falls gradually with age. The older heart has a *reduced myocardial reserve* and is unable to respond to a demand for an increased cardiac output as efficiently as a young heart, and cardiac failure is more easily induced (e.g. by over-zealous intravenous therapy).

With ageing, arteries become less elastic because the muscle fibres of the tunica media are replaced by collagen fibres. The gradual development of arteriosclerosis may cause a further decrease in the distensibility of the arteries. These changes cause an increase in peripheral resistance which results in an increase in blood pressure, especially the systolic pressure.

The respiratory system

Chest movement diminishes because of *increasing rigidity* of the thoracic cage, and vital capacity is reduced. Alveoli become less elastic and enlarge as the alveolar walls become weakened. Bronchioles and alveolar ducts also lose their elasticity and become dilated. A decrease in ciliary action combined with the structural changes increases the susceptibility of the elderly to respiratory infection. Ageing also results in *less efficient diffusion* and tissue utilization of oxygen. It is not suprising that elderly people more easily become short of breath than do young people.

The urinary system

In the ageing kidney the number of functioning nephrons is reduced and the kidneys become smaller. The basement membrane of Bowman's capsule becomes thicker leading to a diminished permeability of the membrane. These changes result in a *reduction in the glomerular filtration rate*. Additionally, the excretory and reabsorptive capacities of the renal tubules decrease as age increases. Therefore the kidneys do not concentrate urine as efficiently in the elderly, although under normal circumstances renal function remains adequate.

Bladder capacity diminishes, especially in the female, and can result in an immediate, urgent need to urinate. It often becomes necessary for the older person to void urine during the night.

The digestive system

Some characteristic changes occur in the anatomy and physiology of the digestive system with ageing.

Teeth become darker with age and the permeability of the enamel decreases. The enamel at the contact points of teeth is gradually reduced and the grinding surfaces of the molars wear down. Changes in the formation of dentine result in a gradual reduction in the size of the pulp cavity. Gum margins tend to recede leading to exposure of the roots.

Dryness of the mouth may be noted due to a reduction in mucus secretion. Since mucus secretion is necessary for the retention of complete dentures an elderly person may experience considerable difficulty in keeping their dentures in place, especially when dentures are new.

A reduction in the flow of saliva diminishes the moisture added to food in the mouth and may affect an individual's enjoyment of particular foods.

Atrophy and diminished enzyme secretion mean that the digestion of food takes longer and its absorption may be impaired. The *reduced gasro-intestinal movement* may lead to constipation, which is exacerbated by a reduction in mucous secretion by the mucosal cells lining the large intestine due to atrophy.

The endocrine system

The secretion of sex hormones declines sharply in women at the menopause (p. 289), while in men there is a slower progressive decrease. Thyroxine secretion and hence *basal metabolic rate (BMR) gradually diminishes* with age. Glucose tolerance also diminishes.

The reproductive system

In the male, changes in the reproductive system occur gradually. The testes continue to produce spermatozoa well into advanced old age, although there is a decrease in the number of viable spermatozoa. Testosterone secretion tends to diminish with increasing age. Sexual arousal is slower with ageing, the number and volume of ejaculations is reduced but feelings of stimulation and satisfaction remain. *The prostate gland usually increases in size* and may cause urinary difficulties.

Following the menopause, the female is no longer able to reproduce because of *atrophy of the ovaries*. The uterus and cervix shrink. The walls of the vagina become less elastic and the mucosal rugae flatten. *Vaginal secretions diminish* and become less acid. There is a loss of subcutaneous fat in the external genitalia; pubic and vulval hair becomes thinner. Due to lack of ovarian hormones there is atrophy of the glandular tissue of the breasts. However, despite the changes occurring in the genital tract there is no loss of female libido.

The nervous system

The brain

Each day approximately 1000 neurons are lost from the brain. Fortunately there is a large reserve and many elderly people never show obvious senile mental changes. The velocity of transmission of nerve impulses is reduced and reaction time is slower in the elderly.

The autonomic nervous system

Vasomotor control is often impaired in elderly people. Diminished blood supply to the brain may cause unsteadiness, faintness and falling after sudden assumption of the erect posture. Nurses have to remember this when getting elderly patients out of bed, and may have to proceed very slowly.

The special senses

Eyes

The lens capsule of the eye becomes less elastic and is unable to accommodate to a shape sufficiently convex to focus on near objects. The 'long sight' resulting from this ageing process is known as *presbyopia*. Sometimes the lens becomes opaque and prevents light entering the eye (cataract). There is loss of peripheral vision, and it takes longer to adapt to light changes.

Ears

Hearing impairment increases with ageing, and is associated with increased deposition of wax in the external auditory meatus, rigidity of the ossicles and degeneration of the vestibulocochlear nerve (VIIIth cranial nerve).

Taste

The number of taste buds on the tongue is reduced with ageing. There appears to be a decreased sensitivity to sweet and salty substances, while there is an increased sensitivity to bitter substances. These changes may result in a preference for very sweet foods.

Temperature regulation

With ageing the ability of the temperature-regulating system to maintain core temperature in extreme environmental temperatures appears to be impaired. There is *impairment of the mechanisms for heat production and preservation*. Some people feel the cold more as they age whilst others are uncomplaining in a cold home environment which their younger relatives, accustomed to efficient modern heating, would not tolerate.

Old people commonly fail to produce a pyrexia when suffering from infections such as pneumonia.

Thirst

The sensation of thirst may be diminished in elderly people, who become dehydrated more easily, especially when they are ill.

Pain

Pain is sometimes less keenly appreciated by elderly people. A myocardial infarction or 'acute abdomen' may be painless (silent).

Resistance to infection

There is increased susceptibility to infection in old age because of a diminished antibody response and reduced serum immunoglobulin concentrations.

The relevance of ageing to clinical medicine

Nurses, physiotherapists and other professional people will realise that, because of impaired vasomotor control, elderly people often cannot be got out of their beds hurriedly without making them feel faint and unsteady.

Doctors avoid misguided attempts at treatment of a raised blood pressure by relating the degree of elevation to the patient's age. For example a blood pressure of 180/95 mmHg in a patient aged 70 years would not call for any treatment. Indeed, effective treatment would be likely to induce disabling postural hypotension.

Drug treatment of any condition may be fraught with special problems in old age. The absorption, metabolism and excretion of a drug may all be slowed. Therefore it is often necessary to reduce the standard dosage of some drugs. Failure to do this may result in digitalis intoxication in elderly patients.

Appendix

The scalp

The scalp forms the soft parts covering the vault of the skull or cranium. It consists of

(1) **Skin** containing a very large number of hairs in their follicles and many sebaceous glands.

(2) **Superficial fascia** consisting of a mesh of fibrous tissue enclosing small lobules of fat.

(3) **Deep fascia**, a very strong layer, sometimes called the **epicranial aponeurosis**, which is attached in front to the frontalis muscle and behind to the occipital bone.

(4) The following *muscles* (Fig. 7.6):
    (a) the frontalis;
    (b) the occipitalis;
    (c) the temporal muscle on each side;
    (d) the small muscles attached to the ears.

(5) Superficial or cutaneous *nerves*.

(6) Numerous *arteries*, which supply the scalp plentifully with blood. The following major vessels are branches of the external carotid:
    (a) the superficial temporal;
    (b) the posterior auricular;
    (c) the occipital artery.
Most of the *veins* drain into the external jugular but a few pass directly through the skull bones from the venous sinuses in the interior of the cranium (sometimes a source of danger in infections of the scalp).

(7) The deepest layer of the scalp is the periosteum covering the cranial bones, here called the **pericranium**.

The triangles of the neck

The neck looked at from the side (see Fig. 7.6) is roughly quadrilateral in shape, bounded above by the lower jaw, below by the clavicle, in front by the midline and behind by the trapezius muscle. It will be seen that the sternomastoid muscle divides this area into two (anterior and posterior) triangles. The **anterior triangle** contains, in its upper part, the external carotid artery and its important main branches, various nerves, parts of the pharynx and larynx and some lymphatic glands. The lower part of the anterior triangle is covered by muscles and also contains the trachea and thyroid gland.

The **posterior triangle**, situated between the anterior border of the trapezius and the posterior border of the sternomastoid, contains the occipital artery, various cutaneous nerves and some lymph nodes. Situated deep in its lower part just above the clavicle are parts of the subclavian artery and vein, and the brachial plexus.

The axilla

The axilla is the space or hollow between the upper part of the arm and side of the chest. When the arm is by the side, the space is pyramidal or cone-shaped, having an apex directed upwards towards the root of the neck and situated close to the coracoid process of the scapula.

The *anterior wall* is formed by the pectoralis major and minor muscles.

The *posterior wall* is formed by the subscapularis and latissimus dorsi muscles.

The *medial wall* is formed by the upper four or five ribs, the intercostal muscles and the serratus anterior.

The *lateral wall* consists of the shaft of the humerus and the attached muscles.

The *floor or base* consists of skin which contains hair and many sweat glands and thick fascia – the axillary fascia.

The axilla contains:
(a) the axillary artery and vein;
(b) the important nerves of the arm, including the radial, ulnar and median, which are arranged around the axillary vessels;

(c) lymph nodes which receive lymph from: (i) the arm; and (ii) the breast (especially important in cancer of the breast).

The cubital fossa

The cubital fossa is a small space situated in front of the elbow joint at the bend of the elbow. Superficially are placed the **median basilic and median cephalic veins**; more deeply the **brachial artery** divides into the radial and ulnar branches. To the lateral side of the brachial artery the **tendon of the biceps muscle** passes to be inserted into the tubercle of the radius. The **radial nerve** also lies to the lateral side. On the medial side is the **median nerve**. The boundaries of the space are formed by various muscles (Fig. A.1).

The palm of the hand

The palm of the hand is of considerable practical importance on account of its liability to injury and infection. It may be conveniently considered in layers.

(1) The **skin** and **superficial fascia**.

(2) The deep or **palmar fascia** is a strong fibrous layer consisting of a central and two lateral portions, the latter being spread over (a) the thenar eminence formed by the short muscles of the thumb and (b) the hypothenar eminence formed by the short muscles of the little finger.

(3) (a) The **superficial palmar arch** which is formed by the continuation of the ulnar artery into the hand joining with a branch from the radial artery on the radial side.

(b) The median and ulnar **nerves** and their branches to the fingers.

(c) The **tendons** of the flexor muscles which are enclosed in synovial sheaths, the arrangement of which is shown in Fig. A.2. These are very important in connection with infections of the fingers and hand. It is clear that infection of the tendon sheath of the little finger can spread and involve the main synovial sac in the palm which is called the ulnar bursa. The tendon sheaths of the other fingers extend upwards only as far as the heads of the metacarpal bones.

(4) The deepest part of the palm is formed by the **metacarpal bones** and the short muscles of the fingers (interossei) on which lies the **deep palmar arch** formed by the continuation of the radial artery joining a deep branch of the ulnar artery on the ulnar side.

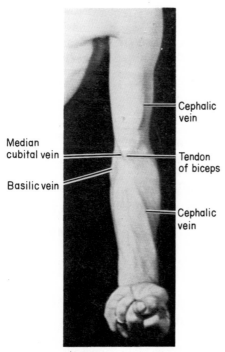

Fig. Al The cubital fossa.

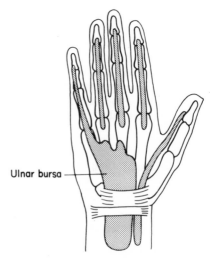

Fig. A2 Diagram illustrating the arrangement of tendon sheaths in the palm.

The umbilicus

The umbilicus or navel is a puckered depression in the skin of the abdominal wall, situated in the midline rather nearer the symphysis pubis than the xiphoid cartilage. It is actually a scar representing the site of the attachment of the umbilical cord which connects the fetus to the placenta and contains blood vessels.

The inguinal canal

The inguinal canal is a passage, 4 cm ($1\frac{1}{2}$ in) long in the lower part of the abdominal wall. It passes obliquely downwards forwards and medially and contains a part of the spermatic cord in the male and the round ligament of the uterus in the female. It is situated immediately above the medial half of the inguinal (Poupart's) ligament. The lateral and deeper end is called the **deep inguinal (internal abdominal) ring**. The superficial end nearer to the midline and just above the spine of the pubis is called the **superficial inguinal (external abdominal) ring**. It has

(*a*) a floor formed mainly by the inguinal ligament;

(*b*) an anterior wall consisting of skin, fascia and the aponeurosis of the external oblique muscle;

(*c*) a posterior wall formed by peritoneum covered by fascia and a tendinous insertion of the internal oblique and transversus muscles called the conjoined tendon.

It is through this canal that the testis travels in its descent from the abdominal cavity to the scrotum, carrying with it the spermatic cord.

Sometimes a rupture or hernia develops in this area. An indirect inguinal hernia consists of a protrusion of the peritoneum through the deep inguinal ring which may extend along the inguinal canal and pass through the superficial inguinal ring to reach the scrotum. This protrusion or peritoneal sac may contain intestine, omentum or some other abdominal viscus. A direct inguinal hernia does not pass through the deep inguinal ring. It bulges directly through the posterior wall of the inguinal canal. It emerges through the superficial inguinal ring but does not usually descend into the scrotum.

A femoral hernia passes down through the femoral ring, which is medial to the femoral vein and behind the inguinal ligament, and emerges at the saphenous opening.

The femoral (Scarpa's) triangle

This is a triangular hollow in the *upper part of the thigh*. Its base is above and is formed by the inguinal ligament. Its lateral boundary is the sartorius muscle, its medial boundary is the adductor longus muscle. The psoas, iliacus and pectineus muscles form the floor.

It contains the **femoral artery** and vein and the femoral nerve which pass vertically downwards from the middle of the inguinal ligament to its apex. Placed superficially are lymph nodes and the saphenous opening through which the long saphenous vein passes to reach the femoral vein (Figs 10.19 and 10.21).

The adductor (Hunter's) canal

After leaving Scarpa's femoral triangle, the **femoral artery** enters Hunter's canal. This is a channel situated deeply in the muscles on the *lower third of the medial side of the thigh*.

The sartorius muscle forms the roof, the adductor longus the posterior wall and the vastus medialis the lateral wall.

The femoral artery and vein leave the lower end of the canal by passing into the popliteal space (Fig 10.19).

The popliteal fossa

The popliteal space or fossa is situated *behind the knee joint*. Its anterior wall or floor is formed from above downwards by

(1) the popliteal surface of the femur;

(2) the posterior ligaments of the knee joint;

(3) the upper part of the head of the tibia and the popliteus muscle which covers it.

The space is diamond-shaped and has the following boundaries.

(1) Upper and medial: the semi-membranosus and semi-tendinosus muscle.

Upper and lateral: the biceps muscle and tendon.

(2) Lower and medial: the medial head of gastrocnemius.

Lower and lateral: the lateral head of gastrocnemius.

Passing vertically downwards through the space are the **popliteal artery** and vein and the medial popliteal nerve.

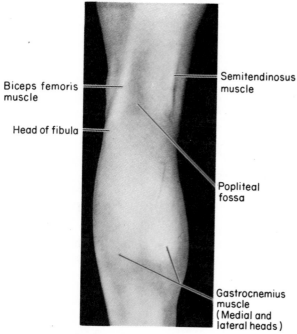

Biceps femoris
muscle

Head of fibula

Semitendinosus
muscle

Popliteal
fossa

Gastrocnemius
muscle
(Medial and
lateral heads)

Fig. A3   Popliteal fossa.

The sole of the foot

This corresponds to the palm of the hand and is noted for the strong plantar fascia which extends throughout its length. The foot is arched both longitudinally and transversely.

The sole of the foot contains the **flexor tendons** of the toes, short muscles passing to the big and little toes, nerves and the arteries which form the plantar arch.

Weights and Measures

1 ounce (oz) = 28.35 grams (g)
1 pound (lb) = 16 ounces = 0.45 kilogram (kg)
1 stone = 14 lb = 6.3 kg

1 fluid ounce = 28.41 ml
1 pint = 560 ml

1 yard = 36 inches (in) = 0.91 metres (m)
1 foot = 30.48 centimetres (cm)
1 inch = 2.54 centimetres

1 kilogram = 2.2 pounds = 35.27 ounces

1 litre = 1.75 pints = 35 fl. oz. = 0.22 gallon
1 ml = 15 minims

1 metre = 1.09 yards = 39.37 inches = 3.28 feet

To convert:
   grams to ounces multiply by 0.03
   ounces to grams multiply by 28.0
   minims to millilitres multiply by 0.06
   pints to litres multiply by 0.57

Fahrenheit to Centrigrade or Celsius, subtract 32 and multiply by $\frac{5}{9}$
Centigrade or Celsius to Fahrenheit, multiply by $\frac{9}{5}$ and add 32

SI units

The International System of Units (Systeme International, SI) is the system used in British Medicine for weights and measures. It is based on seven fundamental units: the metre (m); kilogram (kg); second (s); ampere (A); kelvin (K); candela (cd); and mole (mol). Other units such as those of pressure, the pascal (Pa) and energy, the joule (J), are derived from the basic units. Prefixes are used to indicate fractions or multiples of all these units. The prefixes are as given in the following table:

Prefixes for SI units

Factor by which unit is multiplied	Prefix	Symbol of prefix
10^{12}	tera	T
10^{9}	giga	G
10^{6}*	mega	M
10^{3}*	kilo	k
10^{2}	hecto	h
10^{1}	deca	da
10^{-1}*	deci	d
10^{-2}*	centi	c
10^{-3}*	milli	m
10^{-6}*	micro	m
10^{-9}*	nano	n
10^{-12}*	pico	p
10^{-15}	femto	f
10^{-18}	atto	a

* These are the factors which the nurse is likely to encounter in clinical practice

Index